Springer Undergraduate Mathematics Series

More information about this series at http://www.springer.com/series/3423

Juliusz Brzeziński

Galois Theory Through Exercises

Juliusz Brzeziński
Department of Mathematical Sciences
University of Gothenburg
Sweden

Chalmers University of Technology
Sweden

ISSN 1615-2085 ISSN 2197-4144 (electronic)
Springer Undergraduate Mathematics Series
ISBN 978-3-319-72325-9 ISBN 978-3-319-72326-6 (eBook)
https://doi.org/10.1007/978-3-319-72326-6

Library of Congress Control Number: 2017964367

Mathematics Subject Classification (2010): Primary: 12-01, 12F05, 12F10; Secondary: 11R16, 11R18, 12E05, 12E20, 12F12

Printed on acid-free paper

This Springer imprint is published by the registered company Springer International Publishing AG part of Springer Nature.
The registered company address is: Gewerbestrasse 11, 6330 Cham, Switzerland

In memory of those from whose arms I was torn away, to those whom I had the privilege to carry in mine – Piotr, Jacob and Elsa

Preface

This Book

The present book evolved from different collections of exercises which I wrote when I started my teaching duties at the beginning of the 1960s, first at the University of Warsaw, and 10 years later, at the University of Gothenburg and the Chalmers University of Technology. A few textbooks available at the time, like Emil Artin's seminal book "Galois Theory", which was used frequently, or van der Waerden's "Algebra" (in its part concerned with Galois theory) needed supplementary material in the form of simple examples and some advanced exercises for interested students. Usually, a course in Galois theory followed one or two courses in algebra, which could present some necessary previous knowledge. A common experience of all the courses were the reactions of my students, complaining that it is usually relatively easy to start with a standard exercise in calculus, while an exercise in algebra presents no clues as to how to start solving it, even if you know the relevant theorems. Naturally, the degree of difficulty of the exercises can create an obstacle in any area of mathematics, but in algebra such opinions are often related to rather mundane problems. Therefore, this phenomenon deserves reflection and explanation. There are probably several reasons for such a perception of the subject. One possibility is the character of the tasks related to the same objects in different contexts. For example, in calculus, concrete numbers or functions will have already been encountered in high school and, usually, the first exercises are about concrete objects of this kind. Similar objects appear in algebra, but it is rather a general property of a whole set of numbers or functions which is relevant. Another essential difference is the presence of several new notions related to the sets, which are not necessarily well known from previous encounters with mathematics, e.g. what are the properties of addition or multiplication of different types of numbers. Still another aspect is probably the character of the proofs and the possibility of transferring the arguments from them to concrete situations. The proofs of the theorems in calculus usually give a clue of how to handle concrete objects, while in algebra a proof of a general property of a structure satisfying some conditions usually says very little about suitable examples of such structures and even less about the methods one can use to determine whether or not a concrete structure has a particular property. This is one of the reasons why the adjective "abstract" is used in the names of some algebra courses. Perhaps the only way to address these difficulties, which may appear in connection with courses in algebra, is to expose the students to many concrete examples and to give them the chance to solve many exercises. This is the primary purpose of this text as a book on Galois theory motivated by exercises and examples of their solutions.

A Few Words on the Subject

Galois theory grew from the desire to solve a very concrete mathematical problem related to polynomial equations of degree higher than 4. Already during the 18th century, some mathematicians suspected that the efforts to find general formulae for solutions of quintics (polynomial equations of degree 5) were futile in spite of the existence of such formulae for equations of degree lower than 5. Galois proved that there are quintics with rational coefficients whose zeros cannot be expressed by means of these coefficients using the four arithmetical operations (addition, subtraction, multiplication, division) and extracting roots. A similar result concerning so-called general polynomial equations of degree 5 (polynomials whose coefficients are variables) was published (but not exactly proved) by Ruffini in 1799 and finally proved by Abel independently of Galois. The unsolvability of the quintic is a very concrete, interesting and deep problem, which provides very good reasons to go into the details of a theory which gives its solution. The problem is easy to explain and creates motivation for the students to use a lot of theory from previous courses in algebra as well as to see the need for a suitable language in order to formulate it in mathematical terms.

However, learning Galois theory today is not only motivated by this result concerning polynomial equations. The theory developed by Galois has tremendous value as a path to modern mathematics. In order to solve a concrete algebraic problem, Galois translated it to a new (abstract) language of groups and fields. This was not only a great contribution to these theories (e.g. Galois introduced many notions in group theory), but also gave later ideas about the possibility of using new mathematical structures like groups, rings, fields, linear spaces, topological spaces and many other (abstract) structures in order to reformulate and solve concrete mathematical problems. Thus, Galois theory was a source of future developments of new theories having still broader applicability. Galois theory itself also found many important extensions, generalizations and applications, for example, in number theory, topology and the theory of group representations.

The Structure of the Book

Since it is impossible to exactly define a "standard selection" of notions and theorems which constitute a course in Galois theory, the text is structured in such a way that a student will be able to use it as an exercise book even without previous knowledge of the subject. Chapters 1–15 together with Chap. 17 give more than a very standard course could contain. In the following section, we describe the contents of the book in a way which explains how it may be used as a textbook. The book was often used as a primary source of knowledge and, in several cases, by those students who studied Galois theory on their own with limited assistance from the instructor.

Every chapter, out of the first 15, starts with a presentation of the main notions and theorems followed by a comprehensive set of exercises which explain the fundamental facts about them in a practical way. The exercises graduate from simple, often numerical, to more complex and theoretical. At the same time, they are usually ordered from more standard to more special. As a rule, the exercises given in Chaps. 1–15 should be considered as part of a basic course in the subject. Chapter 16 contains a choice of "one hundred" supplementary problems without hints or solutions.

In contrast, Chap. 18 includes some limited hints and answers (when an answer can be given) to the exercises in Chaps. 1–15. As usual, and in particular in algebraic problems, it is quite often possible to give different solutions. Therefore, the hints cannot be considered as thoughts in "Mao's Little Red Book". This is even more important in Chap. 19, which contains a number of standard examples and complete solutions of more difficult exercises. Let us note that the standard exercises from Chaps. 1 to 15 (in particular, those at the beginning of each chapter) are not solved in the book—almost every such exercise is illustrated in Chap. 19 by an example of a similar nature in order to show how to handle standard problems (often of a computational nature) and to create the possibility of using the exercises in Chaps. 1–15 as homework. Reading the examples with solutions is a part of the learning process and is usually an ingredient in any textbook. On the other hand, complete solutions of some more special problems, as presented in Chap. 19, have at least three different functions. First of all, they contain a number of useful auxiliary results, which are usually proved in the main texts of more standard textbooks. Then, some solutions are examples of how to work with the notions and handle similar problems (there is a rich choice of problems without solutions in Chap. 16). Finally, some of the solutions presented in this chapter may be regarded as the last resort, when serious attempts to solve a problem were fruitless, or perhaps in order to compare one's own solution to a different one suggested in the book.

Chapter 17 contains detailed proofs of all the theorems used in Chaps. 1–15. Each of these chapters starts with an introductory part in which all relevant notions and theorems are formulated and discussed. The presentation of the proofs in a separate chapter is a convenient technical solution. In particular, it makes the presentation of the theoretical background of each chapter more transparent and clear. It also lays stress on the importance of examples and exercises in the learning process. However, the proofs should be studied in direct connection with each chapter, in particular, in those cases when the proof contains clues to practical constructions of those objects whose existence is stated in the theorem. A few of the theorems, which are usually a part of previous courses in algebra, are included in the Appendix. Some others, whose proofs can be considered as an application of Galois theory, are included in the exercises. Their proofs can be considered as optional.

Many chapters include a section on "using computers", which gives examples of how to apply the software package Maple to concrete problems in Galois theory. Maple has a very extensive library of algebraic functions, which helps to define fundamental algebraic structures and carry out different computations. Using this package creates many possibilities to supplement a lot of exercises, which contribute to a better understanding and are impossible without the assistance of a computer. There are other computer packages which can be used, like PARI/GP or SAGE, but the choice of Maple is due to the ease of the interface.

The Appendix at the end of the text is relatively long, and contains, in principle, all the necessary material corresponding to a first course in algebra, discussing groups, rings and fields. The purpose is to provide a convenient source for direct references to the background material necessary in any text on Galois theory. However, even if we give some proofs, there are only a few examples and no exercises, so the purpose is rather to recall the necessary definitions and some fundamental results in the form of a glossary for the convenience of the Reader who has forgotten parts of such a course and needs a short reminder. The Appendix also contains a few topics which are essential in this book, but are not a part of standard courses on groups, rings and fields.

Advice to the Reader

We start here with a very informal "introduction to Galois theory" in order to give some of the general ideas behind it. We want to use this as a guide to the text of the book, showing how to follow the book and how to choose among different directions.

Consider the polynomial equation $f(X) = X^4 - 10X^2 + 1 = 0$. It has four solutions $x_{1,2,3,4} = \pm\sqrt{2} \pm \sqrt{3}$ (note that $f(X) = (X^2 + 1)^2 - 12X^2 = (X^2 - 2\sqrt{3}X + 1)(X^2 + 2\sqrt{3}X + 1)$). The solutions define a field extension of the rational numbers $\mathbb{Q}$ (here the coefficients of $f(X)$ are rational). This extension is the smallest set of numbers which contains both the field of the coefficients $\mathbb{Q}$ and the solutions of the equation $f(X) = 0$. It is not difficult to check that it is the field $\mathbb{Q}(\sqrt{2}, \sqrt{3})$. This field is called a splitting field of $f(X)$. The first observation is that being able to express a solution in terms of the numbers from the field of the coefficients of $f(X)$ using the four arithmetical operations (addition, subtraction, multiplication, division) and extracting roots means that there is a formula for a solution (e.g. $x_1 =$

$\sqrt{2}+\sqrt{3}$). In this case, we have such a formula, since the splitting field is obtained from the rational numbers by adjoining roots of numbers belonging to the field containing the coefficients. This process of building "a formula" may be much more involved (it is possible that in a formula there are roots of other roots) but it is rather clear that the existence of an algebraic formula expressing a solution by means of the four arithmetical operations (addition, subtraction, multiplication, division) and extracting roots may be formulated in terms of suitable field extensions (such field extensions are called radical).

Galois studied all permutations of the solutions which do not change all rational relations among them. Rational relations are different polynomial equalities in x_i ($i = 1, 2, 3, 4$) with rational coefficients, which are true. Let $x_1 = \sqrt{2}+\sqrt{3}$, $x_2 = \sqrt{2}-\sqrt{3}$, $x_3 = -\sqrt{2}+\sqrt{3}$, $x_4 = -\sqrt{2}-\sqrt{3}$. As examples of such relations, we have $x_1 + x_4 = 0$, $x_1x_2 = -1$, and similarly, $x_2 + x_3 = 0$, $x_3x_4 = -1$. There are 24 permutations of x_1, x_2, x_3, x_4. They form the group S_4 of all permutations of four different symbols, here, the numbers x_i. But in order to simplify the notation, we will replace a permutation of x_1, x_2, x_3, x_4 by a suitable permutation of the indices $1, 2, 3, 4$. For example, the permutation x_2, x_1, x_4, x_3 will be denoted by

$$\begin{pmatrix} 1\ 2\ 3\ 4 \\ 2\ 1\ 4\ 3 \end{pmatrix}$$

or, shortly, $(1, 2)(3, 4)$. The symbol $(1, 2)$ denotes the permutation replacing x_1, x_2, x_3, x_4 by x_2, x_1, x_3, x_4. So let us take only those permutations which do not change the set of rational relations which we have found. It is rather evident that the permutations of x_1, x_2, x_3, x_4, which do not change the set of relations $x_1 + x_4 = 0$, $x_1x_2 = -1$, $x_2 + x_3 = 0$, $x_3x_4 = -1$ are $(1, 2)(3, 4)$, $(1, 3)(2, 4)$, $(1, 4)(2, 3)$ and of course, the identity permutation, which takes each solutions onto itself. The identity is usually denoted by (1). If we take any other of the 20 remaining permutations, for example, $(1, 2, 3, 4)$, which means that x_1, x_2, x_3, x_4 goes to x_2, x_3, x_4, x_1, then, for example, the relation $x_1 + x_4 = 0$ goes to $x_2 + x_1 = 0$, which is not true. The permutation $(1, 2)$ replaces $x_1 + x_4 = 0$ by $x_2 + x_4 = 0$ which is not true and so on. In fact, a short check shows that the four permutations which we have found are exactly those which take the set of the four relations into itself. They form a group with four elements: (1), $(1, 2)(3, 4)$, $(1, 3)(2, 4)$, $(1, 4)(2, 3)$. This is just the Galois group of the equation $X^4 - 10X^2 + 1 = 0$.

Galois related a finite group to each polynomial equation $f(X) = 0$ with coefficients in an arbitrary field (here the coefficients are in the field of rational numbers $\mathbb{Q}$). Any polynomial defines a field over the field of its coefficients generated by all its zeros (in a field which contains all zeros like the complex numbers when the field of the coefficients is the field of rational numbers). This field, called a splitting field of the polynomial over its field of the coefficients, contains just all information about the relations among the solutions of the equation $f(X) = 0$. In the case of the polynomial $f(X) = X^4 - 10X^2 + 1$, such a field is $\mathbb{Q}(\sqrt{2}, \sqrt{3})$. A splitting field defines the Galois group of the polynomial. The notion of a splitting field and its group makes the study of the allowed permutations (the elements of the Galois group) much more precise than the process described in the introductory example. Galois established a correspondence between the subgroups of the Galois group of a polynomial and the subfields of its splitting field. Using this correspondence, the existence of a formula for a solution of an equation was related to a very distinctive feature of the Galois group, which, in fact, most of the Galois groups do not have (the group is solvable—a property introduced by Galois). In that way, it is possible to give examples of polynomial equations whose solutions cannot be expressed by means of the four arithmetical operations (addition, subtraction, multiplication, division) and extracting roots—it suffices to find a polynomial whose Galois group is lacking the distinctive property of solvable groups.

Now we are in a position to describe the contents of the book and the possible ways through the topics discussed in different chapters. Their interdependencies are shown in the graph below.

As the unsolvability of some quintics and higher degree equations is chosen in this text as the main concrete motivating problem, we discuss polynomial equations in Chap. 1. It explains some methods used for solving polynomial equations of degree lower than 5 and gives a few historical facts concerning them. It is not necessary to pay too much attention to this chapter, which in mathematical sense is loosely related to Chap. 3. Moreover, many computer packages using symbolic algebra make it possible to obtain exact solutions of polynomial equations, when it is possible (at least up to degree 4). But the purpose of this chapter is rather to describe the historical development of Galois theory and to give a feeling of what should be expected from a general formula for the solution of an algebraic equation.

The main contents, covering a standard course in Galois theory, is presented in Chaps. 2–9. In Chap. 2, we fix the terminology concerning field extensions and discuss the notion of the characteristic of a field, which usually does not get so much attention in introductory courses in algebra. Chapter 3 is about polynomials. Essentially, we repeat several known facts about polynomials, but we concentrate on the notion of irreducibility, which is important when we construct splitting fields and want to investigate their properties. Chapter 4 treats fundamental properties of algebraic field extensions. In Chap. 5, we introduce the notion of the splitting fields of polynomials and investigate their most important properties. In Chap. 6, we define and study automorphism groups of field extensions. We call them Galois groups in order to simplify the terminology and formulations of exercises (but, sometimes in the literature, the notion of Galois group is restricted to Galois extensions of fields). A Galois extension is an extension which has two properties—it is normal and separable. These two properties are introduced and explained in Chaps. 7 (normal extensions) and 8 (separable extensions). Chapter 9 contains the main theorems of Galois theory. One of these theorems describes a correspondence between the subgroups of the Galois group of a polynomial (over its field of coefficients) and the subfields of its splitting field (over the same field). This "Galois correspondence" is one of the best known theoretical results of Galois theory, which has many generalizations and counterparts in many other mathematical situations.

After the path starting with Chap. 2 and ending with Chap. 9, there are different ways to continue. The shortest path to the solution of the motivating problem goes to Chap. 13, where there is a proof of the unsolvability of general quintics and the general equations of degree higher than 4. Chapter 13 needs some knowledge of solvable groups discussed in Chap. 12 (this chapter is only concerned with groups). Notice that in Chap. 13, we use Galois theory in order to give one of the proofs that the field of complex numbers is algebraically closed (see Exercise 13.14)—this result is usually called the fundamental theorem of algebra (see Chap. 1). A natural supplement to Chap. 13 is Chap. 15, which presents some practical methods for finding Galois groups of concrete polynomials as well as constructing polynomials with given Galois groups. The theorems on resolvents proved in this chapter are used in computer implementations of algorithms for computations of Galois groups. We also use them in a proof of a theorem of Richard Dedekind **T.15.4**, which gives a way to construct integer polynomials with the largest possible Galois group when the degree is fixed. Let us note that there is a very quick route to the unsolvability of quintics using Nagell's idea (see Exercise 13.6).

Another path from Chap. 9 goes to Chap. 10 and, possibly, to Chap. 11. In Chap. 10, we study cyclotomic fields. These extensions of the rational numbers by a root of unity are very good examples of how Galois theory works in practice. At the same time, the cyclotomic fields are very interesting and important for their applications in number theory. Chapter 10 has a natural extension in Chap. 11, which is the most "theoretical" part of the text. Here we study several interesting results which are related to different applications and extensions of Galois theory. The normal basis theorem (NBT) is about bases of Galois extensions, which are naturally related to the Galois groups and can be best formulated in terms of modules over groups. Hilbert's Theorem 90 is a famous result with many applications—one of them is a description of all cyclic Galois field extensions (over fields containing suitable roots of unity). Kummer's extensions, which are also discussed in this chapter, generalize Galois extensions with cyclic groups to arbitrary finite abelian groups (with similar assumptions about the presence of roots of unity). As one of the applications in Chap. 11, we give a full description of cubic and quartic Galois extensions over the rational numbers.

Finally, the third path from Chap. 9 goes to Chap. 14, in which Galois theory finds applications to classical problems concerning geometric straightedge-and-compass constructions. The truth is that most of this chapter could already be discussed after Chap. 4 with only a very modest knowledge of field extensions. In fact, geometric constructions are often discussed as an application in introductory courses in algebra. But there are some results for which a deeper knowledge of Galois theory is very helpful, in particular, when one wants to prove that some geometric constructions are possible (for example, in Gauss's theorem about regular polygons).

Mölnlycke, Sweden
January, 2018

Juliusz Brzeziński

Interrelations of Chaps. 1–15

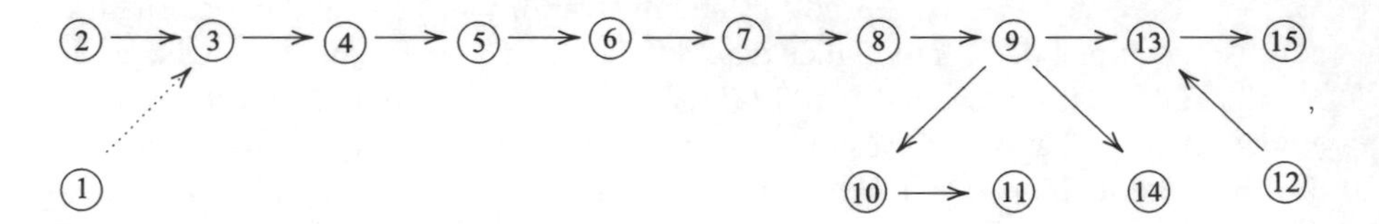

Contents

Chapter 1
Solving Algebraic Equations

Solving algebraic equations of low degrees was for a long time considered to be a very central mathematical activity. Quadratic equations and some very special equations of degree 3 had already been studied in antiquity, even though the notations for variables, numbers and arithmetical operations, as we know them today, were introduced much later (beginning in the middle of the sixteenth century). General equations of degrees 3 and 4 (that is, cubic and quartic equations) were solved during the sixteenth century mainly by Italian mathematicians. They described different methods of finding solutions and also gave formulae expressing the solution in terms of the four arithmetical operations (addition, subtraction, multiplication, division) and extracting roots applied to the coefficients of the equations. The next step was finding similar formulae for the fifth degree equations, which was regarded as an important mathematical problem for almost 300 years. In the eighteenth century, it was more and more clear that the situation in the case of quintic equations (equations of degree 5) may be different. It appeared that for an arbitrary quintic, it was probably not possible to express a solution in terms of the four arithmetical operations and extracting roots applied to the coefficients. These doubts motivated the works of many mathematicians during the eighteenth century and finally led to the results which are a part of the theory developed by Évariste Galois[1] in the beginning of the nineteenth century. One of these shows that indeed, in the general case, the quintic equations cannot be solved in a similar way as the equations of degree less than 5. At the end of this chapter, we add some more remarks about the history of this problem. However, let us note already now that the ideas developed by Galois have many very important consequences and applications in the whole mathematics and the primary cause of his work, that is, expressing solutions of polynomial equations in a particular way, is no longer a central issue.

The purpose of this chapter is to show how to solve equations of degree less than 5. We repeat the well-known formulae for the quadratic equation and show how to

[1] Évariste Galois, 25 October 1811–31 May 1832.

J. Brzeziński, *Galois Theory Through Exercises*, Springer Undergraduate Mathematics Series, https://doi.org/10.1007/978-3-319-72326-6_1

find similar formulae for the cubic and quartic equations. In fact, the description of the algorithms is very simple (and easy to remember), but practical computations may be time consuming if done by hand. There are several computer packages which can manage such computations (e.g. Magma, Maple, Mathematica, Pari/GP, Sage), but the idea here is that solving a few algebraic equations gives some feeling for what should be expected from a "general formula". Moreover, the possibility of finding the solutions will occasionally be useful in some of the exercises in the following chapters.

An **algebraic equation** of **degree** n with complex coefficients is an equation:

$$f(X) = a_nX^n + a_{n-1}X^{n-1} + \cdots + a_1X + a_0 = 0,$$

where $a_i \in \mathbb{C}$, $n \geq 0$ and $a_n \neq 0$ (if $n = 0, f(X)$ is a **constant polynomial**). The **fundamental theorem of** (polynomial) **algebra**, which is often associated with C.F. Gauss,[2] who published its proof in 1799 (but the history of this result is longer), says:

T.1.1 Fundamental Theorem of Algebra. *If $f(X) = a_nX^n + a_{n-1}X^{n-1} + \cdots + a_1X + a_0$ is a nonconstant polynomial of degree n with complex coefficients, then the equation $f(X) = 0$ has n complex solutions (roots) $x_1, \ldots, x_n$ and $f(X) = a_n(X - x_1)\cdots(X - x_n)$.*

Some of the x_i may be equal and the number of occurrences of x_i as a root of $f(X) = 0$ is called its **multiplicity**.

1.1 Quadratic Equations. A quadratic equation $f(X) = 0$ has two roots x_1, x_2 (in $\mathbb{C}$) and $f(X) = a_2X^2 + a_1X + a_0 = a_2(X - x_1)(X - x_2)$. Comparing the coefficients on the left and right in this equality, we get **Vieta's formulae**[3]:

$$x_1 + x_2 = -\frac{a_1}{a_2},$$

$$x_1x_2 = \frac{a_0}{a_2}.$$

With $\frac{a_1}{a_2} = p$ and $\frac{a_0}{a_2} = q$, we get an equivalent equation:

$$X^2 + pX + q = 0$$

and we solve it by transforming to an equation with roots $x_i + \frac{p}{2}$ whose sum is 0, that is, replacing the equation above by the equivalent expression in which $X^2 + pX$

[2]Johann Carl Friedrich Gauss, 30 April 1777–23 February 1855.

[3]François Viète (published his works under the name Franciscus Vieta), 1540 to 23 February 1603.

is completed to a square:

$$\left(X+\frac{p}{2}\right)^2+\left(q-\frac{p^2}{4}\right)=0.$$

Solving, we get the two solutions:

$$x_1=-\frac{p}{2}-\sqrt{\frac{p^2}{4}-q}\quad\text{and}\quad x_2=-\frac{p}{2}+\sqrt{\frac{p^2}{4}-q}.$$

The number $\Delta=p^2-4q=(x_1+x_2)^2-4x_1x_2=(x_1-x_2)^2$ is called the **discriminant** of the polynomial X^2+pX+q. It is nonzero if and only if the roots x_1, x_2 are different.

1.2 Cubic Equations. A cubic equation $f(X)=0$ has three roots x_1, x_2, x_3 (in $\mathbb{C}$) and $f(X)=a_3X^3+a_2X^2+a_1X+a_0=a_3(X-x_1)(X-x_2)(X-x_3)$. Comparing the coefficients on the left and right in this equality, we get Vieta's formulae:

$$\begin{aligned} x_1+x_2+x_3&=-\frac{a_2}{a_3},\\ x_1x_2+x_2x_3+x_1x_3&=\frac{a_1}{a_3},\\ x_1x_2x_3&=-\frac{a_0}{a_3}. \end{aligned}\tag{1.1}$$

The equation with roots $x_i+\frac{a_2}{3a_3}$, where $i=1,2,3$, has the coefficient of X^2 equal to 0, since the sum of these three numbers is zero. Technically, we get such an equation by substituting $X=Y-\frac{a_2}{3a_3}$ into the given cubic equation:

$$a_3X^3+a_2X^2+a_1X+a_0$$

$$=a_3\left(Y-\frac{a_2}{3a_3}\right)^3+a_2\left(Y-\frac{a_2}{3a_3}\right)^2+a_1\left(Y-\frac{a_2}{3a_3}\right)+a_0=a_3(Y^3+pY+q),$$

where p and q are easily computable coefficients (it is not necessary to remember these formulae, since it is easier to remember the substitution and to transform each time, when necessary). Thus, we can start with an equation:

$$X^3+pX+q=0.\tag{1.2}$$

We compare the last equality with the well-known identity:

$$(a+b)^3-3ab(a+b)-(a^3+b^3)=0\tag{1.3}$$

and choose a, b in such a way that:

$$p = -3ab,$$
$$q = -(a^3 + b^3). \tag{1.4}$$

If a, b are so chosen, then evidently $x = a + b$ is a solution of the equation (1.2). Since

$$a^3 + b^3 = -q,$$
$$a^3b^3 = -\frac{p^3}{27},$$

a^3, b^3 are solutions of the quadratic equation:

$$T^2 + qT - \frac{p^3}{27} = 0.$$

Solving this equation gives a^3, b^3. Then we choose a, b which satisfy (1.4) and get $x = a + b$. An (impressive) formula, which hardly needs to be memorized (it is easier to start "from the beginning", that is, from the identity (1.3), than to remember the formula below) is thus the following:

$$x_1 = a + b = \sqrt[3]{-\frac{q}{2} + \sqrt{\frac{q^2}{4} + \frac{p^3}{27}}} + \sqrt[3]{-\frac{q}{2} - \sqrt{\frac{q^2}{4} + \frac{p^3}{27}}}. \tag{1.5}$$

We leave a discussion of different choices of signs of the roots to the exercises, but notice that in order to solve a cubic equation, it is sufficient to find one root x_1 and solve a quadratic equation after dividing the cubic by $X - x_1$. The above expression (in a different notation) was first published by Girolamo Cardano[4] in his book "Ars Magna" in 1545 and is known today as **Cardano's formula**. However, the history surrounding the formula is very involved and several Italian mathematicians knew the method before Cardano (notably, Scipione del Ferro,[5] Antonio Fiore and Niccoló Tartaglia[6] about 20 years earlier than the appearance of Cardano's book).

The **discriminant** of the cubic (1.2) is $\Delta = -(4p^3 + 27q^2) = [(x_1 - x_2)(x_2 - x_3)(x_3 - x_1)]^2$. See Exercise 1.3 below.

1.3 Quartic Equations. Such an equation $f(X) = a_4X^4 + a_3X^3 + a_2X^2 + a_1X + a_0 = a_4(X - x_1)(X - x_2)(X - x_3)(X - x_4) = 0$ has four roots x_1, x_2, x_3, x_4 (in $\mathbb{C}$). Similarly as for quadratic and cubic equations, we can write down Vieta's formulae expressing

[4]Girolamo (Gerolamo, Geronimo) Cardano, 24 September 1501–21 September 1576.

[5]Scipione del Ferro, 6 February 1465–5 November 1526.

[6]Niccoló Fontana Tartaglia, 1499 or 1500 to 13 December 1557.

relations among the roots and the coefficients of the equation (we leave it as an exercise). Solving a quartic equation (we assume $a_4 = 1$ and denote the remaining coefficients avoiding indices):

$$X^4 + mX^3 + pX^2 + qX + r = 0, \tag{1.6}$$

it is not important to assume that $m = 0$ (but it could be achieved by substituting $X - \frac{m}{4}$ for X similarly as in the quadratic and cubic cases). We simply complement $X^4 + mX^3$ in such a way that we get a square of a quadratic polynomial and, when it is done, we find a parameter λ so that the quadratic polynomial in square brackets

$$\left(X^2 + \frac{m}{2}X + \frac{\lambda}{2}\right)^2 - \left[\left(\frac{m^2}{4} + \lambda - p\right)X^2 + \left(\frac{m\lambda}{2} - q\right)X + \left(\frac{\lambda^2}{4} - r\right)\right] = 0 \tag{1.7}$$

is a square of a first degree polynomial. This is achieved when the discriminant of the quadratic polynomial in the square bracket equals 0 (see Exercise 1.8), that is,

$$\left(\frac{m\lambda}{2} - q\right)^2 - 4\left(\frac{m^2}{4} + \lambda - p\right)\left(\frac{\lambda^2}{4} - r\right) = 0. \tag{1.8}$$

This is a cubic equation with respect to λ. Solving it, we get a root λ, which gives a way to factorize the quartic polynomial (1.7) into a product of two quadratic polynomials and then solve two quadratic equations. The method described above was given by L. Ferrari[7] in the middle of the sixteenth century and published in Cardano's book mentioned above.

As we noted earlier, many attempts to solve quintics (that is, fifth degree polynomial equations) were fruitless. During the eighteenth century some mathematicians already suspected that similar formulae as those for equations of degree less than 5 are impossible. It is not too difficult to see that the situation for equations of degree 5 differs from that of equations of lower degree (see Exercise 1.6). J.L. Lagrange[8] started to develop a general theory of algebraic equations. His analysis of different methods used in connection with the solution of algebraic equations led him to the notion of the resolvent polynomial and permutation groups related to it. He was convinced that the "old" methods used for solving equations of degree up to 4 cannot be used for equations of degree 5, which suggested that similar formulae may not exist. The work of Lagrange was a starting point for the work of other mathematicians, who finally proved that it is impossible to express solutions of general quintic equations using the four arithmetical operations and extracting roots

[7] Lodovico Ferrari, 2 February 1522–5 October 1565.

[8] Joseph-Louis Lagrange (born in Italy as Giuseppe Lodovico (Luigi) Lagrangia), 25 January 1736–10 April 1813.

applied to the coefficients. This was proved by Abel[9] in 1823 (earlier an incomplete proof was published by Ruffini[10] in 1799). Galois gave an independent proof of the same result using different methods, which we follow in Chap. 13.

Exercises 1

1.1. Solve the following equations using Cardano's or Ferrari's methods:

(a) $X^3 - 6X + 9 = 0$;
(b) $X^3 + 9X^2 + 18X + 28 = 0$;
(c) $X^3 + 3X^2 - 6X + 4 = 0$;
(d) $X^3 + 6X + 2 = 0$;
(e) $X^4 - 2X^3 + 2X^2 + 4X - 8 = 0$;
(f) $X^4 - 3X^3 + X^2 + 4X - 6 = 0$;
(g) $X^4 - 2X^3 + X^2 + 2X - 1 = 0$;
(h) $X^4 - 4X^3 - 20X^2 - 8X + 4 = 0$.

1.2. Show that the system (1.4) has exactly three solutions (a, b) and find all three solutions of the equation $X^3+pX+q = 0$ following the discussion above concerning Cardano's formula.

1.3. (a) Using Vieta's formulae (1.1) show that the discriminant $\Delta(f) = [(x_1 - x_2)(x_2 - x_3)(x_3 - x_1)]^2$ of $f(X) = X^3 + pX + q$ is equal to $\Delta(f) = -(4p^3 + 27q^2)$ (x_1, x_2, x_3 denote the zeros of $f(X)$). Note that the equation $X^3 + pX + q = 0$ has multiple roots if and only if $\Delta = 0$.
(b) Assume that p, q are real numbers. Show that the equation $f(X) = X^3+pX+q = 0$ has three different real roots if and only if $\Delta(f) > 0$.

1.4. The discriminant of the polynomial $X^3 - 6X + 4$ equals $\Delta = 432$, so the equation $X^3 - 6X + 4 = 0$ has three different real roots (see Exercise 1.3). The formula (1.5) gives

$$x_1 = \sqrt[3]{-2 + 2i} + \sqrt[3]{-2 - 2i}.$$

Find the solutions of the equation without using Cardano's formula (the number 2 is one of them!) and identify these solutions with the expressions given by Cardano's formula.

Remark. The case of the cubic polynomials having three real zeros in the context of Cardano's formula was a serious problem for mathematicians for at least 300 years. As $\Delta = -(4p^3 + 27q^2) > 0$, we have $\frac{q^2}{4} + \frac{p^3}{27} < 0$ in the formula (1.5), so in order to compute the real number x_1, we have to manipulate with complex numbers on the right-hand side. Cardano's formula was considered "not correct" as the real values (even integer values) of the zeros are expressed in terms of complex numbers, which were considered with great suspicion. Mathematician tried to find

[9]Nils Henrik Abel, 5 August 1802–6 April 1829.

[10]Paulo Ruffini, 22 September 1765–10 May 1822.

"better" formulae in which negative numbers under square roots do not appear, in order to eliminate what was known as the "**Casus irreducibilis**". First, during the nineteenth century, it became clear that it is impossible to find "better" formulae—in general, it is impossible to express the real zeros of an irreducible cubic with three real zeros by so-called real radicals, that is, without help of complex numbers. We explain this phenomenon in more detail in Chap. 13 (see **T.13.3**).

1.5. Assume that a cubic equation $X^3 + pX + q = 0$ has three real solutions, that is, $\Delta = -(4p^3 + 27q^2) > 0$. Show that there exists $r \in \mathbb{R}$ and $\varphi \in [0, 2\pi]$ such that the solutions of the cubic equation are

$$x_1 = 2\sqrt[3]{r}\cos\frac{\varphi}{3}, \qquad x_2 = 2\sqrt[3]{r}\cos\frac{\varphi + 2\pi}{3}, \qquad x_3 = 2\sqrt[3]{r}\cos\frac{\varphi + 4\pi}{3}.$$

1.6. (a) Find a cubic equation having a solution $\alpha = \sqrt[3]{a + \sqrt{b}} + \sqrt[3]{a - \sqrt{b}}$, where $a^2 - b = c^3$ and $a, b, c \in \mathbb{Q}$. Write down all solutions of the equation you have found and show that a solution to any cubic equation $X^3 + pX + q = 0$, where p, q are rational numbers, can be written as α above.
(b) Find a quintic equation (that is, an equation of degree 5) having a solution $\alpha = \sqrt[5]{a + \sqrt{b}} + \sqrt[5]{a - \sqrt{b}}$, where $a^2 - b = c^5$ and $a, b, c \in \mathbb{Q}$. Choose a, b in such a way that the polynomial having α as its zero is irreducible over $\mathbb{Q}$. Does every quintic equation with rational coefficients have a solution of the form α above?

Remark. The quintic equation, one of whose zeros is α in (b) above (see p. 152), is sometimes called **de Moivre's quintic**. Of course, we have a formula giving a solution of such an equation, since we construct it starting from a solution. But our objective is to prove that there are quintic equations with rational coefficients for which it is impossible to express a solution using the four arithmetical operations (addition, subtraction, multiplication, division) and extracting roots applied to the coefficients of the equation. Here we see that it is sometimes possible to express a solution in this way even if the polynomial cannot be factored (over the rational numbers) into a product of polynomials of degrees less than 5. In fact, there are very simple examples of polynomials of degree 5, which cannot be factored over the rational numbers but whose zeros can be expressed in terms of radicals, like $X^5 - 2$, but comparing (a) and (b) gives an idea of what can make a difference: there exists a formula for all cubic equations, while there is no such a formula for all quintic equations.

1.7. Solve the binomial equation $X^n - a = 0$ of degree $n > 0$, where $a \in \mathbb{C}$.

1.8. Show that a given trinomial $aX^2 + bX + c \in \mathbb{C}[X]$ $(a \neq 0)$ is a square of a binomial $pX + q \in \mathbb{C}[X]$ if and only if the discriminant of the trinomial $b^2 - 4ac = 0$.

Using Computers 1

Maple makes it possible to find solutions of equations of low degrees in closed form (expressed in terms of arithmetical operations on coefficients and radicals). Thus all equations of degree at most 4 and some other simple equations (like the equations $f(X^n) = 0$ when f has degree at most 3) can be solved using Maple. The solutions are obtained by:

```
>solve(f(X))
```

in order to solve the equation $f(X) = 0$. The numerical real values of zeros can be obtained using `fsolve(f(X))` and all values by `fsolve(f(X),complex)`. Of course, you can use these commands in order to find the zeros of the equations, for example, in Exercise 1.1, but the intention in this chapter is rather to get some experience and feeling of raw computations before the computer era. Nevertheless, if you want to get exact values of all zeros (when such values are possible to give and the suitable procedures are implemented), then it is possible to use the command `solve` as above or the following commands, which give a greater flexibility already for polynomials of degree 4 and which we explain through an example (see Exercise 1.1(h)):

```
>alias(a=RootOf(X^4-4*X^3-20*X^2-8*X+4))
```

$$a$$

```
>allvalues(a)
```

$$1+\sqrt{7}-\sqrt{6+2\sqrt{7}},\ 1+\sqrt{7}+\sqrt{6+2\sqrt{7}},\ 1-\sqrt{7}+\sqrt{6-2\sqrt{7}},\ 1-\sqrt{7}-\sqrt{6-2\sqrt{7}}$$

Notice that the command `alias(a=RootOf(f(X)))` defines one of the zeros of $f(X)$. With the command `allvalues(a)`, we obtain all possible values. If we want to identify the value denoted by a, we can compare it with all possible values. So, for example, `allvalues(a)[1]` gives the first one and we can consider the difference between it and a defined by `alias(a=RootOf(f(X)))`. If it is 0, then a denotes just this value.
It is also possible to obtain the discriminants of the polynomials (see Exercise 1.3, and for general definition A.10.2) identifying the variable with respect to which the discriminant should be computed:

```
>discrim(X^3+pX+q,X)
```

$$-4p^3 - 27q^2$$

but it may be instructive to try to obtain this result for cubic equations in Exercise 1.3 without the use of a computer.

Chapter 2
Field Extensions

In Galois theory, polynomial equations are studied through fields, which they define. Such fields will be discussed in more detail in Chap. 5. In this chapter, we repeat some general notation and properties of fields (like the notion of characteristic), which are important in Galois theory, but are not always sufficiently stressed in introductory courses on fundamental algebraic structures like groups, rings and fields. A short introduction to groups, rings and fields is presented in the Appendix. An important point here are the different notations concerning fields, which we fix in this chapter for the rest of the book.

Fields will be denoted by capital letters, preferably, K, L, M. But on a very few occasions, we also use lower case letters, like k, l, when the need arises. If a field K is a subfield of a field L, then we say that $K \subseteq L$ is a **field extension**. If M is a field such that $K \subseteq M \subseteq L$, then M is called an **intermediate field** (of the field extension $K \subseteq L$). If K_i, $i \in I$ (I an index set) are subfields of L, then the intersection $\cap K_i$, $i \in I$, is also a subfield of L. If $L \supseteq K$ and X is a subset of L, then $K(X)$ denotes the intersections of all subfields of L, which contain both K and X ($K(X)$ is the smallest subfield of L containing both K and X). If $X = \{\alpha_1, \ldots, \alpha_n\}$, then we usually write $K(X) = K(\alpha_1, \ldots, \alpha_n)$ and the elements $\alpha_1, \ldots, \alpha_n$ are called **generators** of this field over K. If $X = K'$ is a subfield of L, then $K(K')$ is denoted by KK' and called the **compositum** of K and K'. If K and K' are two fields and $\sigma : K \to K'$ is a bijective function such that $\sigma(x + y) = \sigma(x) + \sigma(y)$ and $\sigma(xy) = \sigma(x)\sigma(y)$ for $x, y \in K$, then σ is called an **isomorphism** (of fields K, K'). When such a function exists, the fields are called **isomorphic**, which is denoted by $K \cong K'$.

Every field is an extension of the smallest field contained in it—the intersection of all its subfields. Such an intersection does not contain any smaller fields. A field without any smaller subfields (one says often **proper subfields**) is called a **prime field**.

The field of rational numbers $\mathbb{Q}$ is the only infinite prime field (up to isomorphism of fields). For every prime number p, there is a finite prime field $\mathbb{F}_p$ consisting of the residues $0, 1, \ldots, p-1$ of the integers $\mathbb{Z}$ modulo p, which are added and multiplied

J. Brzeziński, *Galois Theory Through Exercises*, Springer Undergraduate Mathematics Series, https://doi.org/10.1007/978-3-319-72326-6_2

modulo p (the field $\mathbb{F}_p$ is the quotient $\mathbb{Z}/p\mathbb{Z}$ of $\mathbb{Z}$ modulo the ideal generated by the prime number p—see section "Fields" in Appendix for more details).

By the **characteristic** of a field K, we mean the number 0 if the prime subfield of K is infinite and the number of elements in the prime subfield of K if it is finite. The characteristic of K is denoted by $\mathrm{char}(K)$. In this chapter, we record only one fact on the characteristic of fields, the proof of which is given on p. 256 (see A.5.1):

T.2.1. (*a*) *The characteristic of a field is* 0 *or a prime number.*
(*b*) *Any field K of characteristic* 0 *contains a unique subfield isomorphic to the rational numbers $\mathbb{Q}$, and any field of characteristic p contains a unique subfield isomorphic to $\mathbb{F}_p$.*

Exercises 2

2.1. Which of the following subsets of $\mathbb{C}$ are fields with respect to the usual addition and multiplication of numbers:

(a) $\mathbb{Z}$;
(b) $\{0, 1\}$;
(c) $\{0\}$;
(d) $\{a + b\sqrt{2},\ a, b \in \mathbb{Q}\}$;
(e) $\{a + b\sqrt[3]{2},\ a, b \in \mathbb{Q}\}$;
(f) $\{a + b\sqrt[4]{2}, a, b \in \mathbb{Q}\}$;
(g) $\{a + b\sqrt{2},\ a, b \in \mathbb{Z}\}$;
(h) $\{z \in \mathbb{C} : |z| \leq 1\}$?

2.2. Show that every subfield of $\mathbb{C}$ contains $\mathbb{Q}$.

2.3. Give an example of an infinite field of characteristic $\neq 0$.

2.4. Show that the characteristic of a field K is the least positive integer n such that $na = 0$ for each $a \in K$ or it is 0 if such n does not exist.

2.5. (a) Let $L \supseteq K$ be a field extension and let $\alpha \in L \setminus K$, $\alpha^2 \in K$. Show that

$$K(\alpha) = \{a + b\alpha, \text{ where } a, b \in K\}.$$

(b) Let $K(\sqrt{a})$ and $K(\sqrt{b})$, where $a, b \in K, ab \neq 0$, be two field extensions of K. Show that $K(\sqrt{a}) = K(\sqrt{b})$ if and only if ab is a square in K (that is, there is a $c \in K$ such that $ab = c^2$).

2.6. Give a description of all numbers belonging to the fields:

(a) $\mathbb{Q}(\sqrt{2})$; (b) $\mathbb{Q}(i)$; (c) $\mathbb{Q}(\sqrt{2}, i)$; (d) $\mathbb{Q}(\sqrt{2}, \sqrt{3})$.

2.7. Show that:

(a) $\mathbb{Q}(\sqrt{5}, i\sqrt{5}) = \mathbb{Q}(i, \sqrt{5})$;
(b) $\mathbb{Q}(\sqrt{2}, \sqrt{6}) = \mathbb{Q}(\sqrt{2}, \sqrt{3})$;

(c) $\mathbb{Q}(\sqrt{5}, \sqrt{7}) = \mathbb{Q}(\sqrt{5} + \sqrt{7})$;
(d) $\mathbb{Q}(\sqrt{a}, \sqrt{b}) = \mathbb{Q}(\sqrt{a} + \sqrt{b})$, when $a, b \in \mathbb{Q}$, $\sqrt{a} + \sqrt{b} \neq 0$.

2.8. Give a description of the following subfields of $\mathbb{C}$:

(a) $\mathbb{Q}(X)$, where $X = \{\sqrt{2}, 1 + 2\sqrt{8}\}$;
(b) $\mathbb{Q}(i)(X)$, where $X = \{\sqrt{2}\}$;
(c) K_1K_2, where $K_1 = \mathbb{Q}(i)$, $K_2 = \mathbb{Q}(\sqrt{5})$;
(d) $\mathbb{Q}(X)$, where $X = \{z \in \mathbb{C} : z^4 = 1\}$.

2.9. Let K be a field.
(a) Show that all the matrices $\begin{bmatrix} a & b \\ -b & a \end{bmatrix}$, where $a, b \in K$, form a field L with respect to matrix addition and matrix multiplication if and only if the equation $X^2 + 1 = 0$ has no solutions in K.
(b) Show that L contains a field isomorphic to K.
(c) Use (a) in order to construct a field with nine elements and find its characteristic.

Remark. We discuss finite fields in Chap. 5, where the general construction of finite fields is given.

2.10. Let K be a field.
(a) Formulate a suitable condition such that all matrices $\begin{bmatrix} a & b \\ -b & a-b \end{bmatrix}$, $a, b \in K$, form a field with respect to matrix addition and multiplication.
(b) Use (a) in order to construct a field with four elements and write down the addition and the multiplication tables for the elements in this field.

2.11. In a field K the equality $a^4 = a$ is satisfied for all $a \in K$. Find the characteristic of the field K.

2.12. Let K be a field of characteristic p.
(a) Show that $(a+b)^{p^m} = a^{p^m} + b^{p^m}$ when $a, b \in K$ and m is a natural number.
(b) Let p divide a positive integer n. Show that $a^n + b^n = (a^{\frac{n}{p}} + b^{\frac{n}{p}})^p$ when $a, b \in K$.
(c) Define $\varphi : K \to K$ such that $\varphi(x) = x^p$. Show that the image of φ, which we denote by K^p, is a subfield of K.

2.13. Let $K \subseteq L$ be a field extension and let M_1, M_2 be two fields containing K and contained in L.
(a) Show that M_1M_2 consists of all quotients of finite sums $\sum \alpha_i\beta_i$, where $\alpha_i \in M_1$ and $\beta_i \in M_2$;
(b) Show that for each $x \in M_1M_2$ there are $\alpha_1, \ldots, \alpha_m \in M_1$ and $\beta_1, \ldots, \beta_n \in M_2$ such that $x \in K(\alpha_1, \ldots, \alpha_m, \beta_1, \ldots, \beta_n)$.

Chapter 3
Polynomials and Irreducibility

In this chapter, we gather a few facts concerning zeros of polynomials and discuss some simple methods which help to decide whether a polynomial is irreducible or reducible. In many practical situations, this information is essential in the study of field extensions. A short presentation of polynomial rings can be found in the Appendix (see section "Polynomial Rings").

Let K be a field and let $f(X) = a_nX^n + \cdots + a_1X + a_0$ be a polynomial with coefficients $a_i \in K$. If $a_n \neq 0$, then n is called the **degree** of $f(X)$ and is denoted by $\deg f$. A polynomial of degree 0 is called a **constant polynomial**. The coefficient a_n is called the leading coefficient and $f(X)$ is called **monic** if $a_n = 1$. The **zero polynomial** is the polynomial with all coefficients equal to 0. Its degree is usually not defined, but sometimes it is defined as -1 or $-\infty$. For practical reasons (in proofs by induction), we shall assume that the degree of the zero polynomial is -1. The set of all polynomials with coefficients in K is denoted by $K[X]$. It is a ring with respect to addition and multiplication of polynomials (see section "Polynomial Rings" in Appendix).

If L is a field containing K, then we say that $\alpha \in L$ is a **zero** of $f \in K[X]$ if $f(\alpha) = 0$.

T.3.1 Factor Theorem. *Let $K \subseteq L$ be a field extension.*

(*a*) *The remainder of $f \in K[X]$ divided by $X - a$, $a \in L$, is equal to $f(a)$.*
(*b*) *An element $a \in L$ is a zero of $f \in K[X]$ if and only if $X - a | f(X)$ (in $L[X]$).*

A polynomial $f \in K[X]$ is **reducible** (in $K[X]$ or over K) if $f = gh$, where $g, h \in K[X]$, $\deg g \geq 1$ and $\deg h \geq 1$. A nonconstant polynomial which is not reducible is called **irreducible**. Of course, every polynomial of degree 1 is irreducible.

J. Brzeziński, *Galois Theory Through Exercises*, Springer Undergraduate Mathematics Series, https://doi.org/10.1007/978-3-319-72326-6_3

In the polynomial rings over fields, every nonconstant polynomial is a product of irreducible factors and such a product is unique in the following sense:

T.3.2. *Let K be a field. Every polynomial of degree ≥ 1 in $K[X]$ is a product of irreducible polynomials. If*

$$f = p_1 \cdots p_k = p'_1 \cdots p'_l,$$

where p_i and p'_i are irreducible polynomials, then $k = l$ and with a suitable numbering of the factors p_i, p'_j, we have $p'_i = c_i p_i$, where $c_i \in K$.

We gather a few facts about irreducible polynomials in different rings. More detailed information is given in the exercises.

In the ring $\mathbb{C}[X]$, the only irreducible polynomials are exactly the polynomials of degree 1. This is the content of the fundamental theorem of (polynomial) algebra (see **T.1.1**), which implies that if $f(X) \in \mathbb{C}[X]$ is a polynomial of degree $n > 1$, then $f(X)$ is reducible. In fact, we have $f(X) = c(X - z_1) \cdots (X - z_n)$, where $c \in \mathbb{C}$ and $z_i \in \mathbb{C}$ are all zeros of $f(X)$ (with suitable multiplicities).

In the ring $\mathbb{R}[X]$ the irreducible polynomials are those of degree 1 together with the polynomials of degree 2 of the form $c(X^2 + pX + q)$ such that $\Delta = p^2 - 4q < 0$ and $c \in \mathbb{R}, c \neq 0$. This follows easily from the description of the irreducible polynomials in $\mathbb{C}[X]$ (see Exercise 3.5).

In the ring $\mathbb{Q}[X]$ of polynomials with rational coefficients, there are irreducible polynomials of arbitrary degree. For example, the polynomials $X^n - 2$ are irreducible for every $n \geq 1$ (see Exercise 3.7). In order to prove that a polynomial with integer coefficients is irreducible in $\mathbb{Q}[X]$ (which is a common situation in many exercises), it is convenient to study the factorization of this polynomial in $\mathbb{Z}[X]$ in combination with Gauss's Lemma or reductions of the polynomial modulo suitable prime numbers (see Exercise 3.8).

T.3.3 Gauss's Lemma. *A nonconstant polynomial with integer coefficients is a product of two nonconstant polynomials in $\mathbb{Z}[X]$ if and only if it is reducible in $\mathbb{Q}[X]$. More precisely, if $f \in \mathbb{Z}[X]$ and $f = gh$, where $g, h \in \mathbb{Q}[X]$ are nonconstant polynomials, then there are rational numbers r, s such that $rg, sh \in \mathbb{Z}[X]$ and $rs = 1$, so $f = (rg)(sh)$.*

Remark. Gauss's Lemma is true for every principal ideal domain R and its quotient field K instead of $\mathbb{Z}$ and $\mathbb{Q}$. The proof of this more general version is essentially the same as the proof of the above version on p. 110 (for a general version and its proof, see [L], Chap. IV, §2).

If $f(X) = a_nX^n + \cdots + a_1X + a_0$ is a polynomial with integer coefficients a_i for $i = 0, 1, \ldots, n$, then its **reduction modulo a prime number** p is the polynomial $\bar{f}(X) = \bar{a}_nX^n + \cdots + \bar{a}_1X + \bar{a}_0$ whose coefficients are the residues of a_i modulo p. We denote the reduction modulo p by $f(X) \pmod{p}$ (or simply write $\bar{f}$ when p is clear from the context). The reduction modulo p is a ring homomorphism of the polynomial ring $\mathbb{Z}[X]$ onto the polynomial ring $\mathbb{F}_p[X]$.

Exercises 3

3.1. (a) Show that a polynomial $f \in K[X]$ of degree 2 or 3 is reducible in $K[X]$ if and only if f has a zero in K, that is, there is a $x_0 \in K$ such that $f(x_0) = 0$.
(b) Show that the polynomial $X^4 + 2X^2 + 9$ over the rational numbers $\mathbb{Q}$ has no rational zeros, but is reducible in $\mathbb{Q}[X]$ (notice that $\mathbb{Q}$ may be replaced by $\mathbb{R}$).

3.2. Make a list of all irreducible polynomials of degrees 1 to 5 over the field $\mathbb{F}_2$ with 2 elements.

3.3. Show that if a rational number $\frac{p}{q}$, where p, q are relatively prime integers, is a solution of an equation $a_nX^n + \cdots + a_1X + a_0 = 0$ with integer coefficients a_i, $i = 0, 1, \ldots, n$, then $p|a_0$ and $q|a_n$.

3.4. Factorize the following polynomials as a product of irreducible polynomials:
(a) $X^4 + 64$ in $\mathbb{Q}[X]$; (e) $X^3 - 2$ in $\mathbb{Q}[X]$;
(b) $X^4 + 1$ in $\mathbb{R}[X]$; (f) $X^6 + 27$ in $\mathbb{R}[X]$;
(c) $X^7 + 1$ in $\mathbb{F}_2[X]$; (g) $X^3 + 2$ in $\mathbb{F}_3[X]$;
(d) $X^4 + 2$ in $\mathbb{F}_{13}[X]$; (h) $X^4 + 2$ in $\mathbb{F}_3[X]$.

3.5. (a) Show that if a real polynomial $f(X) \in \mathbb{R}[X]$ has a complex zero z, then the conjugate $\bar{z}$ is also a zero of $f(X)$.
(b) Show that a real polynomial of degree at least 1 is a product of irreducible polynomials of degrees 1 or 2.

3.6. Using Gauss's Lemma show that the following polynomials are irreducible in $\mathbb{Q}[X]$:

(a) $X^4 + 8$; (b) $X^4 - 5X^2 + 2$; (c) $X^4 - 4X^3 + 12X^2 - 16X + 8$.

3.7. (a) Prove **Eisenstein's**[1] **criterion**: Let $f(X) = a_nX^n + \cdots + a_1X + a_0 \in \mathbb{Z}[X]$ and let p be a prime number such that $p|a_0, p|a_1, \ldots, p|a_{n-1}, p \nmid a_n$ and $p^2 \nmid a_0$. Show that $f(X)$ is irreducible in $\mathbb{Z}[X]$.
(b) Show that the polynomials $X^n - 2$ for $n = 1, 2, \ldots$ are irreducible over $\mathbb{Q}$.
(c) Let p be a prime number. Show that the polynomials $f(X) = p^{n-1}X^n + pX + 1$ are irreducible over $\mathbb{Q}$ for every positive integer n.
(d) Show that the polynomial $f(X) = \frac{X^p-1}{X-1} = X^{p-1} + \cdots + X + 1$ is irreducible over $\mathbb{Q}$ for every prime number p.

3.8. (a) Show that if a reduction of a monic polynomial $f(X) \in \mathbb{Z}[X]$ modulo a prime number p is irreducible as a polynomial in $\mathbb{F}_p[X]$, then $f(X)$ is irreducible in $\mathbb{Q}[X]$.
(b) Show that if a monic polynomial $f(X) \in \mathbb{Z}[X]$ has reductions modulo two prime numbers p, q such that $f(X) \pmod{p}$ is a product of two irreducible factors of

[1] Ferdinand Gotthold Max Eisenstein, 16 April 1823–11 October 1852.

degrees d_{1p}, d_{2p} and $f(X) \pmod q$ is a product of two irreducible factors of degrees d_{1q}, d_{2q} and $\{d_{1p}, d_{2p}\} \neq \{d_{1q}, d_{2q}\}$, then $f(X)$ is irreducible in $\mathbb{Q}[X]$.
(c) Using (a) or (b) show that the following polynomials are irreducible in $\mathbb{Q}[X]$:
(c_1) $X^4 + X + 1$; (c_2) $X^5 + X^2 + 1$; (c_3) $X^4 + 3X + 4$;
(c_4) $X^5 + 3X + 1$; (c_5) $X^6 + 5X^2 + X + 1$; (c_6) $X^7 + X^4 + X^2 + 1$.

3.9. (a) Let $f(X) = X^4 + pX^2 + qX + r$ be a polynomial with coefficients in a field K. Show that $f(X) = (X^2 + aX + b)(X^2 + a'X + b')$, where a, b, a', b' are in some field containing K, if and only if a^2 is a solution of the equation $r(f)(T) = T^3 + 2pT^2 + (p^2 - 4r)T - q^2 = 0$.
(b) Show that $f(X)$ in (a) is a product of two quadratic polynomials over K if and only if
($*$) $q \neq 0$ and the resolvent $r(f)$ has a zero, which is a square in K or
($**$) $q = 0$ and the resolvent $r(f)$ has two zeros, which are squares in K or $\delta = p^2 - 4r$ is a square in K.

Remark. The polynomial $r(f)(T) = T^3 + 2pT^2 + (p^2 - 4r)T - q^2$ is usually called the **resolvent** of $f(X) = X^4 + pX^2 + qX + r$ and, in fact, it is a resolvent in the general sense of this notion in (15.2) (see Exercise 15.4). This exercise gives a somewhat different method of solving quartic equations than the method described in Chap. 1 on p. 4: It is possible to factorize the polynomial $f(X)$ by solving the cubic equation $r(f)(T) = 0$ (this gives a, a', b, b'). The zeros of $f(X)$ are the zeros of the two quadratic factors of $f(X)$.

3.10. Show that over every field there exists infinitely many irreducible polynomials.

3.11. Show that if a monic polynomial with rational coefficients divides a monic polynomial with integer coefficients, then it also has integer coefficients.

Using Computers 3

Many program packages provide implementation of algorithms giving a way to factorize polynomials over the rational numbers $\mathbb{Q}$ or over finite fields $\mathbb{F}_p$. In Maple, it is possible to construct finite field extensions of $\mathbb{Q}$ and $\mathbb{F}_p$ and to factor polynomials over them, but we postpone our general discussion of field extensions to the next chapter. The command `factor(f(X))` gives a way to factor a polynomial $f(X) \in \mathbb{Q}[X]$ over $\mathbb{Q}$ or to establish its irreducibility. If we want to know the same over the real or complex numbers, we can use `factor(f(X),real)`, respectively `factor(f(X),complex)`. The command `irreduc(f(X))` gives `true` or `false` depending on whether $f(X)$ is irreducible or not over the field of rational numbers. The same over the real numbers is achieved by `irreduc(f(X),real)`.

We use the command `Factor(f(X))` **mod** p, where $f(X)$ is any polynomial with integer coefficients, in order to factorize polynomials over a finite field $\mathbb{F}_p$. Observe the usage of the capital letter F in the last command. For example,

```
>Factor(X^21+X+1) mod 2
```

$$(X^7 + X^5 + X^3 + X + 1)(X^{14} + X^{12} + X^7 + X^6 + X^4 + X^3 + 1)$$

The command `Irreduc(f(X))` **mod** p tests whether $f(X)$ is irreducible over the finite field $\mathbb{F}_p$. For example, the following command lists all quadratic irreducible polynomials over the field $\mathbb{F}_5$:

```
>for i from 0 to 4 do for j from 0 to 4 do
  if Irreduc(x^2+i*x+j) mod 5 = true then
  print(x^2+i*x+j) fi od od
```

3.12. List all monic irreducible polynomials of degree less than 4 over the field $\mathbb{F}_3$.

3.13. Verify the results of Exercise 3.1(b), 3.4 and 3.8(c) using Maple.

Chapter 4
Algebraic Extensions

The purpose of this chapter is to describe a kind of field extension which is commonly used in Galois theory. These are the algebraic extensions, that is, the extensions $K \subseteq L$ in which each element $\alpha \in L$ is a zero of a nontrivial polynomial with coefficients in K. We relate the elements of such extensions to the corresponding polynomials and look at the structure of the simplest extensions $K(\alpha)$ of K. We also introduce the notion of the degree of a field extension and prove some of its properties. The notions introduced here and the theorems of this chapter are very important and we will often refer to them in the book.

Let $K \subseteq L$ be a field extension. We say that an element $\alpha \in L$ is **algebraic** over K if there is a nonzero polynomial $f \in K[X]$ such that $f(\alpha) = 0$. If such a polynomial does not exist, α is called **transcendental** over K. If $\alpha \in \mathbb{C}$ is algebraic over $\mathbb{Q}$, then we say that α is an **algebraic number**. A number $\alpha \in \mathbb{C}$ which is not algebraic is called **transcendental**. If $\alpha \in L$ is algebraic over K, then any nonzero polynomial of the least possible degree among all the polynomials $f \in K[X]$ such that $f(\alpha) = 0$ is called a **minimal polynomial** of α over K. The degree of f is called the **degree** of α over K. In the first theorem of this chapter, we characterize all polynomials having α as its zero and show, as a consequence, that there exists one polynomial with this property, which is in a sense the best one—it is **the minimal polynomial** of an algebraic element over K:

T.4.1. *Let $\alpha \in L \supseteq K$ be algebraic over K.*
(a) Any minimal polynomial of α over K is irreducible and divides every polynomial in $K[X]$ which has α as its zero.
(b) An irreducible polynomial $f \in K[X]$ such that $f(\alpha) = 0$ is a minimal polynomial of α over K.
(c) All minimal polynomials of α over K can be obtained by multiplying one of them by nonzero elements of K.

Using the last statement, we usually choose **the minimal polynomial**, which is the unique minimal polynomial whose leading coefficient equals 1. The degree of any

J. Brzeziński, *Galois Theory Through Exercises*, Springer Undergraduate Mathematics Series, https://doi.org/10.1007/978-3-319-72326-6_4

minimal polynomial of α over K is called the **degree** of α over K. An extension $K(\alpha)$ of K is called **simple** and α is called its **generator**.

Every field L containing a given field K can be considered as a vector space over K. By the **degree** $[L : K]$, we mean the dimension of the linear space L over K. If this dimension is not finite, we write $[L : K] = \infty$. Our next theorem gives a complete description of the simple extensions of a given field K. The theorem shows that such an extension $K(\alpha)$ of K has a finite dimension over K exactly when α is algebraic over K.

T.4.2 Simple Extension Theorem. *(a) If $\alpha \in L \supseteq K$ is algebraic over K, then each element in $K(\alpha)$ can be uniquely represented as $b_0 + b_1\alpha + \cdots + b_{n-1}\alpha^{n-1}$, where $b_i \in K$ and n is the degree of the minimal polynomial $p(X)$ of α over K. Thus $[K(\alpha) : K] = n$ and $1, \alpha, \ldots, \alpha^{n-1}$ is a basis of $K(\alpha)$ over K.*
(b) If $\alpha \in L \supseteq K$ is transcendental over K, then $K[\alpha]$ is isomorphic to the polynomial ring $K[X]$ by an isomorphism mapping X to α.

Notice that the first part of **T.4.2** says that $[K(\alpha) : K] = \deg(p)$, when α is algebraic and p its minimal polynomial. The second part says that when α is not algebraic, then the powers of α correspond to the powers of a variable X generating a polynomial ring over K. This means that these powers are linearly independent and the dimension of $K(\alpha)$ over K as a vector space is infinite. As we noted before, in this book we are principally interested in algebraic extensions of fields.

The next theorem gives a relation between the degrees of consecutive finite field extensions. It is very useful and we often refer to it as the "Tower Law":

T.4.3 Tower Law. *Let $K \subseteq L$ and $L \subseteq M$ be finite field extensions. Then $K \subseteq M$ is a finite field extension and $[M : K] = [M : L][L : K]$.*

We meet many types of field extensions in this book. It is very important to understand what all of the adjectives related to the extensions mean. In this chapter, we have three types of extensions: An extension $L \supseteq K$ is called **algebraic** if every element in L is algebraic over K. It is called **finite** if the dimension of L over K is finite (that is, $[L : K] \neq \infty$). The extension $L \supseteq K$ is **finitely generated** if $L = K(\alpha_1, \ldots, \alpha_n)$, where $\alpha_i \in L$, that is, there is a finite number of elements of L which generate it over K (see the notation on p. 9). A relation between these three notions is described by the following theorem:

T.4.4. *A field extension $L \supseteq K$ is finite if and only if it is algebraic and finitely generated.*

Very often it is interesting to know what happens to a property of field extensions when an extension of some type is followed by one more extension of the same type. From this point of view, Theorem **T.4.3** says that if $K \subseteq L$ is finite and $L \subseteq M$ is finite, then $K \subseteq M$ is also finite. A similar transitivity property is true for algebraic extensions. The next theorem states this property but it is formulated in a slightly more general form suitable for an application in the last theorem of this chapter:

T.4.5. *If $K \subseteq M \subseteq L$ are field extensions such that M is algebraic over K and $\alpha \in L$ is algebraic over M, then it is algebraic over K.*

Usually, the elements of a field extension L of a field K may be of both kinds—algebraic or transcendental. In particular, in **T.4.5**, some elements of L may be transcendental, but if they are algebraic over M, then they are algebraic over K. This gives a way to prove a very important property of all elements in a field extension which are algebraic over a ground field:

T.4.6. *If $L \supseteq K$ is a field extension, then all elements in L algebraic over K form a field.*

If $L \supseteq K$, then the field of all elements of L algebraic over K is called the **algebraic closure of** K **in** L. If $K = \mathbb{Q}$ and $L = \mathbb{C}$, then the field of algebraic elements over $\mathbb{Q}$ will be denoted by $\mathbb{A}$ and called **the field of algebraic numbers**. A field K is called **algebraically closed** if it does not posses algebraic extensions, that is, for every field $L \supseteq K$, if $\alpha \in L$ is algebraic over K, then $\alpha \in K$. This means that K is algebraically closed if and only if all irreducible polynomials over K are of degree 1. The fields $\mathbb{C}$ and $\mathbb{A}$ are algebraically closed (see Exercise 13.14 for $\mathbb{C}$ and Exercise 4.19 for $\mathbb{A}$). The fact that $\mathbb{C}$ is algebraically closed is exactly what the fundamental theorem of algebra says (see **T.1.1**).

Exercises 4

4.1. Which of the following numbers are algebraic?

(a) $1 + \sqrt{2} + \sqrt{3}$;
(b) $\sqrt{3} + \sqrt[4]{3}$;
(c) $\sqrt[3]{3} + \sqrt{2}$;
(d) $\sqrt{\pi} + 1$;
(e) $\sqrt{\pi} + \sqrt{2}$;
(f) $\sqrt[3]{1 + \sqrt{e}}$.

4.2. Show that if $f \in K[X]$ is irreducible over K and $L \supseteq K$ is a field extension such that the degree of f and the degree $[L : K]$ are relatively prime, then f is irreducible over L.

4.3. Find the minimal polynomial and the degree of α over K when:

(a) $K = \mathbb{Q}$, $\alpha = \sqrt[3]{\sqrt{3} + 1}$;
(b) $K = \mathbb{Q}$, $\alpha = \sqrt{2} + \sqrt[3]{2}$;
(c) $K = \mathbb{Q}$, $\alpha^5 = 1$, $\alpha \neq 1$;
(d) $K = \mathbb{Q}(i)$, $\alpha = \sqrt{2}$;
(e) $K = \mathbb{Q}(\sqrt{2})$, $\alpha = \sqrt[3]{2}$;
(f) $K = \mathbb{Q}$, $\alpha^p = 1$, $\alpha \neq 1$, p a prime number.

4.4. Find the degree and a basis of the following extensions $L \supseteq K$:

(a) $K = \mathbb{Q}$, $L = \mathbb{Q}(\sqrt{2}, i)$;
(b) $K = \mathbb{Q}$, $L = \mathbb{Q}(\sqrt{2}, \sqrt{3})$;
(c) $K = \mathbb{Q}$, $L = \mathbb{Q}(i, \sqrt[3]{2})$;
(d) $K = \mathbb{Q}$, $L = \mathbb{Q}(\sqrt[3]{2} + 2\sqrt[3]{4})$;
(e) $K = \mathbb{R}(X + \frac{1}{X})$, $L = \mathbb{R}(X)$;
(f) $K = \mathbb{Q}$, $L = \mathbb{Q}(\sqrt{2}, \sqrt{3}, \sqrt{5})$;
(g) $K = \mathbb{Q}(\sqrt{3})$, $L = \mathbb{Q}(\sqrt[3]{1 + \sqrt{3}})$;
(h) $K = \mathbb{F}_2$, $L = \mathbb{F}_2(\alpha)$, where $\alpha^4 + \alpha + 1 = 0$;
(i) $K = \mathbb{F}_3$, $L = \mathbb{F}_3(\alpha)$, where $\alpha^3 + \alpha^2 + 2 = 0$;
(j) $K = \mathbb{R}(X^2 + \frac{1}{X^2})$, $L = \mathbb{R}(X)$.

4.5. Let K be a quadratic field extension of the rational numbers $\mathbb{Q}$. Show that there is a unique square-free integer $d \neq 1$ such that $K = \mathbb{Q}(\sqrt{d})$.

4.6. Show that a complex number $z = a + bi$ is algebraic (over $\mathbb{Q}$) if and only if a and b are algebraic.

4.7. Show that the numbers $\sin r\pi$ and $\cos r\pi$ are algebraic (over $\mathbb{Q}$) if r is a rational number.

4.8. Let $L = \mathbb{Q}(\sqrt[3]{2})$. Find $a, b, c \in \mathbb{Q}$ such that $x = a + b\sqrt[3]{2} + c\sqrt[3]{4}$ when:

$$\text{(a) } x = \frac{1}{\sqrt[3]{2}}; \quad \text{(b) } x = \frac{1}{1+\sqrt[3]{2}}; \quad \text{(c) } x = \frac{1+\sqrt[3]{2}}{1+\sqrt[3]{2}+\sqrt[3]{4}}.$$

4.9. Let $L = \mathbb{Q}(\sqrt{2}, \sqrt{3})$. Find $a, b, c, d \in \mathbb{Q}$ such that $x = a + b\sqrt{2} + c\sqrt{3} + d\sqrt{6}$ when:

$$\text{(a) } x = \frac{1}{\sqrt{2}+\sqrt{3}}; \quad \text{(b) } x = \frac{1}{1+\sqrt{2}+\sqrt{3}}; \quad \text{(c) } x = \frac{\sqrt{2}+\sqrt{3}}{1+\sqrt{2}+\sqrt{3}+\sqrt{6}}.$$

4.10. Let $L = \mathbb{F}_2(\alpha)$, where $\alpha^4 + \alpha + 1 = 0$. Find $a, b, c, d, \in \mathbb{F}_2$ such that $x = a + b\alpha + c\alpha^2 + d\alpha^3$ when:

$$\text{(a) } x = \frac{1}{\alpha}; \quad \text{(b) } x = \alpha^5; \quad \text{(c) } x = \alpha^{15}; \quad \text{(d) } x = \frac{1}{\alpha^2+\alpha+1}.$$

4.11. (a) Show that if $\alpha \in K(X) \smallsetminus K$, $\alpha = \frac{p(X)}{q(X)}$, where $p, q \in K[X]$ and $\gcd(p, q) = 1$, then $[K(X) : K(\alpha)] \leq n$, where $n = \max(\deg p, \deg q)$.
(b) Show that if $\alpha \in K(X) \smallsetminus K$, then α is transcendental over K.
(c) Show that using the notation in (a), we have $[K(X) : K(\alpha)] = n$.

Remark. A famous theorem of Lüroth[1] says that all subfields L of $K(X)$ containing K are of the form $K(\alpha)$, where $\alpha \in K(X)$. The results of the exercise may be used to prove Lüroth's theorem (for a proof see [C], Theorem 2.4, p. 173 or [J], 8.14, p. 522).

4.12. Let M_1, M_2 be two fields between K and L.

(a) Prove that if M_1 and M_2 are algebraic extensions of K, then M_1M_2 is also an algebraic extension of K.
(b) Prove that if $[M_1 : K] \neq \infty$ and $[M_2 : K] \neq \infty$, then $[M_1M_2 : K] \neq \infty$.
(c) Show that if $[M_1 : K] = r$ and $[M_2 : K] = s$, where $\gcd(r, s) = 1$, then $[M_1M_2 : K] = rs$ and $M_1 \cap M_2 = K$.
(d) Is there any "general" relation between $[M_1 : K]$, $[M_2 : K]$ and $[M_1M_2 : K]$?

4.13. Prove **Nagell's lemma**[2]: Let $L \supset K$ be a field extension of prime degree q. If $f \in K[X]$ has prime degree p and is irreducible over K, but reducible over L, then $q = p$.

[1] Jacob Lüroth, 18 February 1844–14 September 1910.
[2] Trygve Nagell (Nagel), 13 July 1895–24 January 1988.

Remark. We use Nagell's lemma in Exercise 13.6 where, following Nagell, we prove, in a very simple way, unsolvability in radicals of some quintic equations.

4.14. Show that if $[K(\alpha) : K]$ is odd, then $K(\alpha) = K(\alpha^2)$. Is it true that $K(\alpha) = K(\alpha^2)$ implies that $[K(\alpha) : K]$ is odd?

4.15. (a) Show that if L is a field containing $\mathbb{C}$ and $[L : \mathbb{C}] \neq \infty$, then $L = \mathbb{C}$.
(b) Show that if L is a field containing $\mathbb{R}$ and $[L : \mathbb{R}] \neq \infty$, then $L = \mathbb{R}$ or $L \cong \mathbb{C}$.

4.16. Is it true that for each divisor d of $[L : K]$ there exists a field M between K and L such that $[M : K] = d$?

4.17. It is (well-)known that the numbers e and π are transcendental.[3] It is (really) not known whether $e + \pi$ and $e\pi$ are transcendental. Show that at least one of the numbers $e + \pi$ or $e\pi$ must be transcendental.

4.18. (a) Let $K \subseteq L$ be a field extension and let $f(\alpha) = 0$, where $\alpha \in L$ and $f(X)$ is a polynomial whose coefficients are algebraic over K. Prove that α is also algebraic over K.
(b) Assume that the number α is algebraic. Prove that the following numbers are also algebraic:
(b_1) α^2; (b_2) $\sqrt{\alpha}$; (b_3) $\sqrt[3]{1 + \sqrt{\alpha}}$.

4.19. Show that the algebraic closure of a field K in an algebraically closed field L is also algebraically closed (for example, since $\mathbb{A}$ is the algebraic closure of $\mathbb{Q}$ in $\mathbb{C}$, it is algebraically closed—see Exercise 13.14).

Using Computers 4

Finite algebraic extensions of rational numbers $\mathbb{Q}$ and of finite fields $\mathbb{F}_p$ can be constructed in Maple using the command `RootOf` like:

```
>alias(a = RootOf(X^2+1))
```

$$a$$

which defines the extension $\mathbb{Q}(i)$ with i denoted here by a. If a field $K = \mathbb{Q}(a)$, where a is defined by `RootOf` (see p. 8), then we can both factorize a polynomial $g(X)$ over K by `factor(g(X),a)` or ask whether it is irreducible using `irreducible(g(X),a)`. For example, the polynomial $X^4 + 1$ over the field $\mathbb{Q}(i)$ is factored by the command:

```
>factor(X^4+1,a)
```

$$(X^2 - a)(X^2 + a)$$

[3] For a proof of transcendence of these numbers see [L], Appendix 1.

If we want to define a finite field as an extension of $\mathbb{F}_p$ for a prime number p, then we use similar commands as above extended by **mod** p, for example:

```
>alias(b = RootOf(X^3+X+1 mod 2))
```

We have defined the field $\mathbb{F}_2(b)$ (we use b if it is the same session in which we used a above), whose degree is 8 over $\mathbb{F}_2$ (the polynomial $X^3 + X + 1$ is irreducible over $\mathbb{F}_2$) and we can make computations with it. For example:

```
>simplify(b^7) mod 2
```

$$1$$

which could be expected in the group of order 7 of all nonzero elements in the field $\mathbb{F}_2(b)$. It is possible to solve problems like Exercises 4.8, 4.9 or 4.10 finding, for example, `simplify(1/(b^2+b+1)) mod 2` in the field $\mathbb{F}_2(b)$.
In Maple, it is possible to find the minimal polynomial of an algebraic number over $\mathbb{Q}$ and over finite algebraic extensions of it (even if these tools are far from being perfect as yet) though the results of computations should be always checked. For example, take $\alpha = \sqrt[3]{1+\sqrt{3}}$. We guess that α has degree 6 over $\mathbb{Q}$ (one can try with bigger values than the expected degree of the minimal polynomial but one has to be cautious). We use the following commands:

```
>with(PolynomialTools)
>r:= evalf((1+sqrt(3))^(1/3))
                    1.397964868
>MinimalPolynomial(r,6)
```

$$-2 - 2X^3 + X^6$$

One can check the irreducibility of this polynomial using `factor(X^6-2*X^3-2)` (of course, the polynomial is irreducible by Eisenstein's criterion) and to check that α really is its zero:

```
>f:=X→X^6-2*X^3-2
```

$$X \to X^6 - 2X^3 - 2$$

```
>a:=(1+sqrt(3))^(1/3)
>simplify (f(a))
```

$$0$$

In general, we can find the minimal polynomial of α over a finite extension K of $\mathbb{Q}$ by finding such polynomial over $\mathbb{Q}$ and then factoring this polynomial over K. We can check which of the factors has α as its zero. If we continue with α above and if we want to get the minimal polynomial of this number over $K = \mathbb{Q}(\sqrt{3})$, then we note easily that $X^3 - (1+\sqrt{3})$ is a polynomial with coefficients in K of degree 3 having α as its zero. Since we already know that $[\mathbb{Q}(\alpha) : \mathbb{Q}] = 6$ and $[K : \mathbb{Q}] = 2$, we have $[\mathbb{Q}(\alpha) : K] = 3$, which means that the polynomial $X^3 - (1 + \sqrt{3})$ is minimal for α over K by **T.4.2**. In Maple, we can use the command:

```
>factor(X^6-2*X^3-2,sqrt(3))
```

$$(X^3 - 1 + \sqrt{3})(-X^3 + 1 + \sqrt{3})$$

which says that the two polynomials of degree 3 are irreducible (the second factor is the minimal polynomial of α). Observe that instead of `sqrt(3)` one could define the field K using construction with `RootOf`.

Working in a field generated by algebraic numbers, it is possible to express the elements of the field in terms of the elements of a basis using the command `rationalise`, as the following example shows. We work in the field $\mathbb{Q}(\sqrt{5}, \sqrt{7})$ and we want express a number in this field in terms of the powers of the generators $\sqrt{5}, \sqrt{7}$:

```
>a:= sqrt(5); b := sqrt(7)
```

$$5^{\frac{1}{2}}$$

$$7^{\frac{1}{2}}$$

```
>expand(rationalize(1/(a+b)))
```

$$\frac{3}{19}\sqrt{5} + \frac{11}{19} - 1/19\sqrt{7} - 2/19\sqrt{7}\sqrt{5}$$

4.20. Solve some of the Exercises 4.8, 4.9 or 4.10 using Maple.

4.21. Find the minimal polynomials of the following numbers α over the given fields:

(a) $\alpha = \sqrt[4]{2} + \sqrt{2} + 1$ over $\mathbb{Q}$;
(b) $\alpha = \sqrt[3]{2} + i$ over $\mathbb{Q}$;
(c) $\alpha = \sqrt[3]{2} + \sqrt{2} + 1$ over $\mathbb{Q}(\sqrt{2})$;
(d) $\alpha = \sqrt[4]{2} + i + 1$ over $\mathbb{Q}(i)$.

4.22. Factorize the given polynomial $f(X) \in \mathbb{Q}[X]$ into irreducible factors over $K = \mathbb{Q}(a)$, where a is a zero of $f(X)$:

(a) $f(X) = X^6 - 2$;
(b) $f(X) = X^5 - 5X + 12$;
(c) $f(X) = X^6 + 3X^3 + 3$;
(d) $f(X) = X^6 - 3X^2 - 1$;
(e) $f(X) = X^6 + 2X^3 - 2$;
(f) $f(X) = X^7 + 7X^3 + 7X^2 + 7X - 1$.

Chapter 5
Splitting Fields

In this chapter, we consider field extensions which are defined by all zeros of a given polynomial over a field containing its coefficients. Thus, if K is a field and $f(X)$ is a polynomial with coefficients in K, then we study the field generated over K by all zeros of $f(X)$. Such a field is called a splitting field of $f(X)$ over K (see below for its definition). Splitting fields of polynomials play a central role in Galois theory. We study them still more in Chap. 7. It is intuitively clear that a splitting field of a polynomial is a very natural object containing a lot of information about it. This will be confirmed in the following chapters, in particular, when we come to the main theorems of Galois theory in Chap. 9 and to its applications in the following chapters (e.g. in Chap. 13, where we study solvability of equations by radicals). As an important example, we study in this chapter finite fields, which can be easily described as splitting fields of very simple polynomials over the finite prime fields. As another example, we consider the notion of an algebraic closure of a field K, which is a minimal (in suitable sense) field extension of K, containing a splitting field of every polynomial with coefficients in K.

If K is a field and $f \in K[X]$, then we say that L is a **splitting field** of f over K if f has all its zeros in L and these zeros generate L over K, that is, $L = K(\alpha_1, \ldots, \alpha_n)$, where $f(X) = a(X-\alpha_1)\cdots(X-\alpha_n)$ and $a \in K$. Sometimes, a splitting field of f over K is denoted by K_f. If $\tau : K \to K'$ is an **embedding** of the field K into a field K', that is, τ is a injective function such that $\tau(x+y) = \tau(x) + \tau(y)$ and $\tau(xy) = \tau(x)\tau(y)$ for $x, y \in K$, then τ can be extended to an embedding of the polynomial rings $\tau : K[X] \to K'[X]$ (denoted by the same letter): $\tau(f(X)) = \tau(a_n)X^n + \cdots + \tau(a_1)X + \tau(a_0)$ when $f(X) = a_nX^n + \cdots + a_1X + a_0 \in K[X]$ (one can say that we apply τ to the coefficients of $f(X)$). Very often, we consider the case when $\tau(K) = K'$, that is, the function τ is an **isomorphism** of the fields K and K'.

Splitting fields are constructed using the theorem below, usually attributed to **Kronecker**.[1] Kronecker's Theorem says that for every polynomial with coefficients

[1] Leopold Kronecker, 7 December 1823–29 December 1891.

J. Brzeziński, *Galois Theory Through Exercises*, Springer Undergraduate Mathematics Series, https://doi.org/10.1007/978-3-319-72326-6_5

in a field K, there exists a field extension of K in which the polynomial has a zero. It gives the first step, which can be repeated using an inductive argument and leads to a construction of splitting fields in the next theorem. Both Kronecker's Theorem and the general result about the existence and uniqueness of splitting fields are the main results of this chapter and are used frequently in the rest of the book.

T.5.1 Kronecker's Theorem. *(a) If f is an irreducible polynomial over K, then there exists a field $L \supseteq K$ such that $L = K(\alpha)$ and $f(\alpha) = 0$.*
(b) If $\tau : K \to K'$ is a field isomorphism, f an irreducible polynomial over $K, L = K(\alpha)$, where $f(\alpha) = 0$ and $L' = K'(\alpha')$, where $\tau(f)(\alpha') = 0$, then there is an isomorphism $\sigma : K(\alpha) \to K'(\alpha')$ such that in the diagram:

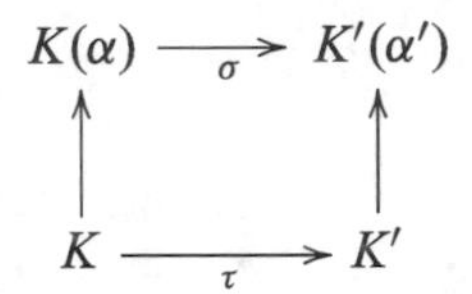

we have $\sigma(\alpha) = \alpha'$ and $\sigma|_K = \tau$.
In particular, if $K = K'$ and $\tau = id$, then σ is an isomorphism over K (that is, the isomorphism σ maps each element in K to itself) of the two simple extensions of K by two arbitrary roots of $f(X) = 0$.

Using Kronecker's Theorem, it is not difficult to use induction in order to obtain the following general result about splitting fields:

T.5.2. *(a) Every polynomial $f \in K[X]$ has a splitting field over K.*
(b) If $\tau : K \to K'$ is an isomorphism of fields, L is a splitting field of a polynomial $f \in K[X]$ and L' is a splitting field of the polynomial $\tau(f) \in K'[X]$, then there exists an isomorphism $\sigma : L \to L'$

$$\begin{array}{ccc} L & \xrightarrow[\sigma]{} & L' \\ \uparrow & & \uparrow \\ K & \xrightarrow[\tau]{} & K' \end{array}$$

which extends τ (that is, $\sigma|_K = \tau$). In particular, if $K = K'$ and $\tau = id$, then two splitting fields for f over K are K-isomorphic (that is, the isomorphism σ maps each element in K to itself).
Moreover, if f has no multiple zeros, then there are exactly $[L : K]$ different possibilities for σ when τ is given.

If $f(X)$ is a polynomial with coefficients in a number field K, that is, a subfield of the complex numbers $\mathbb{C}$, then by the splitting field of $f(X)$ over K, we mean the splitting field generated over K by the zeros of $f(X)$ in $\mathbb{C}$.

As an illustration and application of Theorem **T.5.2** about splitting fields, we describe all finite fields. We have already met finite fields in Chap. 2, where we considered finite prime fields $\mathbb{F}_p$ defined for each prime number p. Here we show that the number of elements in a finite field is always a power of a prime number and that such a field with p^n elements is a splitting field of a polynomial $X^{p^n} - X$ over $\mathbb{F}_p$. In the proof, we need a technical result which is very useful when we want to decide whether a given polynomial has multiple zeros. It uses the notion of the derivative of a polynomial. The definition is exactly as known from calculus, but is given purely formally (without using the notion of limit).

If $f(X) = a_0 + a_1X + \cdots + a_nX^n \in K[X]$, then we define the **derivative** f' of f as the polynomial $f'(X) = a_1 + 2a_2X + \cdots + na_nX^{n-1}$. Exactly as for the polynomials in $\mathbb{R}[X]$, we have $(f_1 + f_2)' = f_1' + f_2'$ and $(f_1f_2)' = f_1'f_2 + f_1f_2'$, which can be checked by an uncomplicated computation (see Exercise 5.10). We will use the derivative in order to determine whether a polynomial has multiple zeros.

T.5.3. *A polynomial $f \in K[X]$ has no multiple zeros in any extension $L \supseteq K$ if and only if* $\gcd(f,f') = 1$.

For the definition of the greatest common divisor of two polynomials, see p. 249 in the Appendix. Using the theorem above and the construction of splitting fields, we have a description of all finite fields:

T.5.4. (*a*) *The number of elements in a finite field is a power of a prime number.*
(*b*) *If p is a prime number and $n \geq 1$, then any splitting field of $X^{p^n} - X$ over $\mathbb{F}_p$ is a finite field with p^n elements.*
(*c*) *Two finite fields with the same number of elements are isomorphic. More precisely, every finite field with p^n elements is a splitting field of $X^{p^n} - X$ over $\mathbb{F}_p$.*

Finite fields are often called **Galois fields** (in particular, in applications to coding theory and cryptography), since Galois gave a construction of such fields for $n \geq 1$ in one of his papers published in 1830. A finite field with p^n elements is often denoted by $GF(p^n)$ following E.H. Moore[2] who proved in 1893 that every such field is isomorphic to a quotient $\mathbb{F}_p[X]/(f(X))$, where $f(X) \in \mathbb{F}_p[X]$ is irreducible of degree n (see Exercise 5.6(a)). In this book, we use notation $\mathbb{F}_{p^n}$ to denote a finite field with p^n elements.

In Chap. 4 (see p. 21), we defined algebraically closed fields. Recall that a field K is algebraically closed if all polynomials with coefficients in K split completely already in K, that is, for every polynomial $f(X) \in K[X]$, we have $f(X) = a_n(X - \alpha_1)\cdots(X - \alpha_n)$, where $a_n, \alpha_i \in K$ (see p. 21). The field of all complex numbers and the field of algebraic numbers have this property (see Exercise 13.14 for $\mathbb{C}$ and Exercise 4.19 for $\mathbb{A}$). By an **algebraic closure** of a field K, we mean a field $\overline{K}$ which is algebraic over K and algebraically closed. This means that such a field $\overline{K}$ contains a splitting field of every polynomial with coefficients in K. We have:

T.5.5. *For every field K there exists an algebraic closure $\overline{K}$ and two algebraic closures of the same field K are K-isomorphic.*

---

[2]Eliakim Hastings Moore, 26 January 1862–30 December 1932.

Exercises 5

5.1. Find the degree and a basis of the splitting field over K for $f \in K[X]$ when:
(a) $K = \mathbb{Q}, f = (X^2 - 2)(X^2 - 5)$;
(b) $K = \mathbb{Q}, f = X^3 - 2$;
(c) $K = \mathbb{Q}, f = X^4 - 2$;
(d) $K = \mathbb{Q}, f = X^4 + X^2 - 1$;
(e) $K = \mathbb{Q}, f = X^4 + 1$;
(f) $K = \mathbb{Q}(i), f = X^4 - 2$;
(g) $K = \mathbb{Q}(i), f = (X^2 - 2)(X^2 - 3)$;
(h) $K = \mathbb{Q}, f = X^p - 1$, p a prime number.

5.2. Decide whether the following pairs of fields are isomorphic:
(a) $\mathbb{Q}(\sqrt[4]{2})$ and $\mathbb{Q}(i\sqrt[4]{2})$;
(b) $\mathbb{Q}(\sqrt[3]{1+\sqrt{3}})$ and $\mathbb{Q}(\sqrt[3]{1-\sqrt{3}})$;
(c) $\mathbb{Q}(\sqrt{2})$ and $\mathbb{Q}(\sqrt{3})$.

5.3. Let L be a splitting field of a polynomial $f(X)$ of degree n with coefficients in a field K. Show that $[L : K] \leq n!$.

5.4. Write down a multiplication table for the elements of the finite field K if

(a) $K = \mathbb{F}_2(\alpha), f(\alpha) = 0$, where $f(X) = X^2 + X + 1$;
(b) $K = \mathbb{F}_2(\alpha), f(\alpha) = 0$, where $f(X) = X^3 + X + 1$;
(c) $K = \mathbb{F}_3(\alpha), f(\alpha) = 0$, where $f(X) = X^2 + 1$.

5.5. Prove that a field with p^n elements contains a field with p^m elements if and only if $m|n$.

5.6. (a) Let $f(X)$ be an irreducible polynomial of degree n over a field $\mathbb{F}_p$. Show that $\mathbb{F}_p[X]/(f(X))$ is a field with p^n elements, which is isomorphic to a splitting field of the polynomial $X^{p^n} - X$ and $f(X)$ divides $X^{p^n} - X$.
(b) Let $f(X)$ be an irreducible polynomial in $\mathbb{F}_p[X]$. Show that $f(X)|X^{p^n} - X$ if and only if $\deg(f(X))|n$.

5.7. (a) Let $v_p(n)$ denote the number of irreducible polynomials of degree n in $\mathbb{F}_p[X]$. Prove that

$$\sum_{d|n} dv_p(d) = p^n \quad \text{and} \quad v_p(n) = \frac{1}{n}\sum_{d|n} p^d \mu\left(\frac{n}{d}\right),$$

where $\mu(n)$ is the **Möbius function**,[3] that is, $\mu(n) = 0$ if there is a prime number p whose square divides n, $\mu(n) = (-1)^k$ if n is a product of k different primes, and $\mu(1) = 1$.
(b) Find the numbers $v_2(n)$ for $n = 1, \ldots, 10$ (compare your result with Exercise 3.2).
(c) Let q be a prime number. What is the number of irreducible polynomials of degree q over $\mathbb{F}_p[X]$? Do you associate the answer with a known theorem?

[3] August Ferdinand Möbius, 17 November 1790–26 September 1868.

Remark. In connection with this exercise, it may be interesting to look at Exercise 5.18.

5.8. (a) Prove that a finite subgroup of the multiplicative group K^* of a field K is cyclic.
(b) Prove that if $L \supseteq K$ are finite fields, then there is an element $\gamma \in L$ such that $L = K(\gamma)$.

5.9. (a) An irreducible polynomial $f \in \mathbb{F}_p[X]$ of degree n is called **primitive** if $f \nmid X^m - X$ when $m < p^n$. Prove that the number of primitive polynomials of degree n is equal to $\frac{1}{n}\varphi(p^n - 1)$, where φ is Euler's[4] function ($\varphi(n)$ equals the number of $0 < k < n$ such that $\gcd(k, n) = 1$—for Euler's function, see p. 279).
(b) Show that for every finite field $\mathbb{F}$ and for every n there exist irreducible polynomials of degree n over $\mathbb{F}$.

5.10. Show that $(fg)' = f'g + fg'$ when $f, g \in K[X]$.

5.11. (a) Let L be a splitting field of the polynomial $X^n - 1$ over a field K. Show that the zeros of this polynomial, that is, the n-th roots of unity, form a finite cyclic group and that $L = K(\varepsilon)$, where ε is a generator of this group.
(b) Show that the number of the n-th roots of unity in a splitting field of $X^n - 1$ over a field K is n if and only if the characteristic of K is 0 or it is relatively prime to n (that is, does not divide n).
(c) Let L be a splitting field of a polynomial $X^n - a$ over a field K ($a \in K^*$). Show that $L = K(\varepsilon, \alpha)$, where ε generates the group of solutions of $X^n - 1 = 0$ in L and α is any fixed solution of the equation $X^n - a = 0$ in L. Show that all solutions of $X^n - a = 0$ are of the form $\eta\alpha$, where η are all different n-th roots of unity.

5.12. (a) Let $a \in K$, where K is a field and let p be a prime number. Show that the polynomial $X^p - a$ is irreducible over K or has a zero in this field.
(b) Show that (a) is not true in general for binomials $X^n - a$ when n is not a prime number.

Remark. The result in (a) was first proved by Abel (see [Tsch], p. 287). A more general result on reducibility of binomial polynomials $X^n - a$ is known as Capelli's Theorem[5] (see [Tsch], p. 294): The polynomial $X^n - a$, where $a \in K$ (K of characteristic 0), is reducible over K if and only if there is a prime p dividing n and $b \in K$ such that $a = b^p$ or $n = 4$ and $a = -4b^4$ (for a proof see [Tsch], IV §2, or [Sch], Thm. 19, p. 92). A field extension $L = K(\alpha)$, where α is a zero of a binomial polynomial $f(X) = X^n - a \in K[X]$, will be called a **simple radical**.

5.13. The polynomial $X^4 + 1$ is irreducible over $\mathbb{Q}$. Show that any reduction of $X^4 + 1$ modulo any prime number p gives a reducible polynomial over $\mathbb{F}_p$.

---

[4]Leonhard Euler, 15 April 1707–18 September 1783.

[5]Alfredo Capelli, 5 August 1855–28 January 1910.

5.14. Let $\overline{K}$ be an algebraic closure of a field K and let L be an extension of K contained in $\overline{K}$. Show that $\overline{K}$ is also an algebraic closure of L.

Using Computers 5

It is possible to construct a splitting field of a polynomial (of not too high degree) using the command `RootOf` (see p. 8) and the command `allvalues`, which gives values of all zeros of the polynomial whose zero is defined by the command `RootOf`, for example:

```
>alias(a=RootOf(X^4-10*X^2+1))
```

$$a$$

```
>allvalues(a)
```

$$\sqrt{3}-\sqrt{2}, \sqrt{3}+\sqrt{2}, -\sqrt{3}+\sqrt{2}, -\sqrt{3}-\sqrt{2}$$

```
>factor(X^4-10*X^2+1,a)
```

$$(-X-10a+a^3)(X-10a+a^3)(X+a)(-X+a)$$

Thus the field $K=\mathbb{Q}(\sqrt{2}+\sqrt{3})$ is a splitting field of $f(X)=X^4-10X^2+1$, since it splits into linear factors when one of its zeros is adjoined to the field $\mathbb{Q}$ (in fact, we have `allvalues(a)[2]` equal to $\sqrt{2}+\sqrt{3}$, but all zeros give, of course, the same field).

The command `factor(f(X),{a,b})` gives a way to check whether the extension by the chosen zeros of $f(X)$ (here a, b) really gives a splitting field (in its splitting field, the polynomial $f(X)$ should split completely into linear factors). Unfortunately, the command `alias` cannot be defined in terms of another alias, which sometimes means that the output in terms of different `RootOf` expressions is not easy to grasp. For example, we find that the splitting field of $f(X)=X^3-2$ is generated by two of its zeros (we choose a simple example in order to save space—see Exercise 5.17):

```
>a:=RootOf(X^3-2)
```

$$RootOf(\_Z^3-2)$$

```
>factor(X^3-2,a)
```

$$-(-X+RootOf(\_Z^3-2))(X^2+RootOf(\_Z^3-2)X+RootOf(\_Z^3-2)^2)$$

```
>b:=RootOf(X^2+a*X+a^2)
```

$$RootOf(\_Z^2 + RootOf(\_Z^3 - 2)\_Z + RootOf(\_Z^3 - 2)^2)$$

```
>factor(X^3-2, {a, b})
```

$$(-X + RootOf(\_Z^2 + RootOf(\_Z^3 - 2)\_Z + RootOf(\_Z^3 - 2)^2))$$

$$(X + RootOf(\_Z^2 + RootOf(\_Z^3 - 2)\_Z + RootOf(\_Z^3 - 2)^2) + RootOf(\_Z^3 - 2))$$

$$(-X + RootOf(\_Z^3 - 2))$$

Using a, b the output is $(-X + b)(X + a + b)(-X + a)$ (which is $X^3 - 2$—use the command `simplify` to obtain the product). It is easy to check that $a = \sqrt[3]{2}$ and $b = \varepsilon\sqrt[3]{2}$, where $\varepsilon = \frac{-1+i\sqrt{3}}{2}$ is a 3rd root of unity.

5.15. Check using Maple that the fields you constructed in Exercise 5.1 really are splitting fields of the given polynomials f.

5.16. Find the degree of the splitting field over the rational numbers for polynomials f in Exercise 4.22.

5.17. Find the smallest number of zeros of the polynomial $f(X)$ which generate its splitting field over $\mathbb{Q}$:

(a) $f(X) = X^5 - 2$; (b) $f(X) = X^5 - 5X + 12$; (c) $f(X) = X^5 - X + 1$.

5.18. Find all irreducible polynomials of degrees $n = 5, 6$ over $\mathbb{F}_2$ using Exercise 5.6(b), that is, factorizing $X^{2^n} - X$.

Chapter 6
Automorphism Groups of Fields

In this chapter, we study automorphism groups of fields and introduce Galois groups of finite field extensions. The term "Galois group" is often reserved for automorphism groups of Galois field extensions, which we define and study later in Chap. 9. The terminology which we use is very common and has several advantages in textbooks (e.g. it is easier to formulate exercises). A central result of this chapter is Artin's[1] Lemma, which is a key result in the modern presentation of Galois theory. It was published in the first edition (from 1942) of Artin's celebrated lectures on Galois theory given at the University of Notre Dame (see [A]). Artin's Lemma is part of Theorem **T.6.2**, which is one of the most important results in this book. In the exercises, we find Galois groups of many field extensions, and we use this theorem for a great variety of problems on field extensions and their automorphism groups.

Let L be a field. An **automorphism** of L is a bijective function $\sigma : L \to L$ such that

(a) $\sigma(x + y) = \sigma(x) + \sigma(y)$
(b) $\sigma(xy) = \sigma(x)\sigma(y)$

for arbitrary $x, y \in L$. If $L \supseteq K$ is a field extension, then an automorphism $\sigma : L \to L$ is called a K**-automorphism** if

(c) $\sigma(x) = x$ for every $x \in K$.

It is not difficult to check that:

T.6.1. *All K-automorphisms of L form a group with respect to the composition of automorphisms.*

The group of all K-automorphisms of L is denoted by $G(L/K)$ and called the **Galois group** of L over K. If G is an arbitrary group which consists of

[1] Emil Artin, 3 March 1898–20 December 1962.

J. Brzeziński, *Galois Theory Through Exercises*, Springer Undergraduate Mathematics Series, https://doi.org/10.1007/978-3-319-72326-6_6

automorphisms of L (e.g. $G = G(L/K)$, where $L \supseteq K$), then we define

$$L^G = \{x \in L : \forall_{\sigma \in G}\, \sigma(x) = x\}.$$

It is easy to check that L^G is a subfield of L. It is called the **fixed (sub)field** of G (see below **T.6.2**).

Sometimes, we want to construct some elements of the field L^G. The two most common constructions are given by the **trace** (denoted Tr_G) and the **norm** (denoted Nr_G) with respect to a finite group G:

$$\mathrm{Tr}_G(\alpha) = \sum_{\sigma \in G} \sigma(\alpha), \quad \text{and} \quad \mathrm{Nr}_G(\alpha) = \prod_{\sigma \in G} \sigma(\alpha), \tag{6.1}$$

since clearly $\mathrm{Tr}_G(\alpha), \mathrm{Nr}_G(\alpha) \in L^G$ for $\alpha \in L$. Notice that $\mathrm{Tr}_G(\alpha + \beta) = \mathrm{Tr}_G(\alpha) + \mathrm{Tr}_G(\beta)$ and $\mathrm{Nr}_G(\alpha\beta) = \mathrm{Nr}_G(\alpha)\mathrm{Nr}_G(\beta)$ for $\alpha, \beta \in L$. The norm and the trace are often denoted by Tr_{L/L^G} and Nr_{L/L^G}.

T.6.2. *If G is a group of automorphisms of L (finite or infinite), then L^G is a subfield of L and $[L : L^G] = |G|$.*

Here $|G|$ denotes the number of elements in G. If G is infinite, we use the notation $|G| = \infty$. Almost all exercises will focus on this important theorem, the part of which claiming the inequality $[L : L^G] \leq |G|$ is often called **Artin's Lemma**. The theorem is a consequence of the following result:

T.6.3 Dedekind's[2] Lemma. *If $\sigma_1, \sigma_2, \ldots, \sigma_n$ are different automorphisms of a field L and the equality $a_1\sigma_1(x) + a_2\sigma_2(x) + \cdots + a_n\sigma_n(x) = 0$, where $a_i \in L$, holds for every $x \in L$, then $a_1 = a_2 = \ldots = a_n = 0$.*

Dedekind's Lemma can be also formulated as a statement saying that different automorphisms of a field L are linearly independent over L (e.g. in the vector space over L consisting of all functions $f : L \to L$).

By the **Galois group** over K of an equation $f(X) = 0$ or the polynomial $f(X)$, where $f(X) \in K[X]$, we mean the Galois group of any splitting field K_f over K. Instead of the notation $G(K_f/K)$, we sometimes use G_f (when K is clear from the context) or $G_f(K)$.

If $K_f = K(\alpha_1, \ldots, \alpha_n)$, where α_i, $i = 1, \ldots, n$, are all zeros of $f(X) = 0$ and $\sigma \in G(K_f/K)$, then $\sigma(\alpha_i)$ is also a zero of $f(X) = 0$ (see Exercise 6.1), so the Galois group $G(K_f/K)$ acts on the set $X_f = \{\alpha_1, \ldots, \alpha_n\}$ of the zeros of $f(X)$. We write $\sigma(\alpha_i) = \alpha_{\sigma(i)}$, that is, we use the same symbol σ in order to denote the permutation $\sigma : \{1, \ldots, n\} \to \{1, \ldots, n\}$ of the indices $1, \ldots, n$ of the zeros of $f(X)$ corresponding to σ (that is, $\sigma(i) = j$ if and only if $\sigma(\alpha_i) = \alpha_j$). It is an easy exercise (see Exercise 6.3) to show that the permutations σ of $1, \ldots, n$ corresponding to the automorphisms in the Galois group $G(K_f/K)$ form a subgroup of the symmetric group S_n (this subgroup depends of course on the numbering

[2] Julius Wilhelm Richard Dedekind, 6 October 1831–12 February 1916.

of the zeros of $f(X) = 0$). We shall sometimes denote any permutation group corresponding to $f(X) = 0$ by $\mathrm{Gal}(K_f/K)$ or $\mathrm{Gal}_f(K)$. In fact, Galois worked with his groups just in this way, considering them as permutation groups. We try to write "Gal" when we consider the Galois group of a polynomial as a subgroup of S_n, where n is the degree of the polynomial. This notational distinction is not standard and sometimes it is necessary to be flexible, considering the Galois group both as permutations of the zeros of $f(X)$ and as permutations of S_n (in particular, when theorems on permutation groups are applied to Galois groups considered as permutation groups of the zeros). As regards the general terminology concerning actions of groups on sets, see section "Group Actions on Sets" in Appendix.

Exercises 6

6.1. Let $L \supseteq K$ be a field extension.
(a) Show that if $\alpha \in L$ is a zero of $f \in K[X]$ and $\sigma \in G(L/K)$, then $\sigma(\alpha)$ is also a zero of f.
(b) Show that if $L = K(\alpha_1, \ldots, \alpha_r)$ and two automorphisms $\sigma, \tau \in G(L/K)$ are equal for every generator α_i (that is, $\sigma(\alpha_i) = \tau(\alpha_i)$ for each i), then $\sigma = \tau$ (that is, $\sigma(\alpha) = \tau(\alpha)$ for every $\alpha \in L$).

6.2. Find the Galois groups $G = G(L/K)$ for the following extensions $L \supseteq K$:
(a) $L = \mathbb{Q}(\sqrt{2})$, $K = \mathbb{Q}$;
(b) $L = \mathbb{Q}(\sqrt[3]{2})$, $K = \mathbb{Q}$;
(c) $L = \mathbb{Q}(\sqrt[4]{2})$, $K = \mathbb{Q}$;
(d) $L = \mathbb{Q}(\sqrt{2}, \sqrt{3})$, $K = \mathbb{Q}$;
(e) $L = \mathbb{F}_2(X)$, $K = \mathbb{F}_2(X^2)$;
(f) $L = \mathbb{F}_5(X)$, $K = \mathbb{F}_5(X^4)$.

6.3. (a) Show that any automorphism of any field restricts to the identity on its prime field. In particular, the identity is the only automorphism of a prime field (that is, $\mathbb{Q}$ or $\mathbb{F}_p$).
(b) Let K be a finite field of characteristic $p \neq 0$. Show that $\sigma(x) = x^p$, $x \in K$, is an automorphism of K.
(c) Use (a) and (b) in order to prove **Fermat's Little Theorem**[3]: If a is an integer and p a prime number, then p divides $a^p - a$.

6.4. Show that the field of real numbers $\mathbb{R}$ has no nontrivial automorphisms.

6.5. Show that if $[L : K] < \infty$, then the order of the Galois group $G(L/K)$ divides the degree $[L : K]$.

6.6. (a) Show that if σ is a K-automorphism of the field $K(X)$, then $\sigma(X) = \frac{aX+b}{cX+d}$, where $a, b, c, d \in K$ and $ad - bc \neq 0$.
(b) Show that all functions on $K(X)$ of the form defined in (a) form the group of all automorphism of $K(X)$ over K (the functions of this form are often called **Möbius functions** or **Möbius transformations** by analogy with the cases $K = \mathbb{C}$).

[3]Pierre de Fermat, 17 August 1601–12 January 1665.

(c) Show that the group of automorphism of $K(X)$ over K is isomorphic to the quotient of the matrix group $GL_2(K)$ of all nonsingular (2×2)-matrices by the subgroup of scalar matrices aI, where $a \in K$, $a \neq 0$ and I is the (2×2) identity matrix.

6.7. Let G be the group of all $\mathbb{F}_2$-automorphisms of $L = \mathbb{F}_2(X)$ (see Exercise 6.6). Find L^G.

6.8. Let $L = \mathbb{F}_5(X)$ and let G be the group of all automorphisms of L such that $\sigma(X) = aX + b$, where $a, b \in \mathbb{F}_5$ and $a \neq 0$. Find L^G.

6.9. (a) Let $L = \mathbb{R}(X, Y)$ and $G = \{\sigma_1, \sigma_2\}$, where σ_1 is the identity and σ_2 is defined by $\sigma_2(X) = -X$, $\sigma_2(Y) = Y$. Find L^G.
(b) Prove with the help of (a) that if $f(X, Y) \in \mathbb{R}(X, Y)$ is such that $f(-X, Y) = -f(X, Y)$, then $\int f(\sin x, \cos x)dx = \int g(t)dt$ for a function $g \in \mathbb{R}(t)$ and $t = \cos x$.

6.10. (a) Let $L = \mathbb{R}(X, Y)$ and $G = \{\sigma_1, \sigma_2\}$, where σ_1 is the identity and σ_2 is defined by $\sigma_2(X) = -X$, $\sigma_2(Y) = -Y$. Find L^G.
(b) Let $f(X, Y) \in \mathbb{R}(X, Y)$ and $f(-X, -Y) = f(X, Y)$. Using (a) show how to express the integral $\int f(\sin x, \cos x)dx$ as an integral of a rational function.

6.11. Let $L = \mathbb{Q}(X, Y)$ and $G = \{\sigma_1, \sigma_2, \sigma_3, \sigma_4\}$ be the group of automorphisms of L defined by the table:

	X	Y
σ_1	X	Y
σ_2	$-X$	Y
σ_3	X	$-Y$
σ_4	$-X$	$-Y$

Find L^G.

6.12. Let $L = \mathbb{Q}(X)$ and $G = \langle\sigma\rangle$, where $\sigma(X) = X + 1$. Find L^G.

6.13. (a) Let $L = K(X, Y)$ and $G = < \sigma_1, \sigma_2 >$, where σ_1 is the identity and $\sigma_2(X) = Y$, $\sigma_2(Y) = X$. Find L^G.
(b) Let $L = K(X_1, \dots, X_n)$ and $G = S_n$, where for $\sigma \in S_n$, $\sigma(X_i) = X_{\sigma(i)}$. Show that $L^G = K(s_1, \dots, s_n)$, where s_i are the elementary symmetric polynomials of $X_1, \dots, X_n$, that is, $s_1 = \sum_{1 \le i \le n} X_i$, $s_2 = \sum_{1 \le i < j \le n} X_iX_j$, $s_3 = \sum_{1 \le i < j < k \le n} X_iX_jX_k$, $\dots$, $s_n = X_1X_2 \cdots X_n$.

6.14. Let $K \subseteq L$ be a field extension and let G be an automorphism group L over K. Show that if N is a normal subgroup of G, then the quotient group G/N is an automorphism group of the field L^N when $(\sigma N)(x) = \sigma(x)$ for $x \in L^N$ and any coset σN, $\sigma \in G$. Moreover, we have $\left(L^H\right)^{G/H} = L^G$.

6.15. (a) Let $L = \mathbb{C}(X_1, \dots, X_n)$ and let $H = \langle\sigma\rangle$ be a cyclic group of order m such that $\sigma(X_i) = \varepsilon_i X_i$, where ε_i for $i = 1, \dots, n$ are m-th roots of unity and ε_1 is a primitive root. Show that $L^H = \mathbb{C}(X_1^m, X_1^{m_2}X_2, \dots, X_1^{m_n}X_n)$ for suitable positive integers $m_2, \dots, m_n \le m$.

(b) Consider the following generalization of (a). Let G be an arbitrary finite abelian group of automorphisms of $L = \mathbb{C}(X_1, \dots, X_n)$ over $\mathbb{C}$ such that for every $\sigma \in G$ and every $i = 1, \dots, n$, we have $\sigma(X_i) = \varepsilon_{\sigma,i} X_i$, where $\varepsilon_{\sigma,i}$ is an m-th root of unity for some positive integer m. Show that there are $Y_i \in L$ such that $L^G = \mathbb{C}(Y_1, \dots, Y_n)$ and each Y_i $(i = 1, \dots, n)$ is a product of suitable powers of X_j $(j = 1, \dots, n)$.

Remark. Let $L = K(X_1, \dots, X_n)$ be the field of rational functions in n variables over a field K, and M a subfield of L containing K such that the extension $L \supseteq M$ is finite. A general **rationality question** is whether there are elements $Y_1, \dots, Y_n \in L$ such that $M = K(Y_1, \dots, Y_n)$. A special case of this situation occurs when G is a finite group of automorphisms of L over K and $M = L^G$ is the fixed field of G. In the exercises above, we have seen several examples of this situation and the last exercise shows that $L^G = \mathbb{C}(Y_1, \dots, Y_n)$ is the field of rational functions over the complex numbers when the group G is abelian and acts as diagonal matrices on the vector $(X_1, \dots, X_n)$ in the vector space $V = \mathbb{C}X_1 + \cdots + \mathbb{C}X_n$. In fact, this proves the same property for any linear action $\sigma(X_i) = \sum a_{ij} X_j$ on V of a finite abelian group G, since for any finite abelian group there is a basis of V in which the action of all the elements of G is diagonal (like in the exercise above)—see [F]. The result of the last exercise was proved by E. Fischer[4] in 1913. Later, in 1918, Emmy Noether asked whether L^G is a field of rational functions over K when $G \subseteq S_n$ is a group of permutations of X_i, that is, if $\sigma \in G$, then $\sigma(X_i) = X_{\sigma(i)}$ (the permutation σ is a bijection on the set of indices $\{1, \dots, n\}$). This question, known as **Noether's[5] problem**, was solved by Emmy Noether for $n = 2, 3, 4$ (for $n = 1$ the much more general Lüroth's theorem is true—see Exercise 4.11), but in 1969, it was proved by Richard Swan that the field L^G is not a field of rational functions when G is a cyclic group of order 47 and $K = \mathbb{Q}$ (see [F]). For more information about rationality questions in function fields, see the very nice survey [F].

6.16. Is it true that if $[L : K_1] \neq \infty$ and $[L : K_2] \neq \infty$, then $[L : K_1 \cap K_2] \neq \infty$, where K_1, K_2 are subfields of the field L?

Using Computers 6

It is not reasonable to expect that there is a general procedure to find the automorphism group of any given field extension $L \supset K$. In some cases, it is possible to describe such a group using Exercise 6.1 when the extension is represented as

[4]Ernst Sigismund Fischer, 12 July 1875–14 November 1954.

[5]Emmy Noether, 23 March 1882–14 April 1935.

$L = K(\alpha)$ and the minimal polynomial $f(X)$ of α over K is given. Then each automorphism is uniquely determined by the image of α and such an image is a zero of the polynomial $f(X)$ in L. For example, take $K = \mathbb{Q}(\sqrt[4]{2})$. Then we define

```
>alias(a=RootOf(X^4-2))
```

$$a$$

```
>factor(X^4-2,a)
```

$$(X - a)(X + a)(X^2 + a^2)$$

Thus $f(X) = X^4 - 2$ has two zeros $\pm a$ in $K = \mathbb{Q}(a)$, where $a = \sqrt[4]{2}$ and there are two automorphisms $\sigma(a) = a$ (the identity) and $\sigma(a) = -a$.
Take $f(X) = X^3 - 7X + 7$ and define $K = \mathbb{Q}(a)$, where a is a zero of $f(X)$. Thus:

```
>alias(a=RootOf(X^3-7X+7))
```

$$a$$

```
>factor(X^3-7X+7,a)
```

$$(-X - 14 + 4a + 3a^2)(X - 14 + 5a + 3a^2)(-X + a)$$

Thus the zeros of $f(X)$ in K are $a, -14 + 4a + 3a^2, 14 - 5a - 3a^2$ and there are three automorphisms $\sigma_0(a) = a, \sigma_1(a) = -14 + 4a + 3a^2, \sigma_2(a) = 14 - 5a - 3a^2$. We can check that $\sigma_1^2(a) = \sigma_2(a)$ (that is, $\sigma_1^2 = \sigma_2$, which is evident, since $G(K/\mathbb{Q}) = \{\sigma_0, \sigma_1, \sigma_2\}$ is a cyclic group of order 3). In Maple, this can be done in the following way:

```
>σ:=a → -14+4*a+3*a^2
```

$$a \to -14 + 4a + 3a^2$$

```
>simplify(σ(σ(a)))
```

$$14 - 5a - 3a^2$$

Thus $\sigma_1^2(a) = \sigma_2(a)$.
Let us consider one more example. Let $K = \mathbb{Q}(i)$ and let $f(X) = X^4 - 2$, so that $L = K(\sqrt[4]{2}) = \mathbb{Q}(i, \sqrt[4]{2})$. We find all automorphisms of L over K defining:

```
>alias(c=RootOf(X^2+1))
```

$$c$$

```
>alias(d=RootOf(X^4-2))
```

$$c, d$$

```
>factor(X^4-2,{c,d})
```

$$(-X + cd)(X + cd)(X + d)(-X + d)$$

Thus, we have four automorphisms of $L = \mathbb{Q}(i, \sqrt[4]{2})$ over $K = \mathbb{Q}(i)$ given by $\sigma(d) = \pm d, \pm cd$, where $c = i, d = \sqrt[4]{2}$.

6.17. Find the orders of the groups $G = G(K/\mathbb{Q})$ and find all automorphisms of the field K when:

(a) $K = \mathbb{Q}(a)$, a is a zero of $f(X) = X^6 - 2X^3 - 1$;
(b) $K = \mathbb{Q}(a)$, a is a zero of $f(X) = X^8 - 2X^4 + 4$;
(c) $K = \mathbb{Q}(a)$, a is a zero of $f(X) = X^{12} - 10X^6 + 1$.

Chapter 7
Normal Extensions

In this chapter, we look at the splitting fields of polynomials, emphasizing one very important property of such fields. This property is contained in the definition of a normal extension. An extension $L \supseteq K$ is **normal** if every irreducible polynomial $f(X) \in K[X]$ which has one zero in L has all its zeros in L (that is, L contains a splitting field of $f(X)$). This definition works equally well for any field extensions (also infinite), but we restrict our interests here to finite extensions and find that normal extensions and splitting fields of polynomials form exactly the same class. In the next chapter, we discuss so-called separable extensions, which will give the last ingredient leading to our final goal—the Galois extensions (exactly those which are both normal and separable).

T.7.1. *A finite extension $L \supseteq K$ is normal if and only if L is a splitting field of a polynomial with coefficients in K.*

For every finite field extension, it is possible to find a kind of minimal normal extension containing it. Moreover, it is unique in a suitable sense. The definition is as follows: N is called a **normal closure** of $L \supseteq K$ if $N \supseteq L$ is a field extension such that $N \supseteq K$ is normal and if $N \supseteq N' \supseteq L$, where N' is a normal extension of K, then $N' = N$.

T.7.2. *Let $L = K(\alpha_1, \ldots, \alpha_n)$ be a finite extension. Then a normal closure of $L \supseteq K$ is unique up to a K-isomorphism. More precisely, every normal closure of $L \supseteq K$ is a splitting field over K of $f = f_1 \cdots f_n$, where f_i is the minimal polynomial of α_i over K.*

Exercises 7

7.1. Which of the following extensions $L \supset K$ are normal?

(a) $L = \mathbb{Q}(\sqrt[4]{2})$, $K = \mathbb{Q}$;
(b) $L = \mathbb{Q}(\sqrt[3]{2})$, $K = \mathbb{Q}$;
(c) $L = \mathbb{Q}(\sqrt{2}, \sqrt{3})$, $K = \mathbb{Q}$;
(f) $L = \mathbb{Q}(\sqrt[4]{2})$, $K = \mathbb{Q}(\sqrt{2})$;
(g) $L = \mathbb{Q}(\sqrt[4]{2}, i)$, $K = \mathbb{Q}$;
(h) $L = \mathbb{Q}(X)$, $K = \mathbb{Q}(X^3)$;

J. Brzeziński, *Galois Theory Through Exercises*, Springer Undergraduate Mathematics Series, https://doi.org/10.1007/978-3-319-72326-6_7

(d) $L = \mathbb{C}$, $K = \mathbb{R}$;
(e) $L = \mathbb{Q}(\sqrt[3]{2}, i)$, $K = \mathbb{Q}$;
(i) $L = \mathbb{C}(X)$, $K = \mathbb{C}(X^3)$;
(j) $L = \mathbb{F}_3(X)$, $K = \mathbb{F}_3(X^2)$.

7.2. Find a normal closure of the following field extensions $L \supset K$:

(a) $L = \mathbb{Q}(\sqrt[4]{2})$, $K = \mathbb{Q}$;
(b) $L = \mathbb{Q}(\sqrt{2}, \sqrt[3]{2})$, $K = \mathbb{Q}$;
(c) $L = \mathbb{Q}(\varepsilon), \varepsilon^5 = 1$, $\varepsilon \neq 1$, $K = \mathbb{Q}$;
(d) $L = \mathbb{Q}(X)$, $K = \mathbb{Q}(X^3)$;
(e) $L = \mathbb{Q}(X)$, $K = \mathbb{Q}(X^4)$;
(f) $L = \mathbb{F}_3(X)$, $K = \mathbb{F}_3(X^4)$.

7.3. Let $L \supseteq M \supseteq K$ be field extensions.

(a) Let $L \supseteq M$ and $M \supseteq K$ be normal extensions. Is $L \supseteq K$ normal?
(b) Let $L \supseteq K$ be normal. Is $L \supseteq M$ normal?
(c) Let $L \supseteq K$ be normal. Is $M \supseteq K$ normal?

7.4. (a) Let $L \supseteq K$ be a normal extension and $\alpha, \beta \in L$ two zeros of an irreducible polynomial with coefficients in K. Show that there is an automorphism $\sigma \in G(L/K)$ such that $\sigma(\alpha) = \beta$.
(b) Show that if L is a finite normal extension of K and $\tau : M \to M'$ is a K-isomorphism of two subfields M and M' of L both containing K, then there is an automorphism $\sigma \in G(L/K)$ whose restriction to M is equal to τ (in particular, every automorphism of M over K has an extension to an automorphism of L over K).

7.5. Let $L \supseteq M \supseteq K$ be field extensions.
(a) Show that if $M \supseteq K$ is normal and $\sigma \in G(L/K)$, then $\sigma M = M$.
(b) Show that if $L \supseteq K$ is normal, then $M \supseteq K$ is normal if and only if $\sigma M = M$ for each $\sigma \in G(L/K)$.

7.6. Let $L \supseteq K$ and M_1, M_2 be two fields between K and L. Prove that
(a) if $M_1 \supseteq K$ and $M_2 \supseteq K$ are normal, then $M_1M_2 \supseteq K$ and $M_1 \cap M_2 \supseteq K$ are normal;
(b) if $M_1 \supseteq K$ is normal, then $M_1M_2 \supseteq M_2$ is normal.

7.7. Let $L \supseteq K$ be a field extension and $\alpha, \beta \in L$. Show that if $K(\alpha) \supseteq K$ and $K(\beta) \supseteq K$ are normal extensions and $K(\alpha) \cap K(\beta) = K$, then $[K(\alpha, \beta) : K] = [K(\alpha) : K][K(\beta) : K]$.

7.8. Let $K \subset L$ be a normal extension and let $f(X)$ be a monic polynomial irreducible over K but reducible over L. Show that for every two monic irreducible factors $f_i(X)$ and $f_j(X)$ of $f(X)$ in $L[X]$, there is an automorphism $\sigma_{ij} : L \to L$ such that $\sigma_{ij}(f_i(X)) = f_j(X)$ (σ_{ij} maps the coefficients of $f_i(X)$ onto the corresponding coefficients of $f_j(X)$). In particular, all irreducible factors of $f(X)$ in $L[X]$ have the same degree.

7.9. An irreducible polynomial $f \in K[X]$ is called **normal** if a splitting field K_f of f over K can be obtained by extending K by only one of the zeros of f, that is, $K_f = K(\alpha)$, where $f(\alpha) = 0$.
(a) Show that an irreducible polynomial $f \in K[X]$ is normal if and only if its splitting field over K has degree equal to the degree of $f(X)$.
(b) Show that the binomials $X^n - 2$ over $\mathbb{Q}$ are not normal when $n > 2$.

Using Computers 7

In the exercises below you can use the command `galois(f(X))`, which gives the Galois group of the polynomial $f(X)$ over $\mathbb{Q}$ and the factorization of $f(X)$ over the extension of $\mathbb{Q}$ by a zero of $f(X)$ using `factor(f(X),a)`.

7.10. For the given polynomial $f(X)$ and its zero a find the degree over $\mathbb{Q}$ of a normal closure L of $K = \mathbb{Q}(a)$ and the minimal number of zeros of $f(X)$ which generate L over $\mathbb{Q}$:

(a) $f(X) = X^6 - 2X^3 - 1$;
(b) $f(X) = X^6 - X^3 + 1$;
(c) $f(X) = X^6 + 2X^3 - 2$;
(d) $f(X) = X^6 + 3X^2 + 3$;
(e) $f(X) = X^7 - X + 1$;
(f) $f(X) = X^7 + 7X^3 + 7X^2 + 7X - 1$.

7.11. Decide whether the polynomial $f(X)$ is normal over $\mathbb{Q}$ (see Exercise 7.9):

(a) $f(X) = X^5 - X^4 - 4X^3 + 3X^2 + 3X - 1$;
(b) $f(X) = X^5 - 5X + 12$;
(c) $f(X) = X^7 + X^6 - 12X^5 - 7X^4 + 28X^3 + 14X^2 - 9X + 1$;
(d) $f(X) = X^8 - 72X^6 + 180X^4 - 144X^2 + 36$ (see [D]).

Chapter 8
Separable Extensions

As we already noted in the previous chapter, our final goal is Galois extensions. The last property of field extensions which we need is separability. This property is rather common, for example, all extensions of fields of characteristic zero (thus, in particular, all number fields) have this property. All finite extensions of finite fields also have this property. This is a reason why there are sometimes "simplified presentations" of Galois theory in which one studies only fields of characteristic zero and finite fields. In that case it is not necessary to mention separability and the theoretical background necessary for the main theorems of Galois theory is more modest. We choose to discuss separability, since there are many branches of mathematics in which non-separable field extensions are important. However, in the sequel, those who wish to may disregard the word "separable", accepting that the results are formulated for fields of characteristic zero or finite fields. In fact, Theorem **T.8.1** says that all extensions of fields of characteristic 0 or algebraic extensions of finite fields are separable. In this chapter, we characterize separable extensions and prove a well-known "theorem on primitive elements" which says that a finite separable extension can be generated over its ground field by only one element. Since in this book the exercises not related to the theorem on primitive elements are often of a theoretical nature, we choose to present them with more detailed solutions (in order to not oversize the role of this chapter).

A polynomial $f \in K[X]$ is called **separable** over K if it has no multiple zeros (in any extension $L \supseteq K$—see **T.5.3**). We say that $\alpha \in L \supseteq K$ is a **separable element** over K if it is algebraic and its minimal polynomial over K is separable. We say that $L \supseteq K$ is a **separable extension** if every element of L is separable over K. A field K is called **perfect** if every algebraic extension of it is separable (that is, every irreducible polynomial in $K[X]$ is separable).

J. Brzeziński, *Galois Theory Through Exercises*, Springer Undergraduate Mathematics Series, https://doi.org/10.1007/978-3-319-72326-6_8

T.8.1. *(a) All fields of characteristic* 0 *and all finite fields are perfect.*
(b) If char $(K) = p$*, then an irreducible polynomial* $f \in K[X]$ *is not separable if and only if* $f' \equiv 0$*, which is equivalent to* $f(X) = g(X^p)$*, where* $g \in K[X]$*.*

We say that $\gamma \in L$ is a **primitive element** (sometimes called a **field primitive element**) of the extension $L \supseteq K$ if $L = K(\gamma)$.

T.8.2 Primitive Element Theorem. *If* $L = K(\alpha_1, \ldots, \alpha_n)$*, where* $\alpha_1, \ldots, \alpha_n$ *are algebraic and all with at most one exception are separable over K, then there is a primitive element of L over K. In particular, every finite separable extension has a primitive element.*

Recall from Chap. 4 that a field extension $K \subseteq L$ is called **simple** if it has a (field) primitive element, that is, there is a $\gamma \in L$ such that $L = K(\gamma)$.

Exercises 8

8.1. (a) Show that $\mathbb{F}_2(X)$ is not separable over the field $\mathbb{F}_2(X^2)$.
(b) For every prime number p give an example of a non-separable extension of a suitable field of characteristic p.
(c) Show that the field extension in (a) is normal.

8.2. Let $K \subset L$ be a quadratic field extension. Show that $L = K(\alpha)$, where $\alpha^2 = a \in K$ unless the characteristic of K is 2 and L is separable over K. Show that then we have $L = K(\alpha)$, where $\alpha^2 = \alpha + a$ for some $a \in K$. Show that $K \subset L$ is not separable if and only if the characteristic of K is 2 and $L = K(\alpha)$, where $\alpha^2 = a \in K$.

8.3. Let K be a field of finite characteristic p.
(a) Let $\alpha \in L \supseteq K$, where L is a finite extension of K. Show that α is separable over K if and only if $K(\alpha^p) = K(\alpha)$.
(b) Show that $K \subseteq L$ is separable if and only if $KL^p = L$ (see Exercise 2.12(c)).
(c) Show that K is perfect if and only if $K^p = K$.
(d) Show that if L is an algebraically closed field and K its subfield such that $[L : K] < \infty$, then L is both separable and normal over K.

8.4. (a) Show that $L = K(\alpha_1, \ldots, \alpha_n) \supseteq K$ is separable if and only if the elements $\alpha_1, \ldots, \alpha_n$ of L are separable over K.
(b) Show that a normal closure of a finite separable extension $L \supseteq K$ (see p. 43) is a separable extension of K.

8.5. Let $L \supseteq K$ be a field extension. Show that all elements $\alpha \in L$ separable over K form a subfield L_s between K and L.

Remark. The degree $[L_s : K]$ is called the **separable degree** of L over K and is denoted by $[L : K]_s$ (if $[L : K] < \infty$, then this means that $[L : K]_s$ divides $[L : K]$ and $[L : K]_s = [L : K]$ when $L \supseteq K$ is separable).

8.6. Let $L \supseteq K$ and char $(K) = p$. Show that for each $\alpha \in L$ there exists an exponent p^r such that α^{p^r} is separable over K (that is, $\alpha^{p^r} \in L_s$, where L_s is defined in Exercise 8.5).

8.7. (a) Let $N \supseteq L \supseteq K$, where N is a finite normal extension of K. Show that the number of different restrictions $\sigma|_L$, where $\sigma \in G(N/K)$ equals $[L : K]_s$ (see Exercise 8.5).
(b) Let $K \subseteq L$ be a finite field extension and let N be an algebraically closed field (see the definition on p. 21). Let $\tau : K \to N$ be an embedding of fields. Show that $[L : K]_s$ is equal to the number of different embeddings $\sigma : L \to N$ whose restriction to K equals τ.

8.8. Let L be a finite extension of K and let $L \supseteq M \supseteq K$. Show that the extension $L \supseteq K$ is separable if and only if the extensions $L \supseteq M$ and $M \supseteq K$ are separable.

8.9. Let L be a finite extension of K and $L \supseteq M \supseteq K$. Show that $[L : K]_s = [L : M]_s[M : K]_s$ (see Exercise 8.5).

8.10. Find a primitive element of $L \supseteq K$ when:

(a) $L = \mathbb{Q}(\sqrt{2}, \sqrt{3})$, $K = \mathbb{Q}$;
(b) $L = \mathbb{Q}(\sqrt{2} - i, \sqrt{3} + i)$, $K = \mathbb{Q}$;
(c) $L = \mathbb{Q}(\sqrt{2} + \sqrt{3}, \sqrt{2} + i, \sqrt{3} - i)$, $K = \mathbb{Q}$;
(d) $L = \mathbb{R}(X, Y)$, $K = \mathbb{R}(X^2, Y^2)$;
(e) $L = \mathbb{Q}(X, Y)$, $K = \mathbb{Q}(X + Y, XY)$;
(f) $L = \mathbb{Q}(\sqrt{2}, \sqrt[3]{2})$, $K = \mathbb{Q}$.

8.11. Show that if $L = \mathbb{F}_2(X, Y)$, $K = \mathbb{F}_2(X^2, Y^2)$, then the extension $L \supset K$ is not simple, that is, one cannot find a $\gamma \in L$ such that $L = K(\gamma)$.

8.12. (a) Let $K \subset L = K(\gamma)$ be a finite field extension and let M be a field such that $K \subseteq M \subseteq L$. Show that M is generated over K by the coefficients of the minimal polynomial of γ over M.
(b) Show that a field extension $L \supseteq K$ is simple and algebraic if and only if the number of fields between K and L is finite.

8.13. Give an example of a finite extension $L \supseteq K$ such that there are infinitely many fields between K and L.

8.14. Let $L \supseteq \mathbb{Q}$ be a finite and normal extension of odd degree. Show that $L \subseteq \mathbb{R}$.

8.15. Let $K \subseteq L$ be an extension of finite fields.
(a) Show that $L = K(\gamma)$ for any generator of the cyclic group L^* (see Exercise 5.8).
(b) Is it true that $L = K(\gamma)$ implies that γ is a generator of the group L^*?

Remark. It follows from (a) that any generator γ of the cyclic group L^* is a (field) primitive element of the extension $K \subseteq L$. In the theory of finite fields, such generators are also called primitive elements of this extension. In order to avoid misunderstandings, we shall use the longer term **group primitive elements** (see [R], p. 214).
(c) Let $K = \mathbb{F}_3$ and $L = \mathbb{F}_{3^n}$. Find the numbers of group primitive and field primitive elements in L for $n = 2, 3, 4$.

8.16. Let M_1, M_2 be two fields between K and L. Prove that if M_1 and M_2 are separable extensions of K, then also M_1M_2 and $M_1 \cap M_2$ are separable extensions of K.

Using Computers 8

The proof of **T.8.2** is effective if the zeros of the minimal polynomials of α, β such that $L = K(\alpha, \beta)$ are known (and K is infinite). Then it is possible to find $c \in K$ such that $\gamma = \alpha + c\beta$ is a primitive element of L over K (see the proof of **T.8.2** on p. 122). Another possibility is to try to guess γ and check that $L = K(\gamma) = K(\alpha, \beta)$. This can be done if we have the minimal polynomial of γ and we are able to define L, for example, using the command `RootOf`. For example, we want to check that $\mathbb{Q}(\sqrt[3]{2}, \sqrt[3]{3}) = \mathbb{Q}(\sqrt[3]{2} + \sqrt[3]{5})$. In some way (say, using Maple), we find the minimal polynomial of $\sqrt[3]{2} + \sqrt[3]{5}$, which is $f(X) = X^9 - 15X^6 - 87X^3 - 125$. Then we define:

```
>f:=X->X^9-15X^6-87X^3-125
>alias(c=RootOf(f(X))
```

We can define `r:=`$\sqrt[3]{2} + \sqrt[3]{5}$ and check `simplify(f(r))`, which gives 0. We can also check that c is just r comparing `evalf(r)` and `evalf(c)` (note that f has only one real zero). Now we take

```
>factor(X^3-2,c)
```

$$-\frac{1}{10125}(-225X^2 - 545X - 175Xc^4 + 10Xc^7 - 209c^2 - 130c^5 + 7c^8)(45X - 109c - 35c^4 + 2c^7)$$

which shows that $\sqrt[3]{2} = \frac{1}{45}(109c + 35c^4 - 2c^7) \in \mathbb{Q}(c)$ and similarly for $\sqrt[3]{3}$. Thus $\mathbb{Q}(\sqrt[3]{2}, \sqrt[3]{3}) \subseteq \mathbb{Q}(\sqrt[3]{2} + \sqrt[3]{3})$, and the inverse inclusion is evident. But Maple can be successfully used in a similar way in order to prove that other elements than those chosen in the proof of **T.8.2** are primitive.

8.17. Find a primitive element c of $L \supseteq K$ when:

(a) $K = \mathbb{Q}, L = \mathbb{Q}(\sqrt{2}, \sqrt[3]{2})$ (see Exercise 8.10(f));
(b) $K = \mathbb{Q}, L = \mathbb{Q}(\sqrt[3]{3}, \sqrt[5]{5})$;
(c) $K = \mathbb{Q}(\sqrt[3]{2}), L = K(\sqrt[3]{3}, \sqrt[3]{5})$;
(d) $K = \mathbb{Q}(\sqrt[4]{2}), L = K(\sqrt[5]{2}, \sqrt[6]{2})$.

Chapter 9
Galois Extensions

Now we have all ingredients necessary to define Galois extensions. The main theorems of Galois theory apply to this type of extension and are proved in this chapter. In the exercises, we find many examples and many interesting properties of Galois extensions.

A field extension $L \supseteq K$ is called **Galois** if it is normal and separable.

T.9.1. *Let $L \supseteq K$ be a finite field extension and $G(L/K)$ its Galois group. Then the following conditions are equivalent:*
(a) $[L : K] = |G(L/K)|$;
(b) $L^{G(L/K)} = K$;
(c) There is a group G of K-automorphisms of L such that $K = L^G$ and then, $G = G(L/K)$;
(d) $L \supseteq K$ is normal and separable;
(e) L is a splitting field of a separable polynomial over K.

If $L \supseteq K$ is a field extension, $\mathcal{F}$ the set of all fields between K and L and $\mathcal{G}$ the set of all subgroups of $G(L/K)$, then we define two functions, which give what is called the **Galois correspondence**:

$$f : \mathcal{G} \to \mathcal{F} \quad \text{and} \quad g : \mathcal{F} \to \mathcal{G}$$

in the following way:

$$f(H) = L^H = \{x \in L : \forall_{\sigma \in H} \sigma(x) = x\}$$

and

$$g(M) = G(L/M) = \{\sigma \in G(L/K) : \forall_{x \in M} \sigma(x) = x\}.$$

J. Brzeziński, *Galois Theory Through Exercises*, Springer Undergraduate Mathematics Series, https://doi.org/10.1007/978-3-319-72326-6_9

T.9.2 Fundamental Theorem of Galois Theory. *If $L \supseteq K$ is a finite Galois extension, then f and g are mutually inverse bijections between the set $\mathcal{F}$ of all fields between K and L and the set $\mathcal{G}$ of all subgroups of $G(L/K)$, that is, $f \circ g = id_{\mathcal{F}}$, $g \circ f = id_{\mathcal{G}}$. Moreover, the functions f and g are inclusion-reversing, that is, $f(H_1) \supseteq f(H_2)$ if $H_1 \subseteq H_2$, and $g(M_1) \supseteq g(M_2)$ if $M_1 \subseteq M_2$.*

Sometimes both the last theorem and the following one are called the fundamental theorem of Galois theory:

T.9.3. *Let $L \supseteq K$ be a Galois extension and M a field between K and L.*
(a) The extension $L \supseteq M$ is a Galois extension.
(b) The extension $M \supseteq K$ is a Galois extension if and only if $G(L/M)$ is normal in $G(L/K)$. If this holds, then $G(M/K) \cong G(L/K)/G(L/M)$.

Exercises 9

9.1. Which of the following extensions $L \supseteq K$ are Galois?
(a) $K = \mathbb{Q}$, $L = \mathbb{Q}(\sqrt[3]{2})$;
(b) $K = \mathbb{Q}$, $L = \mathbb{Q}(\sqrt[4]{2})$;
(c) $K = \mathbb{Q}(\sqrt{2})$, $L = \mathbb{Q}(\sqrt[4]{2})$;
(d) $K = \mathbb{Q}(i)$, $L = \mathbb{Q}(i, \sqrt[4]{2})$;
(e) $K = \mathbb{Q}(X^2)$, $L = \mathbb{Q}(X)$;
(f) $K = \mathbb{F}_p(X^2)$, $L = \mathbb{F}_p(X)$, p a prime number;
(g) $K = \mathbb{F}_2(X^2 + X)$, $L = \mathbb{F}_2(X)$;
(h) $K = \mathbb{R}(X^3)$, $L = \mathbb{R}(X)$.

9.2. Find all subgroups of the Galois group $G(L/K)$ of the splitting field L of the polynomial f as well as all corresponding subfields M between K and L when:
(a) $K = \mathbb{Q}, f(X) = (X^2 - 2)(X^2 - 5)$;
(b) $K = \mathbb{Q}, f(X) = (X^4 - 1)(X^2 - 5)$;
(c) $K = \mathbb{Q}, f(X) = X^5 - 1$;
(d) $K = \mathbb{Q}, f(X) = X^4 + 1$;
(e) $K = \mathbb{Q}(i), f(X) = X^4 - 2$;
(f) $K = \mathbb{Q}, f(X) = X^3 - 5$;
(g) $K = \mathbb{Q}, f(X) = X^4 + X^2 - 1$;
(h) $K = \mathbb{Q}(i), f(X) = X^3 - 1$.

9.3. (a) Let $f(X) \in K[X]$ be a polynomial of degree n over a field K and let $K_f = K(\alpha_1, \ldots, \alpha_n)$ be a splitting field of $f(X)$ over K, where α_i are all zeros of $f(X)$ in K_f. Show that the permutations σ of the indices i of the zeros α_i corresponding to the automorphisms $\sigma \in G(K_f/K)$ according to $\sigma(\alpha_i) = \alpha_{\sigma(i)}$ form a subgroup of S_n.
(b) Give a description of the Galois group $G(K_f/K)$ as a permutation subgroup of S_n ($n = \deg f$) for every polynomial $f(X)$ in Exercise 9.2.

9.4. (a) Let $f(X) \in K[X]$ be a separable polynomial and $f(X) = f_1(X) \cdots f_k(X)$ its factorization in $K[X]$ into a product of irreducible polynomials $f_i(X)$. Show that the permutation group $\mathrm{Gal}(K_f/K)$ (with the notations as in Exercise 9.3(a)) consists of permutations which have k orbits on the set $\{1, \ldots, n\}$ (see p. 267).
(b) Show that the permutation group $\mathrm{Gal}(K_f/K)$ in (a) is transitive (that is, for each pair $i, j \in \{1, \ldots, n\}$, there is a $\sigma \in \mathrm{Gal}(K_f/K)$ such that $\sigma(i) = j$) if and only if the polynomial $f(X)$ is irreducible in $K[X]$.
(c) Show that if the Galois group $G(K_f/K)$ of a polynomial $f(X) \in K[X]$ has k orbits in its action on the set $X_f = \{\alpha_1, \ldots, \alpha_n\}$ of the zeros of $f(X)$, then $f(X)$ is a product of k irreducible polynomials in $K[X]$.

9.5. Show that the extension $L \supseteq K$ is Galois, find the Galois group $G(L/K)$, all its subgroups and the corresponding subfields between K and L when:
(a) $K = \mathbb{Q}, L = \mathbb{Q}(\sqrt{2}, i)$;
(b) $K = \mathbb{Q}, L = \mathbb{Q}(\sqrt[3]{2}, \varepsilon), \varepsilon^3 = 1, \varepsilon \neq 1$;
(c) $K = \mathbb{Q}, L = \mathbb{Q}(\sqrt[4]{2}, i)$;
(d) $K = \mathbb{R}(X^2 + \frac{1}{X^2}), L = \mathbb{R}(X)$;
(e) $K = \mathbb{R}(X^2, Y^2), L = \mathbb{R}(X, Y)$;
(f) $K = \mathbb{R}(X^2 + Y^2, XY), L = \mathbb{R}(X, Y)$.

9.6. Show that every quartic Galois extension of $\mathbb{Q}$ whose Galois group is the Klein[1] four-group[2] V (that is, a group isomorphic to the group $\mathbb{Z}_2 \times \mathbb{Z}_2$) has the form $K = \mathbb{Q}(\sqrt{d_1}, \sqrt{d_2})$, where d_1, d_2 are nonequal square-free integers not equal to 1.

9.7. Is it true that if $L \supseteq M$ and $M \supseteq K$ are Galois extensions, then $L \supseteq K$ is a Galois extension?

9.8. Let G be the automorphism group of the field $\mathbb{F}_3(X)$ consisting of all the automorphisms $X \to aX + b$, where $a \neq 0$ (see Exercise 6.8). Find all subgroups of G and the corresponding subfields between the field $\mathbb{F}_3(X)^G$ and $\mathbb{F}_3(X)$.

9.9. Let $L \supseteq K$ be a Galois extension and M a field between K and L. Show that $[M : K] = |G(L/K)|/|G(L/M)|$.

9.10. Let $L \supseteq K$ be a Galois extension and M a field between K and L.
(a) Show that $H(L/M) = \{\sigma \in G(L/K) | \sigma(M) = M\}$ is a group and $G(L/M)$ its normal subgroup.
(b) Prove that the number of different fields between K and L which are K-isomorphic to M equals

$$\frac{[M : K]}{|G(M/K)|} = \frac{|G(L/K)|}{|H(L/M)|}.$$

(c) Show that $M \supseteq K$ is Galois if and only if every automorphism $\sigma \in G(L/K)$ restricted to M maps M onto M, that is, $H(L/M) = G(L/M)$.
(d) Let $\mathcal{N}(G(L/M))$ be the normaliser of the group $G(L/M)$ in $G(L/K)$ (the normaliser is defined on p. 241). Show that $H(L/M) = \mathcal{N}(G(L/M))$ and $G(M/K) = \mathcal{N}(G(L/M))/G(L/M)$.
(e) Let $\sigma \in G(L/M)$. Show that $G(L/\sigma(M)) = \sigma G(L/M)\sigma^{-1}$.

9.11. Let $L \supseteq K$ be a Galois extension with Galois group $G(L/K) = \{\sigma_1, \ldots, \sigma_n\}$. Let $\alpha \in L$ and let $G(L/K)\alpha$ be the orbit of α under the action of $G(L/K)$, that is, the set of all different images $\sigma_i(\alpha)$ for $\sigma_i \in G(L/K)$. Let $G_\alpha = \{\sigma \in G(L/K) \mid \sigma(\alpha) = \alpha\} = G(L/K(\alpha))$. Show that:

(a) $| G(L/K)\alpha | = \frac{|G(L/K)|}{|G_\alpha|} = [K(\alpha) : K]$;

---

[1] Felix Klein, 25 April 1849–22 June 1925.

[2] The smallest non-cyclic group was named "Viergruppe" by Felix Klein in 1884. It is usually denoted by V or V_4.

(b) The minimal polynomial of α over K is

$$f_\alpha(X) = \prod_{\sigma(\alpha)\in G(L/K)\alpha} (X - \sigma\alpha);$$

(c) $F_\alpha(X) = f_\alpha(X)^{|G_\alpha|}$, where

$$F_\alpha(X) = \prod_{\sigma\in G(L/K)} (X - \sigma\alpha).$$

9.12. Let $K \subseteq L$ be a Galois extension and H a subgroup of the Galois group $G(L/K)$.
(a) Show that $\mathrm{Tr}_H : L \to L^H$ is a surjective linear mapping (see (6.1)).
(b) Let $\alpha_1, \ldots, \alpha_n$ be a basis of L over K. Show that $L^H = K(\mathrm{Tr}_H(\alpha_1), \ldots, \mathrm{Tr}_H(\alpha_n))$.

9.13. Let $L \supseteq K$ be finite fields and let $|K| = q$.
(a) Show that $L \supseteq K$ is a Galois extension.
(b) Show that the Galois group $G(L/K)$ is cyclic and generated by the automorphism $\sigma(x) = x^q$ (called the **Frobenius**[3] **automorphism** of L over K).
(c) Express $\mathrm{Nr}_{L/K}(x)$ as a power of x for $x \in L$ and show that every element of K is a norm of an element of L (see (6.1)).

9.14. Let $L = \mathbb{F}_2(\alpha)$, where α is a zero of the polynomial $X^4 + X^3 + 1 \in \mathbb{F}_2[X]$. Show that $G(L/\mathbb{F}_2)$ contains exactly one subgroup H of order 2 and find $\beta \in L$ such that $L^H = \mathbb{F}_2(\beta)$.

9.15. Let $M_1 \supseteq K$ and $M_2 \supseteq K$ be finite Galois extensions and L a field, which contains both M_1 and M_2. Show that:
(a) $M_1M_2 \supseteq K$ and $M_1 \cap M_2 \supseteq K$ are Galois extensions;
(b) $G(M_1M_2/M_2) \cong G(M_1/(M_1 \cap M_2))$;
(c) $G(M_1M_2/K) \cong G(M_1/K) \times G(M_2/K)$ if $M_1 \cap M_2 = K$.
(d) If there is an isomorphism of M_1 and M_2 over K (that is, there is an isomorphism $\tau : M_1 \to M_2$ which is the identity on K), then the Galois groups $G(L/M_1)$ and $G(L/M_2)$ are conjugate in $G(L/K)$ (that is, there is a $\sigma \in G(L/K)$ such that $G(L/M_1) = \sigma G(L/M_2)\sigma^{-1}$).

9.16. Construct a Galois extension $L \supset \mathbb{Q}$ whose Galois group is:

(a) C_2;	(b) C_3;	(c) C_4;
(d) C_5;	(e) $C_2 \times C_2$	(f) $C_2 \times C_4$;
(g) $C_2 \times C_2 \times C_2$;	(h) $C_3 \times C_3$;	(i) D_4 (the symmetry group of a square);
(j) S_3;	(k) S_4;	(l) A_4.

Remark. **The Inverse Galois Problem** asks whether every finite group is (isomorphic to) the Galois group of some Galois extension of the rational numbers $\mathbb{Q}$. This problem is still unsolved even if for infinitely many finite groups G, it is possible

[3]Ferdinand Georg Frobenius, 26 October 1849–3 August 1917.

to construct suitable Galois extensions $K \supset \mathbb{Q}$ having G as its Galois group. For example, all abelian groups, and even all solvable groups (see Chap. 12) have this property. Also the symmetric groups S_n and alternating groups A_n can be realized as Galois groups over the rational numbers (see Exercises 13.4(c) and 15.11 for S_n and [MM], 9.2 for A_n). Notice that every finite group can be realized as the group of all automorphisms of a finite algebraic extension $K \supseteq \mathbb{Q}$ (see [FK]) when we do not require that K is a Galois extension of $\mathbb{Q}$.
More generally, the same problem can be formulated for other fields K and the corresponding question is whether a given finite group can be realized as the Galois group of a suitable field extension of K. This question has an answer in some cases and remains open in many others (see the presentations of the inverse Galois problem in e.g. [MM]). The inverse Galois problem is closely related to Noether's problem (see p. 39 and [F] for more information).

9.17. Assume that every symmetric group S_p, where p is a prime number, is a Galois group of a Galois extension over the rational numbers (this fact is proved in Exercise 13.4(c)). Show that for every finite group G there exists a Galois extension of number fields whose Galois group is isomorphic to G by embedding G in some group S_p (see A.8.3).

9.18. (a) Let p be a prime number and let ε be a primitive p-th root of unity (e.g. $\varepsilon = e^{\frac{2\pi i}{p}}$). Show that $K = \mathbb{Q}(\varepsilon)$ is a splitting field of $X^p - 1$ and has degree $p - 1$ over $\mathbb{Q}$. Show that the automorphisms of K are $\sigma_k(\varepsilon) = \varepsilon^k$, where $1 \leq k < p$ and $G(K/\mathbb{Q})$ is isomorphic to the cyclic group $(\mathbb{Z}/p\mathbb{Z})^*$ when σ_k corresponds to k.
(b) Prove that every cyclic group is a Galois group of a Galois extension $L \supseteq \mathbb{Q}$.

Remark. We discuss, in more detail, the field extensions of $\mathbb{Q}$ by roots of unity in Chap. 10 (in particular in **T.10.1**).

9.19. Is it true that $L \supset M \supset K$, where $[L : K] < \infty$ and $|G(L/K)| = 1$, implies $|G(M/K)| = 1$?

9.20. Let $L \supseteq K$ be a Galois extension and M_1, M_2 two fields between K and L. Let $G(L/M_1) = H_1$, $G(L/M_2) = H_2$. Find $G(L/M_1M_2)$ and $G(L/M_1 \cap M_2)$.

9.21. Let $L \supseteq K$ be a Galois extension with an abelian Galois group $G(L/K)$. Let f be the minimal polynomial of $\alpha \in L$ over K. Show that f is normal over K (that is, f has all its zeros in $K(\alpha)$—see Exercise 7.9).

9.22. Let $K \subseteq L$ be a Galois extension and let N be a field containing L. Let M be any field such that $K \subseteq M \subseteq N$.
(a) Show that $LM \supseteq M$ is a Galois extension and the natural mapping of $\sigma \in G(LM/M)$ onto its restriction to L is an injection of $G(LM/M)$ into $G(L/K)$.
(b) Consider $G(LM/M)$ as a subgroup of $G(L/K)$ using the restriction in (a) and show that $L^{G(LM/M)} = L \cap M$ (so $[LM : M] = [L : L \cap M]$).
(c) Show that $G(LM/M) = G(L/K)$ if and only if $L \cap M = K$.

Remark. The property of Galois extensions given by (a) and (b) is often called the **theorem on natural irrationalities**.

9.23. Let $K \subseteq L$ be a Galois extension of number fields and assume that K is a real field, that is, $K \subseteq \mathbb{R}$. Show that either L is also real or L contains a subfield M such that $[L : M] = 2$ and M is real.

9.24. Let $K \subseteq L$ be a Galois extension with Galois group $G(L/K) = \{\sigma_1, \dots, \sigma_n\}$. Show that if $\alpha \in L$ has n different images $\sigma_i(\alpha)$, then $L = K(\alpha)$, that is, the element α is primitive for the extension $K \subseteq L$.

9.25. Let K be a field of characteristic different from 2 and $f(X) \in K[X]$ be a separable monic polynomial of degree n. Denote by $\alpha_1, \dots, \alpha_n$ the zeros of the polynomial $f(X)$ in its splitting field $L = K(\alpha_1, \dots, \alpha_n)$ with (permutation) Galois group $\mathrm{Gal}(L/K)$. Let $\Delta(f) = \prod_{1 \le i < j \le n}(\alpha_i - \alpha_j)^2$ be the discriminant of $f(X)$ (see A.10.2). (a) Show that $\Delta(f) \in K$.
(b) Show that $K(\sqrt{\Delta(f)})$ is the fixed field of the subgroup $\mathrm{Gal}_0(L/K)$ of $\mathrm{Gal}(L/K)$ consisting of all even permutations, in particular, all permutations in $\mathrm{Gal}(L/K)$ are even if and only if $\Delta(f)$ is a square in K.

9.26. (a) Let K be a field of characteristic p. Show that for any $a \in K$, the polynomial $f(X) = X^p - X + a$ has p zeros in K or no zeros in K.
(b) Let K be a field of characteristic p such that $f(X) = X^p - X + 1$ has no zero in K. Show that if α is a zero of $f(X)$ in its splitting field L over K, then all other zeros of this polynomial are $\alpha + i$ for $i = 1, \dots, p-1$ and $L = K(\alpha)$. Deduce that L is a Galois extension of $K, f(X)$ is irreducible over K, $[L : K] = p$ and that $G(L/K)$ is a cyclic group of order p (see also Exercise 11.8).
(c) Show that the polynomial $X^p - X + a$ has no zeros in $\mathbb{F}_p$ when $a \in \mathbb{F}_p$ and $a \neq 0$.
(d) Apply (b) and (c) to the polynomial $f(X) = X^p - X + 1$ over the field $\mathbb{F}_p$ and show that it is irreducible in $\mathbb{F}_p[X]$.

Using Computers 9

If $K \subset L$ is a Galois field extension, then a description of its Galois group $G(L/K)$ is easiest if L is represented as a simple extension $L = K(a)$ and the minimal polynomial $f(X)$ of a over K is known. Then it is possible to factorize $f(X)$ over L (using `factor(f(X),a)` where a is defined by `alias(a=RootOf(f(X)))` or the corresponding commands `Factor` and `alias(a=RootOf(f(X)` **mod** `p))` for finite fields) and describe all automorphisms σ of L over K sending a onto all zeros of $f(X)$ in L (which can be obtained from the factorization of $f(X)$ over L). If $L = K(\alpha_1, \dots, \alpha_k)$ for $\alpha_i \in L$ (in Maple it is easier to denote the zeros by, say, *ai*), then as we know from **T.7.2**, the field L is a splitting field of the polynomial $f(X)$, which is the product of the minimal polynomials $f_i(X)$ of α_i. In a particular case when $[L : K] = \prod_{i=1}^{k} \deg f_i(X)$, every automorphism is uniquely given by a mapping of every α_i on a fixed zero of $f_i(X)$. If L is obtained in several steps as a splitting field of a polynomial $f(X)$, then it is also possible to describe this field by gradually adjoining its zeros as in the following example:

```
>a:=RootOf(X^4-2)
```

$$a := RootOf(\_Z^4 - 2)$$

```
>factor(X^4-2,a)
```

$$-(X^2 + RootOf(\_Z^4 - 2)^2)(X + RootOf(\_Z^4 - 2))(-X + RootOf(\_Z^4 - 2))$$

so $\mathbb{Q}(a)$ is not a splitting field. We continue and adjoin another zero:

```
>b := RootOf(X^2+RootOf(_Z^4-2)^2, a)
```

$$b := RootOf(\_Z^2 + RootOf(\_Z^4 - 2)^2, RootOf(\_Z^4 - 2))$$

```
>factor(X^4-2, {a, b})
```

$$(-X + RootOf(\_Z^4 - 2))(-X + RootOf(\_Z^2 + RootOf(\_Z^4 - 2)^2, RootOf(\_Z^4 - 2)))$$

$$(X + RootOf(\_Z^2 + RootOf(\_Z^4 - 2)^2, RootOf(\_Z^4 - 2)))(X + RootOf(\_Z^4 - 2))$$

which using a, b can be rewritten as $X^4 - 2 = (-X + a)(-X + b)(X + b)(X + a)$. Thus $\mathbb{Q}(a, b)$ is a splitting field of $X^4 - 2$.

9.27. Describe the Galois groups of the polynomials in Exercise 7.11(a), (c), (d) using a chosen zero generating the splitting field.

9.28. Show that the polynomials $f(X) \in \mathbb{Q}[X]$ are normal, list all automorphisms of their splitting fields K and describe the Galois groups when:

(a) $f(X) = X^3 - 3X + 1$;
(b) $f(X) = X^6 - X^3 + 1$;
(c) $f(X) = X^6 + 3X^5 + 6X^4 + 3X^3 + 9X + 9$.

9.29. Find the orders of the Galois groups of the following polynomials and determine the minimal number of zeros of each polynomial which generate its splitting field over $\mathbb{Q}$:

(a) $f(X) = X^5 - 5X + 12$; (c) $f(X) = X^5 - X + 1$;
(b) $f(X) = X^5 + 20X + 16$; (d) $f(X) = X^5 + 15X + 12$.

9.30. (a) Construct an irreducible polynomial of degree 10 over $\mathbb{F}_2$ by using irreducible polynomials of degrees 2 and 5 whose splitting fields are subfields of $\mathbb{F}_{2^{10}}$.
(b) Construct an irreducible polynomial of degree 21 by a similar method as in (a) using irreducible polynomials of degrees 3 and 7 whose splitting fields are subfields of $\mathbb{F}_{2^{21}}$.

Remark. It is easy to answer both (a) and (b) using a program like Maple, since testing a few polynomials of degree 10 (in (a)) or 21 (in (b)) gives a way to find irreducible polynomials (the probability that a random polynomial of degree n over $\mathbb{F}_2$ is irreducible is about $1/n$—see Exercise 5.7). In fact, the polynomials $x^{10} + x^3 + 1$ and $x^{21} + x^2 + 1$ are irreducible, which means that their splitting fields over $\mathbb{F}_2$ are just $\mathbb{F}_{2^{10}}$ and $\mathbb{F}_{2^{21}}$, respectively. But the purpose of this exercise is to construct irreducible polynomials using our knowledge of Galois theory (not a random search). Nevertheless, when we present a solution in Chap. 19 (of a similar problem instead of the problem above), we use a computer when the computations are evident, but somewhat cumbersome.

Chapter 10
Cyclotomic Extensions

This chapter can be considered as an illustration of the general theory of Galois extensions in a special case, which plays a central role in number theory. We study field extensions, mostly of the rational numbers, generated by the roots of unity. Even if such fields are simple to describe in purely algebraic terms, they are amazingly rich as mathematical objects. We look at some of their properties, which find different applications in number theory and algebra.

If $\varepsilon \in \mathbb{C}$ is a **primitive** n-th root of unity, that is, $\varepsilon^n = 1$ and $\varepsilon^k \neq 1$, when $0 < k < n$ (for example, $\varepsilon = e^{\frac{2\pi i}{n}}$), then the field $\mathbb{Q}(\varepsilon)$ is called the **n-th cyclotomic field**.

T.10.1. *(a) The degree* $[\mathbb{Q}(\varepsilon) : \mathbb{Q}] = \varphi(n)$, *where* ε *is a primitive n-th root of unity and* φ *is Euler's function.*
(b) Each automorphism σ *in the Galois group* $G(\mathbb{Q}(\varepsilon)/\mathbb{Q})$ *is given by* $\sigma_k(\varepsilon) = \varepsilon^k$, *where* $k \in \{1, \ldots, n\}$ *and* $\gcd(k, n) = 1$. *The mapping* $\sigma_k \mapsto k \pmod n$ *gives an isomorphism* $G(\mathbb{Q}(\varepsilon)/\mathbb{Q}) \cong (\mathbb{Z}/n\mathbb{Z})^*$.

If K is an arbitrary field, then we can also study splitting fields of the polynomials $f_n(X) = X^n - 1$ over it. If K has characteristic $p \neq 0$ and p divides n, then $X^n - 1 = (X^{\frac{n}{p}} - 1)^p$, so $X^n - 1$ and $X^{\frac{n}{p}} - 1$ have the same splitting fields. Thus, in order to investigate the fields K_{f_n}, we can assume that the characteristic of K does not divide n. We always assume this in the exercises below. A consequence of this assumption is that the polynomial $X^n - 1$ has n different zeros (its derivative nX^{n-1} is relatively prime to it—see **T.5.3**). Thus the group of all n-th roots of 1, that is, the group of all solutions of the equation $X^n - 1 = 0$ in K_{f_n} has order n and is cyclic (see Exercise 5.8). The primitive n-th roots of unity over K are the generators of this group. If ε is one of them, then all others are given by ε^j, where $0 < j < n$ and $\gcd(j, n) = 1$ (see A.2.2). Thus the number of such generators is the value of Euler's function $\varphi(n)$ (for the definition of Euler's function, see p. 279).

J. Brzeziński, *Galois Theory Through Exercises*, Springer Undergraduate Mathematics Series, https://doi.org/10.1007/978-3-319-72326-6_10

The **n-th cyclotomic polynomial over** K is the polynomial

$$\Phi_{n,K}(X) = \prod_{\substack{0<k<n,\\ \gcd(k,n)=1}} (X - \varepsilon^k),$$

where ε is any generator of the group of n-th roots of unity in K_{f_n}. If $K = \mathbb{Q}$, then $\Phi_{n,K}(X)$ is usually denoted by $\Phi_n(X)$.

T.10.2. *Let K be a field whose characteristic does not divide n.*
(a) We have:

$$X^n - 1 = \prod_{d|n} \Phi_{d,K}(X) \quad \textit{and} \quad \Phi_{n,K}(X) = \prod_{d|n} (X^d - 1)^{\mu(\frac{n}{d})},$$

where μ denotes the Möbius function (see section "Some Arithmetical Functions" in Appendix *and* Exercise 5.7).
(b) The cyclotomic polynomials $\Phi_{n,K}(x)$ are monic and their coefficients are integer multiples of the identity (that is, the element 1) *of K.*
(c) All irreducible factors of $\Phi_{n,K}(x)$ are of the same degree.
(d) If $K = \mathbb{Q}$, then $\Phi_n(X) = \Phi_{n,\mathbb{Q}}(x)$ is irreducible over $\mathbb{Q}$.

A Galois field extension $K \subseteq L$ is called an **abelian extension** if its Galois group $G(L/K)$ is abelian. Cyclotomic field extensions are examples of abelian extensions.

Exercises 10

10.1. Find the cyclotomic polynomials $\Phi_n(X)$ for $n = 1, 2, \ldots, 10$.

10.2. The **kernel** $r(n)$ of an integer $n > 1$ is the product of all different prime numbers dividing n. Prove the following:
(a) $\Phi_p(X) = X^{p-1} + \cdots + X + 1$, where p is a prime;
(b) $\Phi_n(X) = \Phi_{r(n)}(X^{\frac{n}{r(n)}})$ when $n > 1$;
(c) $\Phi_{pn}(X) = \Phi_n(X^p)/\Phi_n(X)$, where p is a prime not dividing n.
Explain how to use (a)–(c) in order to compute the cyclotomic polynomials $\Phi_n(X)$. Compute $\Phi_{20}(X)$ and $\Phi_{105}(X)$.

10.3. Prove the following identities:
(a) $\Phi_n(X) = \Phi_k(X^{\frac{n}{k}})$, where k is any divisor of n such that $r(k) = r(n)$ (see the definition of $r(k)$ in Exercise 10.2);
(b) If r, s are relatively prime positive integers, then

$$\Phi_{rs}(X) = \prod_{d|r} (\Phi_s(X^d))^{\mu(\frac{r}{d})},$$

where μ denotes the Möbius function (see section "Some Arithmetical Functions" in Appendix).

10.4. Show that $\Phi_{2m}(X) = \Phi_m(-X)$, when $m > 1$ is an odd integer.

10.5. Let m, n be two relatively prime positive integers.
(a) Show that $\mathbb{Q}(\varepsilon_{mn}) = \mathbb{Q}(\varepsilon_m)\mathbb{Q}(\varepsilon_n)$ and $\mathbb{Q}(\varepsilon_m) \cap \mathbb{Q}(\varepsilon_n) = \mathbb{Q}$.
(b) Show that $\Phi_n(X)$ is irreducible over $\mathbb{Q}(\varepsilon_m)$.
(c) What can be said about $\mathbb{Q}(\varepsilon_m)\mathbb{Q}(\varepsilon_n)$ and $\mathbb{Q}(\varepsilon_m) \cap \mathbb{Q}(\varepsilon_n)$ when m, n are not necessarily relatively prime?

10.6. Find η such that $\mathbb{Q}(\eta)$ is the maximal real subfield of the n-th cyclotomic field $\mathbb{Q}(\varepsilon_n)$ $(n > 2)$ and show that $[\mathbb{Q}(\varepsilon_n) : \mathbb{Q}(\eta)] = 2$.

10.7. Let $K = \mathbb{Q}(\varepsilon_p)$ be p-th cyclotomic field, where p is a prime number. According to **T.10.1**, the group $G(K/\mathbb{Q})$ is isomorphic to $(\mathbb{Z}/p\mathbb{Z})^*$ and cyclic of order $p-1$ (see Exercise 9.18).
(a) Show that for each divisor d of $p-1$ there exists exactly one $M \subseteq K$ such that $[M : \mathbb{Q}] = d$.
(b) Let $\sigma \in G(K/\mathbb{Q}) = (\mathbb{Z}/p\mathbb{Z})^*$ generate this Galois group and let $\sigma(\varepsilon_p) = \varepsilon_p^g$, where g generates the cyclic group $(\mathbb{Z}/p\mathbb{Z})^*$. Show that if d is a divisor of $p-1$ and $p-1 = dm$, then $G_d = \{\sigma^0, \sigma^m, \ldots, \sigma^{(d-1)m}\}$ is the (unique) subgroup of $G(K/\mathbb{Q})$ of order d and $K^{G_d} = \mathbb{Q}(\theta_d)$, where

$$\theta_d = \varepsilon_p + \sigma^m(\varepsilon_p) + \cdots + \sigma^{(d-1)m}(\varepsilon_p).$$

Remark. The elements θ_d for d dividing $p-1$ are called **Gaussian periods**. Notice that $\theta_1 = \varepsilon_p$.

10.8. Find all quadratic subfields (that is, K with $[K : \mathbb{Q}] = 2$) in the following cyclotomic fields:

(a) $\mathbb{Q}(\varepsilon_8)$; (b) $\mathbb{Q}(\varepsilon_5)$; (c) $\mathbb{Q}(\varepsilon_7)$.

10.9. Using Gaussian periods (see Exercise 10.6) find the description of all subfields of the cyclotomic fields $K = \mathbb{Q}(\varepsilon_p)$ for:

(a) $p = 5$; (b) $p = 7$; (c) $p = 17$.

In each case, explicitly find the corresponding Gaussian periods.

10.10. Let $p > 2$ be a prime. Show that the only quadratic subfield M of $K = \mathbb{Q}(\varepsilon_p)$ is $M = \mathbb{Q}(\sqrt{p})$ if $p \equiv 1 \pmod 4$ and $M = \mathbb{Q}(\sqrt{-p})$ if $p \equiv 3 \pmod 4$.

Remark. The result shows that quadratic fields $\mathbb{Q}\left(\sqrt{(-1)^{(p-1)/2}p}\right)$ (p an odd prime number) are subfields of cyclotomic fields. This is true for all quadratic fields, and in still more generality, for all finite abelian Galois extensions of $\mathbb{Q}$. In fact, the

Kronecker–Weber[1] theorem says that every finite abelian extension of the rational numbers $\mathbb{Q}$ is a subfield of a cyclotomic field (see [K], p. 200).

10.11. Find and factorize the cyclotomic polynomials:
(a1) $\Phi_{7,\mathbb{F}_2}(X)$; (a2) $\Phi_{18,\mathbb{F}_7}(X)$; (a3) $\Phi_{26,\mathbb{F}_3}(X)$.
(b) Show that the orders of irreducible factors of $\Phi_{n,\mathbb{F}_p}(X)$ are equal to the order of p in the group $\mathbb{Z}_n^*$ if $p \nmid n$.

10.12. (a) Let d, n be positive integers and $d \mid n$, $d < n$. Show that if a prime p divides $\Phi_d(x)$ and $\Phi_n(x)$ for an integer x, then $p \mid n$.
(b) Show that if a prime p divides $\Phi_n(x)$ for an integer x, then $p \mid n$ or $p \equiv 1 \pmod{n}$.
(c) Show that $\Phi_1(0) = -1$ and $\Phi_n(0) = 1$ when $n > 1$.
(d) Let n be a positive integer. Show that there exists infinitely many primes p with $p \equiv 1 \pmod{n}$.

Remark. This is a special case of the famous **Dirichlet's[2] Theorem about primes in arithmetic progressions**, which says that in any arithmetical progression $ak + b$, where a, b are relatively prime integers and $k = 1, 2, \ldots$, there are infinitely prime numbers. The last part of this exercise says that there exists infinitely many primes in the progression $nk + 1$ ($a = n, b = 1$). For a proof of Dirichlet's Theorem in its general version, see [S2], Chap. VI. We apply this case of Dirichlet's Theorem in the proof of inverse Galois problem (over $\mathbb{Q}$) for abelian groups (see Exercise 10.13).

10.13. (a) Prove that for any finite abelian group there is a positive integer n and a surjective homomorphism $\varphi : (\mathbb{Z}/n\mathbb{Z})^* \to G$.
(b) Show that every finite abelian group G is a Galois group $G(K/\mathbb{Q})$ for a number field K (see the Remark after Exercise 9.16).

10.14. (a) Let L be a splitting field of a polynomial $X^n - a$ ($a \in K, a \neq 0$) over a field K whose characteristic does not divide n (see Exercise 5.11). Show that the Galois group $G(L/K)$ contains a normal subgroup H isomorphic to a subgroup of $\mathbb{Z}_n$ such that the quotient $G(L/K)/H$ is isomorphic to a subgroup of $\mathbb{Z}_n^*$. What is the maximal possible order of $G(L/K)$?
(b) Show that the Galois group $G(L/K)$ in (a) is isomorphic to a subgroup of the group of matrices $\begin{pmatrix} a & b \\ 0 & 1 \end{pmatrix}$, where $a \in \mathbb{Z}_n^*$, $b \in \mathbb{Z}_n$ (a subgroup of the group $GL_2(\mathbb{Z}_n)$ of all invertible (2×2)-matrices over the ring $\mathbb{Z}_n$).
(c) Give a description of the Galois groups over $\mathbb{Q}$ of the irreducible binomials $X^p - a$, where $a \in \mathbb{Q}$ and p is a prime number.
(d) Give a description of the Galois group of the binomial $T^n - X$ over $\mathbb{C}(X)$.

---

[1]Heinrich Martin Weber, 5 May 1842–17 May 1913.

[2]Johann Peter Gustav Lejeune Dirichlet, 13 February 1805–5 May 1859.

10.15. Let $\varepsilon_k = e^{\frac{2\pi i}{k}}$ for $k = 1, 2, \ldots$.
(a) Show that the group of all roots of unity in $K = \mathbb{Q}(\varepsilon_m)$ is ε_m^k, $k = 1, \ldots, m$, when m is even or $\pm\varepsilon_m^k$, $k = 1, \ldots, m$, when m is odd.
(b) Show that $\mathbb{Q}(\varepsilon_m) = \mathbb{Q}(\varepsilon_n)$, where $n \geq m$, if and only if $n = m$ or $n = 2m$ and m is odd.
(c) Show that $\mathbb{Q}(\varepsilon_m) \subseteq \mathbb{Q}(\varepsilon_n)$ if and only if $m \mid n$ or m is even, n is odd and $\frac{m}{2} \mid n$.

10.16. Find the orders of the Galois groups of the following binomials over the field $\mathbb{Q}$:

(a) $X^4 + 1$; (b) $X^4 - 3$; (c) $X^6 + 3$; (d) $X^6 - 3$;
(e) $X^6 + 4$; (f) $X^8 + 1$; (g) $X^8 + 2$; (h) $X^8 - 2$.

Using Computers 10

It is not difficult to compute the cyclotomic polynomials over the rational numbers, but if Maple is available, we get $\Phi_{n,\mathbb{Q}}(X)$ using the command `cyclotomic(n,x)` preceded by `with(numtheory)`, which loads several other commands related to number theory. Below, we suggest some numerical experiments.

10.17. Look at the coefficients of the cyclotomic polynomials $\Phi_{n,\mathbb{Q}}(X)$ and find the first n for which there is a coefficient of `cyclotomic(n,x)` whose absolute value is bigger than 1. What could be an explanation that such a coefficient finally appears?

10.18. (a) Study the orders of the Galois groups of irreducible binomials $X^6 - a$ for some set of integer values of a. What can be said about the orders of this group for all possible values of a?
(b) Do the same for binomials $X^7 - a$, $X^8 - a$ and $X^9 - a$.

Remark. For general results on the splitting fields of binomials, see [GV].

Chapter 11
Galois Modules

This chapter, like the preceding one, is also intended to illustrate the main theorems of Galois theory in Chap. 9. It has a little more theoretical character as we discuss some general notions and prove three important theorems having several applications in algebra and number theory. All of these theorems are related in one way or another to the Galois field extensions $K \subseteq L$. The Galois group $G = G(L/K)$ acts on the elements of L^+ (the additive group of L) or L^* (the multiplicative group of L) making them into modules over G (see the definition below) and it is just the structure of these modules over G which is essential in all three of these theorems. The first, on the existence of so-called normal bases, is about especially nice bases of L over K and is related to a property of L^+ as a module over G. The second, called Hilbert's Theorem 90, is related to both L^+ and L^* as G-modules, but it is best known for the second structure. This theorem has many applications and we look at some of them in the exercises. One such application is the third theorem in this chapter, in which we discuss so-called Kummer extensions (see below). A particular case of such extensions are cyclic extensions, that is, Galois extensions $K \subseteq L$ with cyclic Galois group $G = G(L/K)$. If the field K contains sufficiently many roots of unity, then Hilbert's Theorem 90 says how to describe such an extension. Hilbert's Theorem 90 may be considered in the context of so-called cohomology groups (defined for G-modules), which we mention very shortly. This chapter needs some more notions, which are presented in the Appendix. In the following chapters, we only use **T.11.3**.

If G is an arbitrary group and A an abelian group, then A is called a G-**module** if to every $a \in A$ and $\sigma \in G$ corresponds an element $\sigma(a) \in A$ in such a way that $\sigma(a+b) = \sigma(a) + \sigma(b)$, $(\sigma\sigma')(a) = \sigma(\sigma'(a))$ and $e(a) = a$ ($a, b \in A$, $\sigma, \sigma' \in G$, e is the identity in G). If $K \subseteq L$ is a Galois extension with $G = G(L/K)$, then both L^+ and L^* can be considered as G-modules, where the action G is given by $(\sigma, x) \mapsto \sigma(x)$ for $\sigma \in G$ and $x \in L^+$ or $x \in L^*$ (L^+ the additive group of L, L^* the multiplicative group of L). These two G-modules play a very important role in many applications of Galois theory in algebra, number theory and algebraic geometry. For the general

J. Brzeziński, *Galois Theory Through Exercises*, Springer Undergraduate Mathematics Series, https://doi.org/10.1007/978-3-319-72326-6_11

terminology concerning G-modules and more examples see section "Modules Over Rings" in Appendix (in particular A.6.4). We explain there the notion of the group ring $K[G]$ of a finite group $G = \{\sigma_1 = 1, \sigma_2, \dots, \sigma_n\}$ over a field (or a commutative ring) K, whose elements are sums $a_1\sigma_1 + a_2\sigma_2 + \cdots + a_n\sigma_n$, where $a_i \in K$. Any G-module A can be considered as a module over this ring and conversely (for details, see A.6.4).

The first result of this chapter is the **normal basis theorem** (**NBT**), which explains the structure of L^+ as a $G(L/K)$-module when $K \subseteq L$ is a Galois extension. It says that L contains elements whose images by all elements in the Galois group form a basis of L over K. The second result, **Hilbert's**[1] **Theorem 90** (which has number 90 in Hilbert's treatise on algebraic number theory published in 1895) gives information about the multiplicative group of L as a $G(L/K)$-module. Hilbert's Theorem 90 is closely related to so-called **Noether's equations**, which showed the way to the general definitions of the **cohomology groups** and found a natural place in their context. Hilbert's Theorem 90 has many applications. We shall apply it to the abelian **Kummer**[2] **theory**, which is concerned with an important class of Galois extensions whose elementary description follows a pattern that is typical in the much deeper Class Field Theory—a fundamental part of the algebraic number theory of abelian Galois extensions.

T.11.1 Normal Basis Theorem. *Let $K \subseteq L$ be a Galois extension and $G(L/K) = \{\sigma_1 = 1, \sigma_2, \dots, \sigma_n\}$ its Galois group. Then the following properties of the extension $K \subseteq L$ hold and are equivalent:*
(a) There exists an $\alpha \in L$ such that $\sigma_1(\alpha), \sigma_2(\alpha), \dots, \sigma_n(\alpha)$ is a basis of L over K.
(b) L^+ is a cyclic $K[G]$-module, that is, there is an $\alpha \in L$ such that $L^+ = K[G]\alpha$ for some $\alpha \in L$.

Any basis $\sigma_1(\alpha) = \alpha, \sigma_2(\alpha), \dots, \sigma_n(\alpha)$ of L over K is called **normal**. An element α defining such a basis as well as its minimal polynomial are also called **normal**.[3] Thus the last theorem says that every finite Galois extension has a normal basis.

T.11.2 Hilbert's Theorem 90. *Let $L \supseteq K$ be a cyclic extension of degree n and let σ be a generator of the Galois group $G = G(L/K)$. If $\alpha \in L$, then*
(a) $\mathrm{Nr}_G(\alpha) = 1$ if and only if there is a $\beta \in L$ such that $\alpha = \frac{\beta}{\sigma(\beta)}$ (multiplicative version);
(b) $\mathrm{Tr}_G(\alpha) = 0$ if and only if there is a $\beta \in L$ such that $\alpha = \beta - \sigma(\beta)$ (additive version).

[1]David Hilbert, 23 January 1862–14 February 1943.

[2]Ernst Eduard Kummer, 29 January 1810–14 May 1893.

[3]This creates a conflict with the historically older use of the notion of a normal polynomial, which we introduced in Exercise 7.9. Usually it is possible to easily recognize which notion is being discussed.

One of the applications of Hilbert's Theorem 90 (in the multiplicative version (a)) is a description of cyclic extensions over a field which contains sufficiently many roots of unity:

T.11.3. *Let K be a field containing n different n-th roots of unity. If L is a cyclic extension of K of degree n, then there exists an $\alpha \in L$ such that $L = K(\alpha)$ and $\alpha^n \in K$.*

The assumption that K contains n different roots of unity says that n is not divisible by the characteristic of K (see Exercise 5.11(b)). For the case when the characteristic of K divides the degree of a cyclic extension, see Exercise 11.8.

Remark 11.1. The last result **T.11.3** is often called Lagrange's Theorem. It was proved by Lagrange about one century earlier than Hilbert's Theorem 90. It is possible to use Hilbert's result in its proof, which gives a simplification of the argument, so Lagrange's Theorem is often considered as a corollary of Hilbert's Theorem 90. The simplification is obtained when one notes that $\mathrm{Nr}(\varepsilon) = \varepsilon^n = 1$, so that there exists an $\alpha \in L$ such that $\varepsilon = \frac{\alpha}{\sigma(\alpha)}$ by **T.11.2** (a). However, it is useful to define α explicitly for practical reasons (even if this could be deduced from the proof of Hilbert's Theorem 90)—see Exercise 11.6.

Remark 11.2. Hilbert's Theorem 90 is a specialization to cyclic groups of a more general result on **cohomology groups**. If G is a group and A is a G-module, then the first cohomology group $H^1(G,A)$ is the group of functions (with respect to the usual addition of functions) $f : G \to A$ such that $f(\sigma\sigma') = f(\sigma) + \sigma(f(\sigma'))$, called **1-cycles** modulo **1-boundaries**, which are 1-cocyles of the form: $f(\sigma) = a - \sigma(a)$ for some $a \in A$. In multiplicative notation, the 1-cocyles are $f : G \to A$ such that $f(\sigma\sigma') = f(\sigma)\sigma(f(\sigma'))$ and the 1-boundaries $f(\sigma) = \frac{a}{\sigma(a)}$ for some $a \in A$.

If we write $f(\sigma) = \alpha_\sigma$, then we recognize that in the proofs of **T.11.2**, we consider 1-cocycles $f : G \to K^*$ (**T.11.2**(a)) and $f : G \to K^+$ (**T.11.2**(b)) with suitable choices of α. The conditions defining a 1-cocyle $a_{\sigma\sigma'} = \alpha_\sigma\sigma(\alpha_{\sigma'})$ are often called **Noether's equations** and Hilbert's theorem 90 can be considered as a statement about solutions of these equations.

Using the notion of the first cohomology group, the statements of Hilbert's Theorem 90 are simply the claims that $H^1(G,K^*)$ and $H^1(G,K^+)$ are trivial (every 1-cocyle is 1-coboundary). In fact, the same arguments as those given in the proof of **T.11.2** show that for any finite automorphism group G of K these two groups are trivial. The cohomology groups $H^n(G,A)$ are defined in a natural way for all integers n and the groups $H^n(G,K^+)$ are trivial for all $n > 0$, while $H^n(G,K^*)$ are very interesting groups related to the field K.

As an illustration and application of some of the results in this chapter, we shall consider a class of field extensions which are splitting fields of arbitrary families of binomials $X^m - a$ where $a \in K$ and m is fixed. We exclude the case $a = 1$, that is, the case of cyclotomic fields, considered in Chap. 10, assuming that the field K contains m *different* m-th roots of unity. This theory is known as (abelian) **Kummer theory**. Notice that the condition concerning roots of unity simply says that the ground field K has characteristic 0 or its characteristic is relatively prime to m (see Exercise 5.11(b)).

The **exponent** of any group G is the least positive integer m such that $\sigma^m = 1$ for all $\sigma \in G$. A Galois extension $K \subseteq L$ is said to be of **exponent** m if the exponent of its Galois group divides m (notice a difference between the exponent of a group and the exponent of a Galois extension). A Galois extension $K \subseteq L$ of exponent m is called a **Kummer extension** if it is abelian (that is, its Galois group is abelian) and the field K contains m different m-th roots of unity. The simplest example is a splitting field of a polynomial $X^m - a$ over such a field K. It is $L = K(\alpha)$, where α is any fixed zero of this polynomial (see Exercise 5.11). The Galois group $G(L/K)$ is abelian and $\sigma^m = 1$ for every automorphism $\sigma \in G(L/K)$ (see Exercise 9.22). A zero α of $X^m - a$ will be denoted by $\sqrt[m]{a}$. Even if this symbol may denote any of the m different zeros $\eta\alpha$, where η is an m-th root of unity, the field $K(\sqrt[m]{a})$ is uniquely defined, since the factor η belongs to K. We denote by K^{*m} the subgroup of the multiplicative group $K^* = K \setminus \{0\}$ consisting of all m-th powers of nonzero elements in K.

The fundamental facts about Kummer extensions of exponent m are contained in the following theorem:

T.11.4. *Let K be a field containing m different m-th roots of* 1.
(*a*) *If $K \subseteq L$ is a Kummer extension of exponent m, then every subextension of fields $M \subseteq N$, where $K \subseteq M \subseteq N \subseteq L$, is also a Kummer extension.*
(*b*) *All Kummer extensions of K of exponent m are exactly the splitting fields of sets of binomial polynomials $X^m - a$ for some $a \in K$. In particular, all finite Kummer extensions of K are $L = K(\sqrt[m]{a_1}, \ldots, \sqrt[m]{a_r})$ for some elements $a_1, \ldots, a_r \in K$.*
(*c*) *There is a one-to-one correspondence between the isomorphism classes of finite Kummer extensions of K of exponent m and the subgroups A of K^* containing K^{*m} such that the index $[A : K^{*m}]$ is finite. In this correspondence, to a Kummer extension L of K corresponds the subgroup A of K^* consisting of all $a \in K$ such that $a = \alpha^m$ for some $\alpha \in L$, and to a subgroup A of K^* corresponds any splitting field over K of all binomials $X^m - a$ for $a \in A$. Moreover,*

$$|G(L/K)| = [L : K] = [A : K^{*m}]. \tag{11.1}$$

Exercises 11

11.1. Construct a normal basis for each of the following field extensions:

(a) $\mathbb{Q}(i)$ over $\mathbb{Q}$;
(b) $\mathbb{Q}(\sqrt{2}, \sqrt{3})$ over $\mathbb{Q}$;
(c) $\mathbb{Q}(\varepsilon)$, $\varepsilon^5 = 1, \varepsilon \neq 1$ over $\mathbb{Q}$;
(d) $\mathbb{F}_2(\gamma)$, $\gamma^3 + \gamma^2 + 1 = 0$ over $\mathbb{F}_2$.

11.2. (a) Show that in a quadratic Galois extension $K \subset L$ an element $\alpha \in L$ is normal if and only if $\alpha \notin K$ and $\mathrm{Tr}(\alpha) \neq 0$ (see (6.1)).
(b) Show that in a cubic Galois extension $K \subset L$, where K is a real number field, an element $\alpha \in L$ is normal if and only if $\alpha \notin K$ and $\mathrm{Tr}(\alpha) \neq 0$.

11.3. (a) Let $K \subseteq L$ be a Galois extension and let $\alpha \in L$ be a normal element. Show that $L = K(\alpha)$, that is, a normal element is field primitive. Is the converse true?
(b) Let $K \subseteq L$ be finite fields. Is it true that the group primitive elements (that is, the generators of the cyclic group L^*) are normal? Is the converse true? (see Exercise 8.15).

11.4. (a) Let X, Y, Z be a Pythagorean triple, that is, an integer solution of the equation $X^2 + Y^2 = Z^2$ with positive and relatively prime X, Y, Z. Consider the complex number $\alpha = x + yi \in \mathbb{Q}(i)$, where $x = X/Z$ and $y = Y/Z$ and notice that $\mathrm{Nr}(\alpha) = 1$ in $\mathbb{Q}(i)$. Use this observation together with Hilbert's Theorem 90 in order to find a formula for X, Y, Z.
(b) Find in a similar way a formula for integer solutions of the equation $X^2 + 2Y^2 = Z^2$.

11.5. (a) Let K be a field containing n different n-th roots of unity. Using Kummer theory (**T.11.4**) show that if $X^n - a$ and $X^n - b$ are irreducible polynomials in $K[X]$, then $K(\sqrt[n]{a}) = K(\sqrt[n]{b})$ if and only if there is an r such that $0 < r < n$, $\gcd(r, n) = 1$ and $ba^{-r} \in K^{*n}$.
(b) Give a description of all quadratic extensions of the rational numbers $\mathbb{Q}$ using (a).

11.6. (a) Show that $E = \mathbb{Q}(\sqrt{-3})$ is the smallest number field containing the 3rd roots of unity, and show that for every cubic Galois extension L of E there is an $\alpha \in E$ such that $L = E(\sqrt[3]{\alpha})$. Show also that one can always choose $\alpha \in \mathbb{Z}[\varepsilon]$, where $\varepsilon = \frac{1+\sqrt{-3}}{2}$, that is, $\alpha = a + b\varepsilon$, $a, b \in \mathbb{Z}$ (that is, α is an Eisenstein integer—see p. 246).
(b) Show that there is a one-to-one correspondence between cyclic cubic extensions $K \supset \mathbb{Q}$ and the cyclic cubic extensions $L = EK$ of E such that $L \supset \mathbb{Q}$ is Galois with cyclic Galois group $\mathbb{Z}_6$.
(c) Show that a cubic extension $L = E(\sqrt[3]{\alpha})$ of E ($\alpha \in E$) is a Galois extension of $\mathbb{Q}$ if and only if $\mathrm{Nr}(\alpha) = \alpha\bar{\alpha} \in \mathbb{Q}^{*3}$ or $\alpha\mathrm{Nr}(\alpha) \in \mathbb{Q}^{*3}$. In the first case, the Galois group $G(L/\mathbb{Q})$ is the cyclic group $\mathbb{Z}_6$, and in the second, it is the symmetric group S_3.
(d) Show that $\alpha = f\bar{f}^2$, where $f \in \mathbb{Z}[\varepsilon]$ satisfies the condition $\mathrm{Nr}(\alpha) = \alpha\bar{\alpha} \in \mathbb{Q}^{*3}$ in (c). Show that $\mathbb{Q}(\gamma)$, where $\gamma = \sqrt[3]{\alpha} + \sqrt[3]{\bar{\alpha}}$, is a cubic Galois extension of $\mathbb{Q}$ and find the minimal polynomial of γ over $\mathbb{Q}$. Give a few examples of cyclic cubic extensions $K \supset \mathbb{Q}$.

11.7. (a) Let $E = \mathbb{Q}(i)$. Show that for every cyclic quartic Galois extension L of E there is an $\alpha \in E$ such that $L = E(\sqrt[4]{\alpha})$. Show also that one can always choose $\alpha \in \mathbb{Z}[i]$, that is, $\alpha = a + bi$, $a, b \in \mathbb{Z}$ (that is, α is a Gaussian integer).
(b) Show that there is a one-to-one correspondence between pairs of cyclic quartic extensions $K_1 \supset \mathbb{Q}$, $K_2 \supset \mathbb{Q}$, where K_1 is real and K_2 is nonreal, such that $EK_1 = EK_2$, and the cyclic quartic extensions L of E such that $L \supset \mathbb{Q}$ is Galois with Galois group $\mathbb{Z}_2 \times \mathbb{Z}_4$.
(c) Show that a quartic extension $L = E(\sqrt[4]{\alpha})$ of E ($\alpha \in E$) is a Galois extension of $\mathbb{Q}$ if and only if $\mathrm{Nr}(\alpha) = \alpha\bar{\alpha} \in \mathbb{Q}^{*4}$ or $\alpha^2\mathrm{Nr}(\alpha) \in \mathbb{Q}^{*4}$. In the first case the Galois

group $G(L/\mathbb{Q})$ is the group $\mathbb{Z}_2 \times \mathbb{Z}_4$, and in the second, it is the dihedral group D_4 (the symmetry group of a square—see A.2.4).
(d) Show that $\alpha = f\bar{f}^3g^2$, where $f \in \mathbb{Z}[i], f = k + li, k, l \in \mathbb{Z}$ and $g \in \mathbb{Q}^*$, satisfies the condition $\mathrm{Nr}(\alpha) = \alpha\bar{\alpha} \in \mathbb{Q}^{*4}$ from (c). Show that $\mathbb{Q}(\gamma)$, where $\gamma = \sqrt[4]{\alpha} + \sqrt[4]{\bar{\alpha}}$, $\sqrt[4]{\alpha}\sqrt[4]{\bar{\alpha}} = \mathrm{Nr}(f)|g|$, is a cyclic quartic extension of $\mathbb{Q}$ (real if $g > 0$, nonreal if $g < 0$) and find the minimal polynomial of γ over $\mathbb{Q}$.
(e) Show that a quartic Galois extension of $\mathbb{Q}$ has cyclic Galois group C_4 if and only if there are nonzero integers k, l, m such that $K = \mathbb{Q}\left(\sqrt{m(\Delta + k\sqrt{\Delta})}\right)$, $\Delta = k^2 + l^2$ and $\sqrt{\Delta} \notin \mathbb{Q}$. Show also that $M = \mathbb{Q}(\sqrt{\Delta})$ is the only quadratic subfield of K. Give a few examples of cyclic quartic extensions $K \supset \mathbb{Q}$ and show that the fields K_1 and K_2 in (b) correspond to the triples (k, l, m) and $(k, l, -m)$.

Remark. This exercise and the previous one can be seen in a broader context: to describe all Galois extensions of a field (here the rational numbers) with given Galois group (the cyclic group C_3 of order 3 in Exercise 11.6 and the cyclic group C_4 of order 4 in this exercise). In general, the idea is that the cyclic Galois extensions of degree n over $\mathbb{Q}$ are easier to handle when they are "lifted" to a bigger field $E = \mathbb{Q}(\varepsilon)$, where ε is a primitive n-th root of unity. Over this bigger field, we have Lagrange's Theorem **T.11.3**, which gives a precise description of cyclic extensions. When we have such a description, we hope to "go back" ("descend") to get a description of the extensions over $\mathbb{Q}$. The methods used in these two exercises are examples of a very general procedure, which is usually studied in the context of Galois cohomology. There are several books on the subject with the first (and one of the best) by J.-P. Serre [S1]. In the very general frame of Galois cohomology, one has a Galois extension $K \subset E$ and one wants to study some kind of objects over K (in our case, the Galois field extensions $K \subset L$ with a cyclic Galois group of order n). One takes a fixed object over K, which one wants to study, and one looks at all objects of the same type which become isomorphic when "lifted" to E. In our case, the lifting means the fields LE and $L'E$ which became isomorphic (here an isomorphism means equality) over E. We use the extensions $K \subset E \subset LE$ in order to "descend" to all extensions $K \subset L'$ for which $L'E$ becomes isomorphic (here equal) to LE over E.
A general procedure to handle such situations goes through suitable cohomology groups (see the definition of $H^1(G, A)$ in Remark 11.2). The set of all extensions $K \subset L'$ which become isomorphic to the given extension $K \subset L$ after lifting to E (that is, LE and $L'E$ are E-isomorphic) is in a one-to-one correspondence with the elements of the cohomology group $H^1(G, A)$, where in our case $G = G(E/K)$ and $A = G(L/K)$. For the details, see, e.g. [S1], Chap. III, Prop. 1. The group $A = G(L/K)$ is considered as a module over $G = G(E/K)$ (see the beginning of this chapter for definitions). In the present case this structure is trivial, that is, $\sigma a = a$ for $\sigma \in G$ and $a \in A$. Hence, the cohomology group $H^1(G, A)$ defined in Remark 11.2 for $G = G(E/\mathbb{Q}) = C_2$ and $A = G(K/\mathbb{Q}) = C_4$ (C_3 in Exercise 11.6) with the trivial action of G on A is equal to $H^1(G, A) = \mathrm{Hom}(C_2, C_4)$ (all group homomorphism from C_2 to C_4 in the present exercise), since the 1-cocyles are

simply homomorphisms of G into A, and the 1-coboundaries are trivial (any 1-coboundary maps G onto the identity in A). But we have exactly two different homomorphisms of C_2 into C_4 (the trivial one, and the embedding sending the element of order 2 in C_2 onto the element of order 2 in C_4). This means that there are exactly two non-isomorphic cyclic extensions of $\mathbb{Q}$ which become isomorphic (here equal) over $E = \mathbb{Q}(i)$. In Exercise 11.6, we have a similar situation, but $H^1(G, A) = \mathrm{Hom}(C_2, C_3)$, which only has the trivial element (the trivial homomorphism). This explains why there is a one-to-one correspondence between cubic Galois extensions of $\mathbb{Q}$ and cubic Galois extensions of $E = \mathbb{Q}(\varepsilon)$. An easy description of such extensions over E is the key to a description of all cubic (or quartic when $E = \mathbb{Q}(i)$) cyclic extensions over $\mathbb{Q}$.

11.8. Let K be a field of prime characteristic p and let L be a Galois field extension of K of degree p. Show that $L = K(\alpha)$, where α is a zero of a polynomial $X^p - X - a$ for some $a \in K$ (compare Exercise 9.26).

Remark. The polynomial $X^p - X - a$ over a field K of characteristic p is often called an **Artin–Schreier[4] polynomial**, and $K \subset L$ is called an **Artin–Schreier extension**. The description of cyclic Galois extensions of degree p given in the exercise above is called the Artin–Schreier theorem. Such extensions play an important role, for example, in Kummer theory and in the study of radical extensions over fields of characteristic p (see Chap. 13 for characteristic 0).

11.9. Let $K \subset L$ be a finite field extension and L an algebraically closed field. Show that $[L : K] = 2$.

Remark. This is a part of the **Artin–Schreier Theorem**, which says that not only an algebraically closed field of finite degree > 1 over its subfield must have degree 2, but also that K must be a real field (that is, -1 is not a sum of squares in K), which is really closed (that is, there is no bigger real field containing it) and $L = K(i)$, where $i^2 = -1$ (see [L], Chap. VI, Thm. 6.4). The best known example of such an extension is $\mathbb{R} \subset \mathbb{C}$.

Using Computers 11

11.10. Use Exercise 11.6, respectively Exercise 11.7, in order to write Maple procedures which list cubic, respectively quartic, polynomials with integer coefficients whose Galois groups are cyclic of order 3, respectively 4.

[4] Otto Schreier, 3 March 1901–2 June 1929.

Chapter 12
Solvable Groups

As preparation for the next chapter, in which we discuss solvability of equations by radicals, we need some knowledge about solvable groups. In this chapter, we define solvable groups, discuss several examples of such groups and prove those of their properties which we shall apply in the next chapter.

A group G is called **solvable** if there exists a chain of groups:

$$G = G_0 \supset G_1 \supset \ldots \supset G_n = \{e\}$$

such that G_{i+1} is normal in G_i and the quotients G_i/G_{i+1} are abelian for $i = 0, 1, \ldots, n-1$.

T.12.1. *(a) If G is solvable and H is a subgroup of G, then H is solvable.*
(b) If N is a normal subgroup of G and G is solvable, then the quotient group G/N is solvable.
(c) If N is a solvable normal subgroup of G such that the quotient group G/N is solvable, then G is solvable.

We look at several examples of solvable and some non-solvable groups in the exercises below. In particular, we look at solvable and non-solvable subgroups of the permutation groups. By the **symmetric group** S_n, we mean the group of all bijective functions on a set X with n elements. Usually, we choose $X = \{1, 2, \ldots, n\}$, but sometimes other choices are more suitable (see section "Permutations" in Appendix). The following result is usually used in the proof that among algebraic equations of degree at least 5, there are equations not solvable by radicals:

T.12.2. *The symmetric group S_n is not solvable when $n \geq 5$.*

J. Brzeziński, *Galois Theory Through Exercises*, Springer Undergraduate Mathematics Series, https://doi.org/10.1007/978-3-319-72326-6_12

Exercises 12

12.1. Show that the following groups are solvable:
(a) Every abelian group;
(b) The symmetric group S_3 (this may be considered as the group of all symmetries of an equilateral triangle—see section "Transitive Subgroups of Permutation Groups" in Appendix);
(c) The symmetric group S_4 (this may be considered as the group of all symmetries of a regular tetrahedron—see section "Transitive Subgroups of Permutation Groups" in Appendix);
(d) The dihedral group D_n (the group of all symmetries of a regular polygon with n sides for $n = 3, 4, \ldots$—see A.2.4);
(e) The group $\mathbb{H}^*(\mathbb{Z})$ of quaternion units $\pm 1, \pm i, \pm j, \pm k$, where $i^2 = j^2 = -1, ij = k$ and $ji = -ij$ (this group is called the **quaternion group** and is often denoted by Q_8).

Remark. According to the famous Feit–Thompson Theorem (proved by Walter Feit and John G. Thompson in 1962 and conjectured by William Burnside[1] in 1911), every finite group of odd order is solvable. As a rather long exercise, one can prove that every group of order less than 60 is solvable. The smallest non-solvable group is the group A_5 (of order 60) of all even permutations of numbers $1, 2, 3, 4, 5$ (see Exercise 12.2).

12.2. The aim of this exercise is to show that every alternating group A_n, where $n \geq 5$, is not solvable (for the definition of A_n, see p. 265).
(a) Let G be a subgroup of the group A_n, where $n \geq 5$, and N a normal subgroup of G such that G/N is abelian. Show that if G contains every cycle (a, b, c), then N also contains every such cycle.
(b) Deduce from (a) that the group A_n, for $n \geq 5$, is not solvable.
(c) Why are the symmetric groups S_n, $n \geq 5$, not solvable?

12.3. Show that a group G is solvable if and only if there exists a chain of groups:

$$G = G_0 \supset G_1 \supset \ldots \supset G_n = \{e\}$$

such that G_{i+1} is normal in G_i and the quotient group G_i/G_{i+1} is cyclic of prime order for $i = 0, 1, \ldots, n-1$.

12.4. Denote by $\mathcal{G}_n$ the set of all functions $\varphi_{a,b} : \mathbb{Z}/n\mathbb{Z} \to \mathbb{Z}/n\mathbb{Z}$, where $\varphi_{a,b}(x) = ax + b$, $a, b \in \mathbb{Z}/n\mathbb{Z}$, $a \in (\mathbb{Z}/n\mathbb{Z})^*$, and by $\mathcal{T}_n$ all translations $\varphi_{1,b}(x) = x + b$.
(a) Show that the functions $\varphi_{a,b}(x)$ are bijections on the set $\{0, 1, \ldots, n-1\}$ and as such can be considered as permutations belonging to the group S_n of all permutations of this set (with n elements).

[1] William Burnside, 2 July 1852–21 August 1927.

(b) Show that $\mathcal{G}_n$ is a subgroup of S_n of order $n\varphi(n)$, where $\varphi(n)$ is Euler's function (for the definition of Euler's function, see p. 279).
(c) Define $\Phi : \mathcal{G}_n \to (\mathbb{Z}/n\mathbb{Z})^*$ such that $\Phi(\varphi_{a,b}) = a$. Show that Φ is a surjective group homomorphism whose kernel is $\mathcal{T}_n$. Show that $\mathcal{G}_n$ is solvable.
(d) Let p be a prime number. Show that $\mathcal{G}_p$ is transitive on the set $\{0, 1, \ldots, p-1\}$ and that every $\varphi_{a,b} \in \mathcal{G}_n$, $\varphi_{a,b} \neq \varphi_{1,0}$, has at most one fixed point $x \in \mathbb{Z}/p\mathbb{Z}$.
(e) If p is a prime number prove that $\mathcal{G}_p$ has exactly one subgroup of order p—the subgroup $\mathcal{T}_p$ consisting of all translations $\varphi_{1,b}$, $b = 0, 1, \ldots, p-1$, which is generated by $\varphi_{1,1}$. Show that $\mathcal{T}_p$ contains all elements of order p in $\mathcal{G}_p$.
(f) Show that $\mathcal{G}_n$ is isomorphic to the group consisting of the matrices

$$\begin{pmatrix} a & b \\ 0 & 1 \end{pmatrix},$$

where $a \in (\mathbb{Z}/n\mathbb{Z})^*$, $b \in \mathbb{Z}/n\mathbb{Z}$ with respect to matrix multiplication. Notice that the determinant:

$$\det : \mathcal{G}_n \to (\mathbb{Z}/n\mathbb{Z})^*$$

is a surjective group homomorphism (in (c) this was denoted by Φ) and that its kernel is the subgroup isomorphic to $\mathcal{T}_n$ consisting of matrices with $a = 1$. Check that the group $\mathcal{G}_n$ acts on the set of the column vectors $[x, 1]^t$, $x \in \mathbb{Z}/n\mathbb{Z}$ (that is, maps by matrix multiplication a vector of this type to a vector of the same type) and, when $n = p$ is a prime, every matrix has at most one fixed vector $[x, 1]^t$ (compare to (d)).

Remark. The subgroup of $\mathcal{G}_n$ ($n > 2$) consisting of matrices with $a = \pm 1$ is isomorphic to the dihedral group D_n—see A.2.4.

12.5. Let G be a transitive subgroup of the symmetric group S_p on a set X with p elements. Show that the following conditions are equivalent:
(a) G is solvable.
(b) Each non-identity element of G fixes at most one element of X.
(c) The group G is conjugate to a subgroup of $\mathcal{G}_p$ containing $\mathcal{T}_p$ (for the notation, see Exercise 12.4).
In order to show (a)–(c) prove the following two facts:
(d) The order of any transitive group G on a set X with p elements is divisible by p;
(e) If H is a subgroup of a transitive group G on a set X with p elements, then H is also transitive on X or every element of H acts as the identity on X.

12.6. Show that every p-group, that is, a group whose order is a power of a prime, is solvable.

Remark. A famous result proved by William Burnside in 1904 says that if the order of a finite group is divisible by at most two prime numbers, then the group is solvable.

Chapter 13
Solvability of Equations

In this chapter, we apply Galois theory to the classical problem of unsolvability by radicals of general polynomial equations of degree at least 5. We restrict to the fields of characteristic zero even if it is possible to prove corresponding results in arbitrary characteristic. Such a theory needs some technical reformulations of the notions to the case of radical extensions of fields of prime characteristic. On the other hand, if one has the more modest goal of proving that the fifth degree general equation over a number field is unsolvable by radicals, then there exists a very simple proof given by Nagell, which needs only very limited knowledge on field extensions and no knowledge of Galois theory (see Exercise 13.6). The questions related to the unsolvability by radicals of polynomial equations gave birth to Galois theory. Now they are mostly of historical value, while Galois theory as such is essential in many parts of mathematics and its different generalizations have developed into very important theories with many applications.

In this chapter, we prove the main theorem about polynomial equations solvable by radicals, that is, we show that equations solvable by radicals are characterized by the solvability of their Galois groups. We use this result in order to prove that the general equation of degree at least 5 is not solvable by radicals. Finally, we discuss the case of irreducible equations of degree 3 with 3 real zeros when Cardano's formula cannot avoid complex numbers in the algebraic expressions of the zeros, even though the coefficients of the equation are real. This so-called "Casus irreducibilis" played an important role in the history of algebraic equations. The last theorem in this chapter gives an explanation of this phenomenon.

Throughout this chapter, we assume that all fields have characteristic 0. A field extension $L \supseteq K$ is called **radical** if there is a chain of fields

$$K = K_0 \subseteq K_1 \subseteq \ldots \subseteq K_n = L \tag{13.1}$$

such that $K_i = K_{i-1}(\alpha_i)$, where $\alpha_i^{r_i} \in K_{i-1}$ and r_i are positive integers for $i = 1, \ldots, n$, that is, each K_i is a simple radical extension (see p. 31) of the preceding

J. Brzeziński, *Galois Theory Through Exercises*, Springer Undergraduate Mathematics Series, https://doi.org/10.1007/978-3-319-72326-6_13

field K_{i-1}. We say that an equation $f(X) = 0, f \in K[X]$, is **solvable by radicals** over K if a splitting field K_f of $f(X)$ over K is contained in a radical extension $L \supseteq K$. In general, we say that an extension $K \subseteq L$ is **solvable** (by radicals) if L is a subfield of a radical extension of K (so $f(X) = 0$ is solvable by radicals over K says that K_f is a solvable extension of K). We say that an equation $f(X) = 0, f \in K[X]$, where K is a subfield of the real numbers $\mathbb{R}$, is **solvable by real radicals** if a splitting field of $f(X)$ over K is contained in a radical extension $L \supseteq K$ such that $L \subset \mathbb{R}$.

T.13.1. *If char $(K) = 0$, then an equation $f(X) = 0, f \in K[X]$, is solvable by radicals if and only if the Galois group of f over K is solvable.*

The **general equation** of degree n over K is

$$f(X) = \prod_{i=1}^{n}(X - X_i) = X^n - s_1X^{n-1} + s_2X^{n-2} + \cdots + (-1)^n s_n = 0,$$

where s_i denote the elementary symmetric functions of $X_1, X_2, \ldots, X_n$, that is, $s_1 = \sum X_i$, $s_2 = \sum X_iX_j, \ldots, s_n = X_1X_2 \cdots X_n$.

T.13.2. *The Galois group over $K(s_1, s_2, \ldots, s_n)$ of the general equation $f(X) = 0$ of n-th degree is S_n, so $f(X) = 0$ is not solvable by radicals when $n \geq 5$.*

This result is a consequence of **T.13.1**, taking into account that the symmetric groups S_n are not solvable for $n \geq 5$ (see Exercise 12.2). It is not difficult to construct a rational polynomial $f(X)$ of degree 5 having the symmetric group S_5 as its Galois group. According to the same result, such an equation $f(X) = 0$ is not solvable by radicals (see several exercises below).

Another famous result is concerned with a surprising phenomenon related to irreducible (rational) cubic polynomials (and also, similar polynomials of higher degrees)—even if such a polynomial has real coefficients and 3 real zeros, the formulae expressing these zeros (like Cardano's formula (1.5) in Chap. 1) cannot contain only real radicals. This follows from the following fact:

T.13.3 Casus Irreducibilis. *Let $f(X)$ be an irreducible polynomial having 3 real zeros and coefficients in a real number field K. Then the equation $f(X) = 0$ is not solvable by real radicals over the field K.*

Exercises 13

13.1. Show that $L \supset \mathbb{Q}$ is a radical extension when:
(a) $L = \mathbb{Q}(\sqrt[5]{1 + \sqrt{3}})$; (b) $L = \mathbb{Q}(\sqrt[3]{1 - \sqrt{5}}, \sqrt[7]{\sqrt{2} + \sqrt{3}})$.

13.2. Argue that the following equations are solvable by radicals over $\mathbb{Q}$ (without solving them):
(a) $X^4 - 4X^2 - 21 = 0$; (b) $X^6 - 2X^3 - 2 = 0$.

13.3. Show that in the definition of an equation solvable by radicals, we can always assume that in the chain (13.1) the degrees r_i of the consecutive extensions $K_{i-1} \subset K_i$ are prime numbers.

13.4. Let p be a prime number and K a real number field.
(a) Prove **Weber's theorem**: If $f \in K[X]$ is an irreducible polynomial of degree p with $p-2$ real and 2 complex nonreal zeros, then its Galois group is S_p.
(b) Show that the equation $X^5 - p^2X - p = 0$, p a prime number, is not solvable by radicals over $\mathbb{Q}$.
(c) Show that the polynomial $f(X) = (X^2+4)(X-2)(X-4)\cdots(X-2(p-2))-2$ is irreducible and has exactly $p-2$ real zeros (hence its Galois group is S_p according to (a)).

Remark. Probably the simplest example of a polynomial over $\mathbb{Q}$ with Galois group S_n $(n > 1)$ is $X^n - X - 1$. This is proved in a paper by H. Osada in J. Number Theory 25(1987), 230–238.

13.5. For every $n \geq 5$ give an example of a polynomial equation $f(X) = 0$ over $\mathbb{Q}$ whose degree is n and which is not solvable by radicals.

13.6. The aim of this exercise is a direct proof (using only the definition of solvability by radicals, but without Galois theory and the relation between solvable equations and solvable groups) that any polynomial equation $f(X) = 0$, where $f(X) \in \mathbb{Q}[X]$ is irreducible and has 3 real and 2 nonreal (conjugate) complex roots, is not solvable by radicals over $\mathbb{Q}$. Let $f(X)$ be such a polynomial and assume that there is a chain of fields

$$\mathbb{Q} = K_0 \subseteq K_1 \subseteq \ldots \subseteq K_n = L$$

such that $K_i = K_{i-1}(\alpha_i)$, where $\alpha_i^{r_i} \in K_{i-1}$, the exponents r_i are prime numbers (see Exercise 13.3) for $i = 1, \ldots, n-1$ and the splitting field of $f(X)$ over $\mathbb{Q}$ is contained in L. Of course, the polynomial $f(X)$ has a factorization over $K_n = L$. Choose the least $i < n$ such that $f(X)$ is irreducible in K_{i-1} but reducible in K_i. In the exercise, we want to show that this is impossible, that is, $f(X)$ must still be irreducible over K_i. This contradiction shows that $f(X) = 0$ cannot be solvable by radicals.
(a) Using Nagell's lemma (see Exercise 4.13) show that $[K_i : K_{i-1}] = 5$, where $K_i = K_{i-1}(\alpha_i)$, $\alpha_i^5 = a_i \in K_{i-1}$.
(b) Show that $f(X)$ is irreducible over K_i.

13.7. Show that every equation $f(X) = 0$ of degree $\deg(f) \leq 4$, where $f \in K[X]$, is solvable by radicals.

13.8. Show that the equation $f(X) = 0$ is solvable in radicals over K if and only if the equation $f(X^n) = 0$ is solvable by radicals over K (n a positive integer).

13.9. Prove **Galois' theorem**: An irreducible polynomial equation of prime degree p over a number field is solvable by radicals if and only if its splitting field is generated by any two of its zeros:
(a) Let $f(X)$ be an irreducible polynomial of degree p over a number field K and let L be its splitting field over K. Show that L is generated by two of the zeros of $f(X)$ if and only if each nontrivial automorphism in $G(L/K)$ has at most one fixed point as a permutation of the zeros of $f(X)$. Show that the group $G(L/K)$ is transitive (see Exercise 6.2);
(b) Prove Galois' theorem by regarding $G(L/K)$ as a subgroup of S_p (see p. 36) and using Exercise 12.5.

13.10. Let $f(X) \in K[X]$ be an irreducible polynomial of prime degree $p > 2$ over a real number field K (that is, $K \subset \mathbb{R}$). Using Galois' theorem from Exercise 13.9 show that if the equation $f(X) = 0$ is solvable by radicals, then $f(X)$ has exactly one or p real zeros. Notice that this result gives an alternative solution of Exercise 13.4.

13.11. Let $f(X)$ be an irreducible polynomial of degree 5 over the rational numbers and $\Delta(f)$ its discriminant (see A.10.2).
(a) Show that if $\Delta(f) < 0$, then the equation $f(X) = 0$ is not solvable by radicals.[1]
(b) Show that there are both solvable and unsolvable equations $f(X) = 0$ over K with $\Delta(f) > 0$.

13.12. Show that the general cubic equation $f(X) = X^3 - s_1X^2 + s_2X - s_3 = 0$ over the field $K = \mathbb{Q}(\varepsilon, s_1, s_2, s_3)$, where $\varepsilon^3 = 1, \varepsilon \neq 1$, is solvable by radicals by constructing a suitable chain of fields (13.1) between $K_0 = K$ and $L = K(X_1, X_2, X_2)$ ($s_1 = X_1 + X_2 + X_3$, $s_2 = X_1X_2 + X_2X_3 + X_3X_1$, $s_3 = X_1X_2X_3$).

13.13. Show that if an equation $f(X) = 0, f \in K[X]$, where K is a subfield of the real numbers $\mathbb{R}$, is solvable by real radicals, then the degree of its splitting field $K_f \subseteq \mathbb{R}$ over K is a power of 2.

13.14. It is well-known (and easy to prove) that any real polynomial of odd degree has a real zero. Using this fact show that the field of complex numbers is algebraically closed (that is, the fundamental theorem of algebra—see **T.1.1**) in the following steps:
(a) If K is a Galois extension of $\mathbb{R}$ and the degree $[K : \mathbb{R}]$ is odd, then $K = \mathbb{R}$;
(b) If K is a quadratic extension of $\mathbb{R}$, then $K = \mathbb{C}$, while there are no quadratic extensions of the field $\mathbb{C}$;
(c) If K is a Galois extension of $\mathbb{C}$ (a splitting field of an irreducible polynomial over $\mathbb{C}$), then there is a field containing it, which is a Galois extension over $\mathbb{R}$. Using (a) and (b) show that this field containing K is equal to $\mathbb{C}$ (so $K = \mathbb{C}$).

[1] I owe this exercise to Erik Ljungstrand—a botanist with a real interest in mathematics.

Chapter 14
Geometric Constructions

In this chapter, we discuss some straightedge-and-compass constructions, in particular, the three classical problems known from antiquity—the impossibility of squaring the circle, doubling the cube and angle trisection. Problems concerning the impossibility (or possibility) of some geometric constructions by various means (like straightedge-and-compass, only compass or otherwise) are usually discussed in courses on Galois theory. But, in fact, the classical impossibility problems mentioned above do not need so much knowledge of Galois groups (if any at all) and can be presented already for those with a rudimentary knowledge of finite field extensions (say directly after Chap. 4 in this book). Some other problems, like Gauss's theorem on straightedge-and-compass constructible regular polygons need a little more knowledge related to Galois groups of field extensions. This chapter contains several exercises concerned with geometric straightedge-and-compass constructions. We prove two theorems—the first is usually used in the proofs of impossibility of some straightedge-and-compass constructions, and the second, usually in the proofs of possibility. Only the second one needs references to Galois groups.

Let X be an arbitrary set of points in the plane containing $(0, 0)$ and $(1, 0)$.

- A line is defined by X if it goes through two points belonging to X.
- A circle is defined by X if its center belongs to X and its radius is equal to the distance between two points belonging to X.

We say that a point $P = (a, b)$ **is directly constructible** from X by using a straightedge and a compass (a straightedge-and-compass construction) if P is an intersection point of two lines or two circles or a line with a circle, which are defined by X. Let X_1 be the set of all points in the plane which can be directly constructed from $X = X_0$, X_2 the set of all points which can be directly constructed from X_1, X_3 the set of all points which can be directly constructed from X_2 and so on. We say that a point $P = (a, b)$ **is constructible from** X by a straightedge-and-compass

J. Brzeziński, *Galois Theory Through Exercises*, Springer Undergraduate Mathematics Series, https://doi.org/10.1007/978-3-319-72326-6_14

construction if $P \in X^\star = \bigcup_{i=0}^{\infty} X_i$ (that is, $P \in X_i$ for some $i \geq 0$). We say shortly that X^* is the set of points constructible from X.
One also defines **real numbers constructible** from X as those $r \in \mathbb{R}$ such that $|r| =$ the distance between two points constructible from X.
Very often one takes $X = \{(0,0),(1,0)\}$. Those numbers which can be constructed from $X = \{(0,0),(1,0)\}$ will be denoted by $\mathbb{K}$. The smallest number field which contains the coordinates of $(0,0)$ and $(1,0)$ is of course $\mathbb{Q}$. The constructible lines through $(0,0),(1,0)$ and $(0,0),(0,1)$ (it is easy to see that the point $(0,1)$ is constructible from $(0,0)$ and $(1,0)$) are, as usual, called the **axes**.

T.14.1. *Let K be the smallest subfield of $\mathbb{R}$ which contains the coordinates of all points belonging to a given set of points X in the plane. A point $P = (a,b)$ can be constructed from X by a straightedge-and-compass construction if and only if there is a chain of fields:*

$$K = K_0 \subset K_1 \subset \ldots \subset K_n = L \subset \mathbb{R} \tag{$*$}$$

such that $a, b \in L$ and $[K_{i+1} : K_i] = 2$ for $i = 0, 1, \ldots, n-1$. In particular, the set of numbers which are constructible from X (like $\mathbb{K}$ when $X = \{(0,0),(1,0)\}$) is a field.

In practice, one uses this theorem when one wants to show that a point $P = (a,b)$ is not constructible—one shows that $[K(a,b) : K]$ is not a power of 2. If one wants to show that a point can be constructed the following theorem is often used:

T.14.2. *Let K be the smallest subfield of $\mathbb{R}$ which contains the coordinates of all points belonging to a point set X in the plane $\mathbb{R}^2$. A point $P = (a,b)$ is constructible from X by a straightedge-and-compass construction if and only if one of the following equivalent conditions hold:*
(a) There exists a Galois extension $L \supseteq K$ such that $a, b \in L$ and $[L : K]$ is a power of 2.
(b) There exists a Galois extension $L \supseteq K$ such that $a + bi \in L$ and $[L : K]$ is a power of 2.

In the following exercises, the terms "to construct" or "can be constructed" should be understood as constructed by using a straightedge and a compass. We always start from a set X which contains $(0,0)$ and $(1,0)$.

Exercises 14

14.1. Show that the following geometric constructions are impossible:
(a) to construct a cube of volume 2 when a cube of volume 1 is given, that is, to construct a segment of length $\sqrt[3]{2}$ when a segment of length 1 is given ("doubling of a cube");
(b) to construct the angle 20° when the angle 60° is given ("trisection of an angle");
(c) to construct a square of area π when a disk of area π (that is, of radius 1) is given ("squaring of a circle").

14.2. Let X be a set of points in the plane. Prove that $P = (a, b)$ can be constructed from X if and only if its coordinates can be constructed from X.

14.3. (a) Is it possible to construct a disk whose area is equal to the sum of the areas of two given disks?
(b) Is it possible to construct a sphere whose volume is equal to the sum of the volumes of two given spheres?

14.4. Is it possible to construct a square whose area is equal to the area of a given triangle?

14.5. Is it possible to construct a cube whose volume is equal to the volume of a regular tetrahedron whose sides are equal to 1?

14.6. Prove the following **theorem of Gauss**: A regular polygon with n sides is constructible (by straightedge-and-compass construction) when a segment of length 1 is given if and only if $n = 2^r$, $r \geq 2$, or $n = 2^r p_1 p_2 \cdots p_s$, $r \geq 0$, $s \geq 1$ and p_i are different Fermat primes (p is a Fermat prime when $p = 2^{2^t} + 1$, $t \geq 0$):
(a) If $k|n$ and a regular polygon with n sides is constructible, then a regular polygon with k sides is constructible;
(b) If $n = kl$, where k and l are relatively prime, then a regular polygon with n sides is constructible if and only if regular polygons with k sides and l sides are constructible;
(c) If $n = 2^r$, $r \geq 2$, then a regular polygon with n sides is constructible;
(d) If $n = p^2$, where p is an odd prime, then a regular polygon with n sides is not constructible;
(e) If $n = p$, where p is an odd prime, then a regular polygon with n sides is constructible if and only if p is a Fermat prime.

14.7. Construct a regular polygon with n sides when $(0, 0)$ and $(1, 0)$ are given if:
(a) $n = 5$; (b) $n = 15$; (c) $n = 20$.

14.8. Which of the following angles α can be constructed when $(0, 0)$ and $(1, 0)$ are given:
(a) $\alpha = 1°$; (b) $\alpha = 3°$; (c) $\alpha = 5°$.

14.9. Give an example of an angle α which is not constructible when $(0, 0)$ and $(1, 0)$ are given, but which can be trisected when this angle is given (an angle is given if three different points $(0, 0)$, $(1, 0)$ and (a, b) are given defining the rays from $(0, 0)$ through $(1, 0)$ and (a, b)).

Chapter 15
Computing Galois Groups

In earlier chapters, we had several opportunities to find Galois groups of specific polynomials. In general, computing the Galois group of a given polynomial over a given field is numerically complicated when the degree of the polynomial is modestly high. For polynomials of (very) low degrees it is possible to specify some simple numerical invariants, which tell us about the isomorphism type of the Galois group depending on the values of these invariants. For arbitrary polynomials there is a variety of numerical methods which, for not too high degrees, make the computational task possible to implement in a more or less effective way. There are several computer packages in which Galois groups of irreducible polynomials up to not too high degrees (sometimes, up to 11) can be computed, notably, Maple, GP/Pari, Sage and Magma. In this chapter, we discuss some theoretical background to these numerical methods and apply them in a few cases. In the exercises, we give several examples how to compute and classify Galois groups for low degree polynomials using different methods. Of particular interest is a very general theorem proved by Dedekind, which in a special case relates the Galois group of an integer irreducible polynomial to Galois groups of its reductions modulo prime numbers. Several exercises are concerned with Dedekind's Theorem.

Let K be a field and $F(X_1, \ldots, X_n) \in K(X_1, \ldots, X_n)$ a rational function in n variables $X_1, \ldots, X_n$. The symmetric group S_n acts as an automorphism group of the field $K(X_1, \ldots, X_n)$ when for $\sigma \in S_n$:

$$\sigma F(X_1, \ldots, X_n) = F(X_{\sigma(1)}, \ldots, X_{\sigma(n)})$$

and σ is a permutation of $\{1, \ldots, n\}$. Let G be a subgroup of S_n. We denote by G_F **the stabilizer of** F **in** G, that is, $G_F = \{\sigma \in G \mid \sigma F = F\}$.

T.15.1. (*a*) *Let G be a subgroup of S_n. Then for every subgroup H of G there exists a polynomial $F \in K[X_1, \ldots, X_n]$ such that $H = G_F$.*

J. Brzeziński, *Galois Theory Through Exercises*, Springer Undergraduate Mathematics Series, https://doi.org/10.1007/978-3-319-72326-6_15

(b) Let $G = \sigma_1 G_F \cup \cdots \cup \sigma_m G_F$ be the presentation of G as a union of different left cosets with respect to G_F. Then $\sigma_i F(X_1, \ldots, X_n)$, for $i = 1, \ldots, m$, are all different images of F under the permutations belonging to G.

Let $f(X) \in K[X]$ be a polynomial of degree n and $L = K_f = K(\alpha_1, \ldots, \alpha_n)$ its splitting field over K, where α_i are all the zeros of $f(X)$ in L taken in an arbitrarily fixed order. Assume that $\mathrm{Gal}(K_f/K) \subseteq G$, where the Galois group of $f(X)$ over K is considered as a group of permutations of $\{1, \ldots, n\}$ in the usual way: $\sigma(\alpha_i) = \alpha_{\sigma(i)}$ (this means that we use the same symbol σ to denote the permutation $\sigma \in \mathrm{Gal}(K_f/K)$ of the zeros of $f(X)$ in K_f and the corresponding permutation on the set of indices of these zeros). From now on, we assume that $F(X_1, \ldots, X_n)$ is a polynomial.

By a **general polynomial** with respect to a subgroup $G \subseteq S_n$ and $F(X_1, \ldots, X_n) \in K[X_1, \ldots, X_n]$, we mean the polynomial:

$$r_{G,F}(T) = \prod_{i=1}^{m} (T - (\sigma_i F)(X_1, \ldots, X_n)), \tag{15.1}$$

where the product is over a set σ_i, $i = 1, \ldots, m$, of representatives of all left cosets of G_F in G. Of course, $\sigma_i F$ does not depend on the choice of a representative of $\sigma_i G_F$, since $\sigma_i \tau F = \sigma_i F$, when $\tau \in G_F$. We usually omit G in $r_{G,F} \in K[X_1, \ldots, X_n, T]$ when $G = S_n$. If $G = S_n$ and $F(X_1, \ldots, X_n) = X_1$, then $r_{G,F}(T)$ is the general polynomial of degree n as defined on p. 78.

Let $f(X) \in K[X]$ be a polynomial with zeros $\alpha_1, \ldots, \alpha_n$ in a field extension of K. By the **resolvent polynomial** of $f(X)$ with respect to a subgroup $G \subseteq S_n$ and $F(X_1, \ldots, X_n) \in K[X_1, \ldots, X_n]$, we mean the polynomial

$$r_{G,F}(f)(T) = \prod_{i=1}^{m} (T - (\sigma_i F)(\alpha_1, \ldots, \alpha_n)). \tag{15.2}$$

The values $(\sigma_i F)(\alpha_1, \ldots, \alpha_n)$ are obtained by a homomorphism of $K[X_1, \ldots, X_n]$ onto $K[\alpha_1, \ldots, \alpha_n]$ sending X_i to α_i. Notice that these values need not be different. They generate over K a subfield of K_f which plays an important role in the study of the Galois group $\mathrm{Gal}(K_f/K)$, in particular, via the following result:

T.15.2. *Let $f(X) \in K[X]$, $F(X_1, \ldots, X_n) \in K[X_1, \ldots, X_n]$ and assume that $\mathrm{Gal}(K_f/K) \subseteq G$, where G is a subgroup of S_n. Then:*
(a) The resolvent polynomial $r_{G,F}(f)$ has its coefficients in K;
(b) If $K = \mathbb{Q}$, $F(X_1, \ldots, X_n) \in \mathbb{Z}[X_1, \ldots, X_n]$ and $f(X)$ has integer coefficients and the leading coefficient is 1, then the same is true of $r_{G,F}(f)$;
(c) If all the zeros of $r_{G,F}(f)$ are different, then $\mathrm{Gal}(K_f/K)$ is conjugate in G to a subgroup of G_F if and only if at least one of the zeros of $r_{G,F}(f)$ belongs to K.
Even if the assumptions above are not satisfied, the splitting field of the resolvent $r_{G,F}(f)$ in K_f and the degrees of irreducible factors of the resolvent very often give information about the Galois group $\mathrm{Gal}(K_f/K)$ (see the exercises below).

There exists a general algorithmic procedure for finding the isomorphism type of the Galois group $\mathrm{Gal}(K_f/K)$ which uses **T.15.2** and was already known to Galois. Unfortunately, the value of this procedure for practical computations is rather limited. We describe it for two reasons. First of all, it shows that the Galois group of any polynomial can be found if its zeros are known (over number fields, it is often sufficient to know the zeros with appropriate precision). On the other hand, we will use it in the proof of Dedekind's Theorem (**T.15.4**), which gives a very good method for computing Galois groups in many situations.

Consider a field k and let $K = k(Y_1, \ldots, Y_n)$ be the field of rational functions of Y_i. Take the polynomial

$$F(X_1, \ldots, X_n) = X_1Y_1 + \cdots + X_nY_n$$

in the polynomial ring $K[X_1, \ldots, X_n]$. Let $f(X) \in k[X]$ and let $k_f = k(\alpha_1, \ldots, \alpha_n)$ be a splitting field of $f(X)$ over k. Then $K_f = k_f(Y_1, \ldots, Y_n)$ is a splitting field of $f(X) \in K[X]$ over K and $\mathrm{Gal}(K_f/K) \cong \mathrm{Gal}(k_f/k)$ (see Exercise 15.9). Using the notation of **T.15.2**, choose $G = S_n$. It is clear that $G_F \subset S_n$ consists of only the identity permutation. The resolvent of $f(X)$ with respect to $G = S_n$ and $F(X_1, \ldots, X_n)$ is:

$$r_{G,F}(f)(T) = \prod_{\sigma \in S_n} (T - (\alpha_{\sigma(1)}Y_1 + \cdots + \alpha_{\sigma(n)}Y_n)).$$

This polynomial has degree $n!$ and is a product of irreducible polynomials $r_i(T, Y_1, \ldots, Y_n)$ in $k[T, Y_1, \ldots, Y_n]$:

$$r_{G,F}(f)(T) = r_1(T, Y_1, \ldots, Y_n) \cdots r_t(T, Y_1, \ldots, Y_n). \tag{15.3}$$

The polynomials r_i are called **Galois resolvents** of $f(X)$. We will assume that r_1 is the polynomial having $\theta = \alpha_1Y_1 + \cdots + \alpha_nY_n$ as its zero.

T.15.3. *Let $K = k(Y_1, \ldots, Y_n)$ be the field of rational functions in the variables Y_i over a field k. Let $f(X) \in k[X] \subset K[X]$ and let $r_{G,F}(f)(T)$ be the resolvent of $f(X)$ with respect to the Galois group* $\mathrm{Gal}(k_f/k) = \mathrm{Gal}(K_f/K) \subseteq S_n$ *and the polynomial*

$$F(X_1, \ldots, X_n) = X_1Y_1 + \cdots + X_nY_n.$$

Let $r_i(T, Y_1, \ldots, Y_n)$, for $i = 1, \ldots, t$, be the irreducible factors of the resolvent $r_{G,F}(f)(T)$ in $k[T, Y_1, \ldots, Y_n]$. Then the Galois group $\mathrm{Gal}(k_f/k)$ *is isomorphic to any group G_{r_i} of those permutations of $Y_1, \ldots, Y_n$ which map $r_i(T, Y_1, \ldots, Y_n)$ onto itself. Moreover, all $r_i(T, Y_1, \ldots, Y_n)$ have the same degree $[k_f : k]$ with respect to T and a common splitting field $K_f = k_f(Y_1, \ldots, Y_n)$ over K.*

Another, often very effective technique for the description of Galois groups of polynomials, in particular, over number fields, uses a result proved by Dedekind. This is also a very general method, but in the case of number fields there is usually a lot of additional arithmetical information, which can be used in order to determine the Galois groups. Therefore, we formulate both the general case and a special case of Dedekind's Theorem over the integers.

Let $\varphi : R \to R^*$ be a ring homomorphism mapping an integral domain R into an integral domain R^*. Let K and K^* be fields of quotients of R and R^*, respectively. Let $f \in R[X]$ be separable (that is, without multiple zeros). Denote by f^* the image of f under the homomorphism extending φ to $R[X]$ by applying φ to the coefficients of the polynomials in $R[X]$. Such an extension is sometimes called **reduction modulo** φ, since the simplest example is the homomorphism $\varphi : \mathbb{Z} \to \mathbb{F}_p$ of reduction modulo a prime p (see p. 14).

T.15.4 Dedekind's Theorem. *(a) Let $f \in R[X]$ be a separable monic polynomial and assume that its image $f^* \in R^*[X]$ is also separable and* $\deg(f) = \deg(f^*) = n$. *Then the Galois group of f^* over K^* as a permutation group of its zeros is a subgroup of the Galois group of f over K as a permutation group of its zeros.*
(b) In (a), let $R = \mathbb{Z}$, $R^* = \mathbb{F}_p$ and let $\varphi : \mathbb{Z} \to \mathbb{F}_p$ be the reduction modulo a prime number p. If $f \in \mathbb{Z}[X]$ and

$$f^* = f_1^* \cdots f_k^*,$$

where f_i^* are irreducible over $\mathbb{F}_p$, then $\mathrm{Gal}(\mathbb{Q}_f/\mathbb{Q})$ *considered as a permutation subgroup of S_n contains a permutation which is a product of cycles of length* $\deg(f_i^*)$ *for $i = 1, \ldots, k$.*

Old proofs of the first part of this theorem use the relation, mentioned earlier, between Galois groups and the Galois resolvents of f and f^*. We choose this "old fashioned" proof using Galois' resolvents in order to avoid introducing several notions, which are outside the scope of this book, and in order to use the technique of resolvents for the computation of the Galois groups (see **T.15.3**). A proof in the modern language can be found in [L, Chap. VII, §2].

Exercises 15

15.1. Consider $f(X) = X^3 + pX + q \in K[X]$ with zeros $\alpha_1, \alpha_2, \alpha_3$ in some extension of the field K, where $\mathrm{char}(K) \neq 2$. Let $\Delta = \Delta(f) = (\alpha_1 - \alpha_2)^2(\alpha_2 - \alpha_3)^2(\alpha_3 - \alpha_1)^2 = -4p^3 - 27q^2$ be the discriminant of $f(X)$ (see Exercise 1.3).
(a) Show that $K_f = K(\alpha_1, \alpha_2, \alpha_3) = K(\sqrt{\Delta}, \alpha_1)$ and if $f(X)$ is irreducible in $K[X]$, then its Galois group is isomorphic to C_3 or S_3, depending on $\sqrt{\Delta} \in K$ or $\sqrt{\Delta} \notin K$.
(b) In the notation of **T.15.2**, choose $G = S_3$ and $F(X_1, X_2, X_3) = (X_1 - X_2)(X_2 - X_3)(X_3 - X_1)$. Show that $G_F = A_3$ and $r_{G,F}(f) = X^2 - \Delta(f)$. Assume that $f(X)$ is irreducible over K and deduce the description of the Galois group of $f(X)$ in (a) from **T.15.2**.

15.2. (a) Let $L = K(X_1, \ldots, X_n)$, $F(X_1, \ldots, X_n) = \prod_{1 \le i < j \le n}(X_i - X_j)$ and $G = S_n$, where K is a field of characteristic different from 2. Show that $r_{G,F}(T) = T^2 - F^2$ and $G_F = A_n$.

(b) Let $f(X) \in K[X]$ be an irreducible polynomial with zeros $\alpha_1, \dots, \alpha_n$ in an extension of K and $\Delta(f) = \prod_{1 \le i < j \le n} (\alpha_i - \alpha_j)^2$ its discriminant (see Exercise 9.25). Use (a) and **T.15.2** in order to show that the Galois group $\mathrm{Gal}(K_f/K)$ is contained in A_n if and only if $\sqrt{\Delta(f)} \in K$.

15.3. Denote by $\alpha_1, \alpha_2, \alpha_3, \alpha_4$ the zeros of a separable polynomial $f(X) = X^4 + pX^2 + qX + r$ in its splitting field K_f over K and let $r(f)(T) = T^3 + 2pT^2 + (p^2 - 4r)T - q^2$ be its resolvent as defined in Exercise 3.9.
(a) Show that the zeros of the resolvent $r(f)(T)$ are $(\alpha_1+\alpha_4)^2, (\alpha_2+\alpha_4)^2, (\alpha_3+\alpha_4)^2$.
(b) Show that the discriminants of the polynomials f and $r(f)$ are equal (see Exercises 15.2 and A.10.2). In particular, if a quartic polynomial $f(X)$ is separable, then its resolvent $r(f)(X)$ is also separable.
(c) Show that a splitting field K_f of f can be obtained from a splitting field $K_{r(f)}$ of its resolvent $r(f)$ over K by adjunction of one arbitrary solution α_i ($i = 1, 2, 3, 4$) of the equation $f(X) = 0$, that is, $K_f = K(\alpha_1, \alpha_2, \alpha_3, \alpha_4) = K_{r(f)}(\alpha_i)$.

15.4. (a) Let K be a field of characteristic different from 2 and $F(X_1, X_2, X_3, X_4) = \frac{1}{2}((X_1 + X_2)^2 + (X_3 + X_4)^2) \in K[X_1, X_2, X_3, X_4]$. Show that $G_F = D_4 = V_4 \cup \{(1,3,2,4), (1,4,2,3), (1,2), (3,4)\}$, where $V_4 = \{(1), (1,2)(3,4), (1,3)(2,4), (1,4)(2,3)\}$ if $G = S_4$ and $G_F = V_4$ if $G = A_4$. Show that $r_{G,F}(T)$ is the same for both $G = S_4$ and $G = A_4$.
(b) Let $f(X) = X^4 + pX^2 + qX + r$ and let $\alpha_1, \alpha_2, \alpha_3, \alpha_4$ be the zeros of $f(X)$ in its splitting field K_f over K. Show that the resolvent defined in Exercise 3.9 equals the resolvent $r_{G,F}(f)(T) = r(f)(T) = T^3 + 2pT^2 + (p^2 - 4r)T - q^2$.

Remark. Very often one meets a somewhat simpler choice $F(X_1, X_2, X_3, X_4) = X_1X_2 + X_3X_4$ for which $G_F = D_4$ (when $G = S_4$). The choice of F in the text of the exercise is adjusted to the "ad hoc" resolvent obtained in a natural way from factorization of a quartic polynomial as a product of two quadrics in Exercise 3.9. The resolvent $r_{G,F}(f)$ for the "simpler" choice of F is: $r_{G,F}(f)(T) = T^3 - pT^2 - 4rT + 4pr - q^2$.

15.5. (a) Assume that $f(X) = X^4 + pX^2 + qX + r$ is irreducible over K. Let

$$\Delta = \Delta(f) = \Delta(r(f)) = -4p^3q^2 - 27q^4 + 16p^4r - 128p^2r^2 + 144pq^2r + 256r^3$$

be the discriminant of f and $r(f)$ (see Exercise 15.3(b)) and $\delta = p^2 - 4r$. Using Exercise 15.3 (with or without Exercise 15.4) show that

$$G(K_f/K) = \begin{cases} S_4 \text{ if } [K_{r(f)} : K] = 6, \\ A_4 \text{ if } [K_{r(f)} : K] = 3, \\ D_4 \text{ if } [K_{r(f)} : K] = 2 \;\text{ and } f \text{ is irreducible over } K_{r(f)}, \\ C_4 \text{ if } [K_{r(f)} : K] = 2 \;\text{ and } f \text{ is reducible over } K_{r(f)}, \\ V_4 \text{ if } [K_{r(f)} : K] = 1. \end{cases}$$

(b) Show that:

$$G(K_f/K) = \begin{cases} S_4 \text{ when } r(f) \text{ does not have zeros in } K \text{ and } \sqrt{\Delta} \notin K, \\ A_4 \text{ when } r(f) \text{ does not have zeros in } K \text{ and } \sqrt{\Delta} \in K, \\ D_4 \text{ when } r(f) \text{ has only one zero } \beta \in K \text{ and } \sqrt{\beta\Delta} \notin K \\ \qquad \text{if } \beta \neq 0 \text{ and } \sqrt{\delta\Delta} \notin K \text{ if } \beta = 0, \\ C_4 \text{ when } r(f) \text{ has only one zero } \beta \in K \text{ and } \sqrt{\beta\Delta} \in K \\ \qquad \text{if } \beta \neq 0 \text{ and } \sqrt{\delta\Delta} \in K \text{ if } \beta = 0, \\ V_4 \text{ when } r(f) \text{ has all its zeros in } K. \end{cases}$$

In each case give an example of a polynomial f over $\mathbb{Q}$ with the corresponding Galois group.

15.6. Let $f(X) = X^4 + pX^2 + qX + r$ be reducible in K but without zeros in K. Then

$$G(K_f/K) = \begin{cases} V \text{ when } r(f) \text{ has only one zero in } K, \\ C_2 \text{ when } r(f) \text{ has all its zeros in } K. \end{cases}$$

15.7. (a) Show that the resolvent $r(f)$ of $f(X) = X^4 + pX^2 + r$ always has one of its zeros in K and it has all three zeros in K if and only if $\sqrt{r} \in K$ (K a field of characteristic $\neq 2$). Check also that the discriminant of $f(X)$ (and $r(f)$—see Exercise 15.3(b)) is

$$\Delta = \Delta(f) = 16r(p^2 - 4r)^2,$$

so $K(\sqrt{\Delta}) = K(\sqrt{r})$.
(b) Show that if $f(X) = X^4 + pX^2 + r$ is irreducible in $K[X]$, where K is a field of characteristic different from 2, then

$$G(K_f/K) = \begin{cases} V_4 \text{ if } r \text{ is a square in } K, \\ C_4 \text{ if } r \text{ is not a square in } K \text{ and } r(p^2 - 4r) \text{ is a square in } K, \\ D_4 \text{ if } r \text{ and } r(p^2 - 4r) \text{ are not squares in } K. \end{cases}$$

15.8. Show that the splitting fields of irreducible trinomials $X^4 + qX + r$ over $\mathbb{Q}$ may give all possible types of Galois groups which appear in Exercise 15.5, that is, S_4, A_4, D_4, C_4 and V_4.

15.9. (a) Let $K = k(Y_1, \ldots, Y_n)$ be the field of rational functions of Y_i over a field k. Let k' be a finite Galois extension of k. Show that $K' = k'(Y_1, \ldots, Y_n)$ is a Galois extension K and $\mathrm{Gal}(K'/K) \cong \mathrm{Gal}(k'/k)$.
(b) Compute the resolvent $r_{G,F}(f)$ in (15.3) when $n = 2$, $k = \mathbb{Q}$ and $f(X) = X^2 + pX + q$ has zeros α_1, α_2 ($F(X_1, X_2) = X_1Y_1 + X_2Y_2$). When is $r_{G,F}(f)$ irreducible?

15.10. Using Dedekind's Theorem **T.15.4** show that the Galois groups over $\mathbb{Q}$ of the given polynomials are:

(a) S_5 for $f(X) = X^5 + 2X^3 + 1$;
(b) S_5 for $f(X) = X^5 - X - 1$;
(c) A_5 for $f(X) = X^5 + 10X^2 + 24$;
(d) A_5 for $f(X) = X^5 - 55X + 88$.

15.11. (a) Using Dedekind's Theorem **T.15.4** show that for every n the symmetric group S_n is the Galois group of a Galois field extension of $\mathbb{Q}$.
(b) Construct polynomials in $\mathbb{Z}[X]$ with Galois groups S_7 and S_9.

Chapter 16
Supplementary Problems

In this chapter, we formulate "100 problems" without solutions, but occasionally with some hints or references.

Field Extensions

16.1. Let $K \subset L$ be a field extension of degree 4. What are the possibilities for the number of quadratic extensions of K contained in L? (Make a distinction between separable and non-separable extensions.)

16.2. Let $K \subset L$ be a separable field extension of degree 4 and N its normal closure. What can be said about the degree of N over K when:

(a) there are no quadratic extensions of K contained in L?
(b) there is exactly one quadratic extension of K contained in L?
(c) there are at least two quadratic extensions of K contained in L?

16.3. (a) Consider $K = \mathbb{Q}(\sqrt{1+\sqrt[3]{2}})$. Show that $\alpha = \sqrt{1+\sqrt[3]{2}}$ is a zero of $f(X) = X^6 - 3X^4 + 3X^2 - 3$, $[K : \mathbb{Q}] = 6$ and K does not contain any quadratic extension of $\mathbb{Q}$.
(b) Let $K = \mathbb{Q}(\sqrt[3]{1+\sqrt{3}})$. Show that K has degree 6 over $\mathbb{Q}$, does not contain any cubic but contains a quadratic extension of $\mathbb{Q}$.
(c) Show that there exist extensions of degree 6 over $\mathbb{Q}$ which do not contain subextensions of degrees 2 and 3 over $\mathbb{Q}$.

16.4. Show that the polynomial $f(X) = X^4 - 20X^2 + 16$ is normal over $\mathbb{Q}$. Let $K = \mathbb{Q}(\alpha)$ be the splitting field of $f(X)$, where α is a zero of it. Find all intermediate fields $M = \mathbb{Q}(\beta) \subseteq K$ with β expressed as a linear combination of powers of α (see the definition of a normal polynomial in Exercise 7.9).

16.5. Let $K \subseteq L$ be a field extension and M_1, M_2 two subfields containing K and contained in L.
(a) Let M_1, M_2 both have one of the properties: separable, normal, Galois over K. What can be said about the same property of M_1M_2 and $M_1 \cap M_2$ over K?

J. Brzeziński, *Galois Theory Through Exercises*, Springer Undergraduate Mathematics Series, https://doi.org/10.1007/978-3-319-72326-6_16

(b) Let L have one of the properties: separable, normal, Galois over both M_1 and M_2. What can be said about the same property of L over M_1M_2 and $M_1 \cap M_2$?

16.6. Let $L_i \supseteq K$, $i = 1, 2$, be finite field extensions. Show that there exists a normal extension L of K such that $L_i \subseteq L$ for $i = 1, 2$. Show also that if $L_i \supseteq K$ are separable, then there exists a Galois extension L of K containing both.

16.7. (a) Let $\alpha = \sqrt{2} + \sqrt{3} + \cdots + \sqrt{n}$ $(n \geq 2)$. Show that $[\mathbb{Q}(\alpha) : \mathbb{Q}] = 2^{\pi(n)}$, where $\pi(n)$ denotes the number of primes less than or equal to n.
(b) Find a primitive element of $\mathbb{Q}(\sqrt{2}, \sqrt{3}, \sqrt{5})$ over $\mathbb{Q}$.
(c) Using (a) show that $\alpha = \sqrt{2} + \sqrt{3} + \cdots + \sqrt{n}$ is a primitive element of $\mathbb{Q}(\sqrt{2}, \sqrt{3}, \ldots, \sqrt{n})$ over $\mathbb{Q}$.
(d) Assume that $[\mathbb{Q}(\sqrt{a_1}, \ldots, \sqrt{a_n}) : \mathbb{Q}] = 2^n$. Show that $\mathbb{Q}(\sqrt{a_1}, \ldots, \sqrt{a_n}) = \mathbb{Q}(\sqrt{a_1} + \cdots + \sqrt{a_n})$ ($a_1, \ldots, a_n$ integers).
(e) Show that $\mathbb{Q}(\sqrt{p_1}, \ldots, \sqrt{p_n}) = 2^n$ when $p_1, \ldots, p_n$ are different prime numbers.

16.8. Let $K = \mathbb{Q}(\sqrt[n]{a})$, where a is a positive integer, be such that $[K : \mathbb{Q}] = n$. Show that if M is a subfield of K and $[M : \mathbb{Q}] = d$, then $E = \mathbb{Q}(\sqrt[d]{a})$.

16.9. Let $f(X) \in K[X]$ be an irreducible polynomial of degree n and let K_f be its splitting field over a field K. By Exercise 5.3, we know that $[K_f : K] \leq n!$. Show that $[K_f : K]$ always divides $n!$.

16.10. Let $K \subseteq L$ be a Galois extension. Show that L is a splitting field of a normal polynomial over K (see the definition of a normal polynomial in Exercise 7.9).

16.11. Let $f(X)$ be an irreducible polynomial of degree n over $\mathbb{Q}$ and let α, β be two of its different zeros such that $\mathbb{Q}(\alpha, \beta) = \mathbb{Q}(\alpha + \beta)$. What can be said about the polynomial $f(X)$ and about the degree of the splitting field $\mathbb{Q}_f$ over $\mathbb{Q}$ when $n = 2, 3, 4, 5$?

16.12. Let $K \subset L$ be a finite field extension and R a subring of L containing K. Show that R is also a field.

16.13. Show that a quartic field with Galois group V (the Klein four-group), that is, $K = \mathbb{Q}(\sqrt{d_1}, \sqrt{d_2})$, where d_1, d_2 are nonequal square-free integers not equal to 1 (se Exercise 9.6), is a splitting field of a trinomial $X^4 + pX + q$, $p, q \in \mathbb{Z}$ if and only if the quadratic form $X_0^2 + d_1X_1^2 + d_2X_2^2$ is isotropic (that is, it has a nontrivial zero $X_0 = x_0, X_1 = x_1, X_2 = x_2$).

16.14. Show that a quartic field with Galois group C_4 (a cyclic group of order 4), that is, $K = \mathbb{Q}\left(\sqrt{m(\Delta + k\sqrt{\Delta})}\right)$, where $\Delta = k^2 + l^2$, and k, l, m are integers and $\sqrt{\Delta} \notin \mathbb{Q}$ (see Exercise 11.7) is a splitting field of a trinomial $X^4 + pX + q$ if and only if the quadratic form $X_0^2 + tX_1^2 + tX_2^2$ is isotropic (that is, it has a nontrivial zero $X_0 = x_0, X_1 = x_1, X_2 = x_2$).

16.15. Let $K = \mathbb{Q}(\sqrt{d})$ be a quadratic extension of the rational numbers. Show that the splitting field of a cubic polynomial $f(X) = X^3 + pX + q \in \mathbb{Z}[X]$ contains K if and only if there is a nontrivial rational point on the surface $4X^3 + 27Y^2 + dZ^2 = 0$ (that is, this Diophantine equation has a nontrivial solution $(X, Y, Z) = (x, y, z) \neq (0, 0, 0)$).

16.16. (a) Let $K \subseteq L$ be a separable field extension and assume that there is an n such that $[K(\alpha) : K] \leq n$ for every $\alpha \in L$. Show that the extension $K \subseteq L$ is finite and $[L : K] \leq n$.
(b) Show that (a) need not be true when $K \subseteq L$ is not separable.

16.17. Let L be a separable extension of K and $[L : K] = n$. Show that there are at most 2^{n-1} fields M such that $K \subseteq M \subseteq L$.

16.18. Let $K \subset M \subset L$ be three fields and let L be algebraic over K. Is it true that the degree of $\alpha \in L$ over M divides the degree of α over K?

16.19. Let $K \subset M \subset L$ be three fields and let L be algebraic over K. Show that if $[M : K]$ and $[M(\alpha) : M]$ are relatively prime for $\alpha \in L$, then the coefficients of the minimal polynomial of α over M are in K.

16.20. Let M_1, M_2 be two finite extensions of a field K contained in a field L. Show that $[M_1M_2 : K] = [M_1 : K][M_2 : K]$ implies that $M_1 \cap M_2 = K$.

16.21. Let $K \subseteq L$ be a finite field extension. Show that K is perfect if and only if L is perfect (see p. 47 and Exercise 8.3).

16.22. Show that a real Galois extension $\mathbb{Q} \subset K$ (that is, $K \subset \mathbb{R}$) of odd degree is not radical (see p. 77).

16.23. Let K be a field of characteristic $p > 0$ and let L be a finite extension of K such that p does not divide $[L : K]$. Show that $K \subseteq L$ is a separable extension.

16.24. Let $L = K(\alpha_1, \ldots, \alpha_r)$ be an algebraic extension of K and let $f_{\alpha_i}(X)$ be the minimal polynomial of α_i over K for $i = 1, \ldots, r$. Show that the order of the Galois group of L over K (that is, the automorphism group of L over K) is at most equal to the product $\prod_1^r \deg f_{\alpha_i}$.

16.25. Let K be field and $f(X), g(X) \in K[X]$ two irreducible polynomials. Let L be a splitting field of $f(X)g(X)$ over K. Let α be a zero of $f(X)$ and β a zero of $g(X)$ in L.
(a) Show that $f(X)$ is irreducible over $K(\beta)$ if and only if $g(X)$ is irreducible over $K(\alpha)$.
(b) Generalize (a): If $f(X) = f_1(X) \cdots f_r(X)$, where $f_i(X)$ are irreducible in $K(\beta)[X]$, and $g(X) = g_1(X) \cdots g_s(X)$, where $g_j(X)$ are irreducible in $K(\alpha)[X]$, then $r = s$. Moreover, it is possible to number f_i, g_j in such a way that if $\alpha_i \in L$ is a zero of $f_i(X)$ and $\beta_i \in L$ is a zero of $g_i(X)$, then $K(\alpha, \beta_i)$ is isomorphic to $K(\beta, \alpha_i)$ for $i = 1, \ldots, r$. Show that $\deg f \deg g_i = \deg g \deg f_i$.

Remark. This correspondence between pairs of irreducible polynomials over fields was first observed by Richard Dedekind in 1855 (see [E, p. 31]) and is sometimes called **Dedekind's duality**.

16.26. Show that if L is a simple and algebraic extension of a field K, then every intermediate field M between K and L is also simple over K (L need not be algebraic over K, in which case the claim is Lüroth's theorem—see Exercise 4.11).

16.27. Let $K \subset L$ be a field extension and $\alpha \in L$. Show that for any nonzero polynomial $f(X) \in K[X]$, the elements α and $f(\alpha)$ are either both algebraic or both transcendental over K.

16.28. Let $L = K(\alpha, \beta)$ be a field extension of K such that the degrees $[K(\alpha) : K] > 1$ and $[K(\beta) : K] > 1$ are finite and relatively prime.
(a) Give examples showing that the fields $K(\alpha, \beta)$ and $K(\alpha\beta)$ may be equal or nonequal.
(b) Is it true that $K(\alpha, \beta) = K(\alpha\beta)$ when $K(\alpha)$ and $K(\beta)$ are Galois extensions of K?

Remark. Compare Exercise 16.100.

16.29. Let $L = K(\alpha, \beta)$ be a field extension of K such that α is algebraic, and β is transcendental over K. Show that L is not a simple extension of K.

Function Fields

16.30. Let k be a field of characteristic 0, $L = k(X_1, \ldots, X_n)$ and $K = k(s_1, \ldots, s_n)$, where s_i are the elementary symmetric polynomials of the variables $X_1, \ldots, X_n$. Show that $\alpha = X_1 + 2X_2 + \cdots + nX_n$ is a primitive element of the extension $K \subset L$.

16.31. Let $\mathbb{F}$ be a finite field of characteristic p. Find the following degrees:

(a) $[\mathbb{F}(X) : \mathbb{F}(X^p)]$; (b) $[\mathbb{F}(X, Y) : \mathbb{F}(X^p, Y^p)]$.

16.32. Show that $\mathbb{C}(X)$ is a Galois extension of $\mathbb{C}(X^n + X^{-n})$, $n \geq 1$. Find the degree and the Galois group of this extension (the Galois group is the dihedral group D_n generated by $\sigma(X) = \frac{1}{X}$ and $\tau(X) = \varepsilon X$, where $\varepsilon = e^{\frac{2\pi i}{n}}$—see Example A.2.4 and the Remark after Exercise 12.4).

16.33. Let $\mathbb{F}$ be a finite field with $q = p^n$ elements. Show that $L = \mathbb{F}(X)$ is a Galois extension of $K = \mathbb{F}(X^q - X)$. Find the degree $[L : K]$ and give a description of the Galois group $G(L/K)$.

16.34. Let $\mathbb{F}$ be a finite field with q elements and let $G = G(\mathbb{F}(X)/\mathbb{F})$ be the group of all Möbius transformations over $\mathbb{F}$ (see Exercise 6.6). Show that:
(a) $|G| = q^3 - q$;
(b) $\mathbb{F}(X)^G = \mathbb{F}(Y)$, where $Y = \frac{(X^{q^2}-X)^{q+1}}{(X^q-X)^{q^2+1}}$;
(c) If H_1 is the subgroup of G consisting of $\sigma(X) = aX + b, a, b \in \mathbb{F}, a \neq 0$, then $\mathbb{F}(X)^{H_1} = \mathbb{F}((X^q - X)^{q-1})$;
(d) If H_2 is the subgroup of G consisting of $\sigma(X) = X + b, b \in \mathbb{F}$, then $\mathbb{F}(X)^{H_2} = \mathbb{F}(X^q - X)$; and
(e) Find all subfields of $\mathbb{F}(X)$ containing $\mathbb{F}(X)^G$ when $\mathbb{F}$ has 2 or 3 elements.
This exercise is partly taken from [V].

16.35. Show that if K is a field and X a variable, then there exist infinitely many intermediate fields M such that $K \subseteq M \subseteq K(X)$.

16.36. (a) Find the degree of the extension $K(X_1, \ldots, X_n) \supseteq K(X_1^{d_1}, \ldots, X_n^{d_n})$, where $X_1, \ldots, X_n$ are variables and $d_1, \ldots, d_n$ are positive integers (K any field).
(b) Let K be a field and let k, l, m, n be nonnegative integers such that $kn - lm \neq 0$. Find the degree $[K(X, Y) : K(X^k Y^l, X^m Y^n)]$, where X, Y are variables.
This exercise is partly taken from [B] and [V].

16.37. (a) Let k, l, m, n be positive integers. Show that $\mathbb{C}(X^k Y^l, X^m Y^n) \subseteq \mathbb{C}(X, Y)$ is a Galois extension and describe its Galois group.
(b) Let m, n be positive integers. Show that $\mathbb{C}(X^n + Y^n, X^m Y^m) \subseteq \mathbb{C}(X, Y)$ is a Galois extension and describe its Galois group.

16.38. Let $K \subseteq L$ be a finite field extension. Show that $K(X) \subseteq L(X)$ is also finite and $[L : K] = [L(X) : K(X)]$. Compare Exercise 15.9.

16.39. Let K be an arbitrary field and let $L = K(X)$ be the field of rational functions over K. Show that the six functions $X \mapsto X, X \mapsto 1-X, X \mapsto 1/X, X \mapsto 1/(1-X)$, $X \mapsto X/(X-1), X \mapsto (X-1)/X$ form an automorphism group G of $L = K(X)$ over K, which is isomorphic to S_3. Show also that $L^G = K(g)$, where

$$g(X) = \frac{(X^2 - X + 1)^3}{X^2(X-1)^2}.$$

Compare Exercises 6.7 and 16.34.

16.40. Let K be a field. Find a rational function $f(X) \in K(X)$ such that $K(X)^G = K(f)$, where $G = \langle \sigma \rangle$ and $\sigma(X) = -\frac{1}{X+1}$ is an automorphism of order 3.

16.41. Consider the following 12 Möbius transformations over the complex numbers $\mathbb{C}$:

$$z \to \pm z, \;\; \pm\frac{1}{z}, \;\; \pm i\frac{z+1}{z-1}, \;\; \pm i\frac{z-1}{z+1}, \;\; \pm\frac{z+i}{z-i}, \;\; \pm\frac{z-i}{z+i}.$$

Show that these transformations form a group G isomorphic to the permutation group A_4 (the tetrahedron rotation group). Considering this group as a group of automorphism of $K = \mathbb{C}(z)$ find a function φ such that $K^G = \mathbb{C}(\varphi)$.

16.42. Let $L = K(X_1, \ldots, X_n)$ and $K = k(s_1, \ldots, s_n)$, where $s_1, \ldots, s_n$ are elementary symmetric polynomials of the variables $X_1, \ldots, X_n$ and k is a field.
(a) Show that two functions $f(X, \ldots, X_n)$ and $g(X_1, \ldots, X_n)$ in L have equal stabilizers (see p. 85) if and only if $k(s_1, \ldots, s_n, f) = k(s_1, \ldots, s_n, g)$ (see Exercise 9.11).
(b) Choose suitable polynomials $f(X_1, X_2, X_3)$ so that $k(s_1, s_2, s_3, f)$ are all subfields of $L = K(X_1, X_2, X_3)$ containing $K = k(s_1, s_2, s_3)$.

16.43. Let $K \subset L$ be an algebraic field extension. An element $\alpha \in L \setminus K$ is called **essentially defined** over K if there is no proper subfield of K containing all coefficients of the minimal polynomial of α over K.

Assume that $L = k(X_1, X_2, X_3)$ and $K = k(s_1, s_2, s_3)$, where s_i are the elementary symmetric polynomials in the variables X_1, X_2, X_3.
(a) Show that $\alpha = X_1$ is essentially defined over K.
(b) Show that $\alpha = X_1X_2$ is not essentially defined over K. Let K_α be the field generated over k by the coefficients of the minimal polynomial of α over K. Find the degree $[K : K_\alpha]$ and show that $K_\alpha \subset L$ is a Galois extension of degree 12.
(c) Is $\alpha = X_1X_2^2$ essentially defined over K? What is the degree $[L : K_\alpha]$, where, as before, K_α denotes the field generated over k by the coefficients of the minimal polynomial of α over K.
(d) Is it true that any primitive element of L over K (that is, any α such that $L = K(\alpha)$) is essentially defined over K?

16.44. As we know (see Exercise 6.13(b)) $[K(X_1, \ldots, X_n) : K(s_1, \ldots, s_n)] = n!$, when K is a field and $s_1, \ldots, s_n$ the elementary symmetric polynomials of the variables $X_1, \ldots, X_n$. Find a basis of $K(X_1, \ldots, X_n)$ as a linear space over $K(s_1, \ldots, s_n)$.

16.45. Let n be a positive integer. Give an example of a Galois field extension $L \supset \mathbb{C}(X)$ whose Galois group is cyclic of order n

16.46. Let K be a field. Give an example of two subfields M, M' of $L = K(X)$ such that $[L : M] < \infty$ and $[L : M'] < \infty$, but $M \cap M' = K$.

Groups and Field Extensions

16.47. Show that a finite group is solvable if and only if every nontrivial subgroup H of G contains a normal subgroup N such that H/N is a nontrivial abelian group.

16.48. There exists five non-isomorphic groups of order 8—the cyclic group $\mathbb{Z}_8$ of order 8, the products $\mathbb{Z}_2 \times \mathbb{Z}_4$, $\mathbb{Z}_2 \times \mathbb{Z}_2 \times \mathbb{Z}_2$, the dihedral group D_4 and the quaternion group Q_8 (see Example A.2.4). For each of these groups, construct a Galois field extension $K \supset \mathbb{Q}$ having it as its Galois group. (All examples can be found in different places in this book.)

16.49. Show that the field $K = \mathbb{Q}(\alpha)$, where $\alpha = \sqrt{(2+\sqrt{2})(3+\sqrt{3})}$, is Galois and show that the Galois group $G(K/\mathbb{Q})$ is the quaternion group of order 8 (see Exercise 12.1(e)). (The minimal polynomial of α is $f(X) = X^8 - 24X^6 + 144X^4 - 288X^2 + 144$.)

16.50. There exists five non-isomorphic groups of order 12—the cyclic group $\mathbb{Z}_{12}$, the product $\mathbb{Z}_2 \times \mathbb{Z}_2 \times \mathbb{Z}_3$ (=$\mathbb{Z}_2 \times \mathbb{Z}_6$) and three nonabelian groups: the alternating group A_4, the dihedral group D_6 (=$\mathbb{Z}_2 \times \mathbb{S}_3$) and the dicyclic group Dic_3 (called also binary dihedral). This last group can be described as the subgroup of quaternions generated by $\varepsilon = e^{\frac{\pi i}{3}}$ and j satisfying relation $\varepsilon^6 = 1, j^2 = -1$ and $\varepsilon j \varepsilon = j$. For each of these groups, construct a Galois field extension $K \supset \mathbb{Q}$ having it as its Galois group.

Remark. Only the last example cannot be found in previous chapters, but observe that $Dic_3 = \langle \varepsilon, j \rangle$, where ε generates a cyclic group of order 6, j a cyclic group of order 4 and $\varepsilon j \varepsilon = j$. The polynomial which we need has a splitting field containing a splitting field of a cubic with group S_3 and a quartic with group C_4, whose intersection is the only quadratic field (all three over $\mathbb{Q}$). Use Exercises 11.7 and 16.15. Notice that for the dihedral group, we have $D_6 = \langle \varepsilon, s \rangle$, $\varepsilon^6 = s^2 = 1$, $\varepsilon s \varepsilon = s$.

16.51. (a) Call a number field (that is, a subfield of the complex numbers) normal if it is normal over the rational numbers (one could say Galois, but this term is used as the name of finite fields—it could be extended to all finite Galois extensions of prime fields). Show that two isomorphic normal subfields of the complex numbers are equal. Generalize to normal intermediate fields in an arbitrary separable field extension.
(b) Show that for every $n > 2$ there exists n isomorphic but different subfields of the complex numbers.

16.52. Let $G = G(L/K)$ be the Galois group of a Galois field extension $K \subseteq L$ and let G' be the commutator subgroup of G (see p. 245). Show that $L^{G'} \supseteq K$ is an abelian Galois extension of K. Show also that for every Galois extension $M \supseteq K$ such that $G(M/K)$ is abelian, we have $M \subseteq L^{G'}$ (thus $L^{G'}$ is the maximal abelian extension of K, which is contained in L).

16.53. Let L be an algebraically closed field and K a subfield of L such that $[L : K] > 1$ is finite. Show that the characteristic of L is 0, $[L : K] = 2$ and $L = K(i)$, where $i^2 = -1$ (use Exercise 11.9).

16.54. Let $K \subseteq L$ be a Galois extension and $G = G(L/K)$ its Galois group. Show that if L is finite, then both functions $\mathrm{Tr}_G : L \to K$ and $\mathrm{Nr}_G : L \to K$ are surjective. Give an example showing that this is not true when L is infinite (see p. 36 and Exercise 9.12).

16.55. Let $K \subseteq L$ be a finite Galois field extension. Show that if a prime number p divides $[L : K]$, then there is a subfield M of L containing K such that $[L : M] = p$.

16.56. Let G be a group and let G' denote the commutator subgroup of G (see p. 245).
(a) Show that a finite group G is solvable if and only if for each subgroup $H \neq 1$ of G, we have $H' \neq H$.
(b) Show that if H is a normal subgroup of a solvable group G, then H' is also normal in G.

Remark. Notice that this exercise shows that in the definition of solvable group it is possible to choose a chain $G = G_0 \supset G_1 \supset \ldots \supset G_n = \{e\}$ such that G_{i+1} is normal in G_i and G_i/G_{i+1} is abelian in such a way that G_{i+1} is normal in G (and not only in G_i) for $i = 0, 1, \ldots, n-1$.

16.57. The **normal core** of a subgroup H in a group G is the biggest subgroup of H which is normal in G.
(a) Show that the normal core of H in G is

$$H_G = \bigcap_{g \in G} gHg^{-1}.$$

(b) Let $K \subseteq M \subseteq L$ be field extensions such that L is a Galois extension of K and let $H = G(L/M)$, so that $M = L^H$. Show that the normal closure M^* of M in L is the fixed field of the normal core H_G.

16.58. Let $K \subseteq L$ be a Galois extension of degree n. Show that L is not a splitting field of a polynomial of degree less than n if and only if the normal core (see Exercise 16.57) of each nontrivial subgroup of $G(L/K)$ is nontrivial.

16.59. Let $K \subseteq L$ be a Galois extension and let $G(L/K) = G$. Show that if G contains a subgroup H such that H does not contain any normal subgroups of G other than the identity subgroup (the core of H is trivial—see Exercise 16.57), then there is a polynomial over K of degree $|G/H|$ for which the field L is its splitting field.

16.60. Let G be a group and K a field. Assume that G is a Galois group of a Galois extension of K and call the **Galois index** of G over K the smallest degree of the polynomials with coefficients in K whose splitting field has G as its Galois group. Show that the Galois index of G over K is equal to the minimal index of subgroups of G whose core is trivial.

16.61. (a) Let G be a group of order n and d a divisor of n. Assuming that $G = G(L/K)$ for some Galois extension $K \subset L$ formulate suitable conditions (in terms of the group G), which assure that there exists an irreducible polynomial in $K[X]$ of degree d whose splitting field is L.
(b) What are the degrees of irreducible polynomials whose splitting field over $\mathbb{Q}$ has Galois group S_4? Answer the same question for S_5.

16.62. (a) Show that there exists a polynomial in $\mathbb{Z}[X]$ of degree 6 whose Galois group is A_5.
(b) Find a polynomial satisfying (a).

Hint Take a Galois extension of $\mathbb{Q}$ with group A_5 and consider the fixed field of a subgroup of order 10 in A_5.

16.63. Let $K \subseteq L$ be a Galois extension of degree n whose Galois group $G(L/K)$ is abelian. What is the degree of irreducible polynomials in $K[X]$ having L as its splitting field? Answer the same question when $G(L/K)$ is the quaternion group (of order 8—see Exercise 12.1(e)).

16.64. Let $K \subseteq L$ be a separable field extension and let N be a normal closure of $K \subseteq L$. Show that the group $G(N/L)$ does not contain nontrivial normal subgroups of the group $G(L/K)$. Notice that $K \subseteq N$ is a Galois extension by Exercise 8.4(b).

Remark. Another way of expressing the situation in the exercise is to say that the core of $G(N/L)$ in $G(N/K)$ is trivial (see Exercise 16.57).

16.65. Let $K \subseteq L$ be a Galois extension and let $\sigma \in G(L/K)$ be an involution of L (that is, an element of order 2 in $G(L/K)$).
(a) Show that $K \subseteq L^\sigma$ is Galois if and only if σ is in the center of $G(L/K)$.
(b) Let $K = \mathbb{Q}$ and $L \subset \mathbb{C}$ in (a). Explain when $L_0 = L \cap \mathbb{R}$ is Galois over $\mathbb{Q}$ (see Exercise 9.23).

16.66. Let $f(X) \in K[X]$ be an irreducible polynomial such that its splitting field K_f is a Galois extension of K with an abelian Galois group. Show that $K_f = K(\alpha)$ for any zero of $f(X)$ in K_f. (Compare Exercise 9.21.)

16.67. Let $K \subset L$ be a Galois extension and $\alpha \in L$ a zero of an irreducible polynomial $f(X) \in K[X]$. Show that if $K(\alpha)$ is a proper subfield of L, then the Galois group $G(L/K)$ is not abelian.

16.68. Let $N \supset L$ be a normal closure of a finite separable field extension $L \supset K$. Show that $G(N/K)$ is not abelian if $L \supset K$ is not normal.

16.69. Let K be a real number field (that is, $K \subseteq \mathbb{R}$) and $L \subset \mathbb{C}$ a Galois extension of K of odd degree. Show that L is also a real field.

Polynomials

16.70. Assume that an irreducible polynomial $f(X) = a_nX^n + \cdots + a_1X + a_0 \in \mathbb{Q}[X]$ ($a_n \neq 0$) has a zero of absolute value 1. Show that the degree of $f(X)$ is even and the polynomial is symmetric, that is, $a_i = a_{n-i}$ for $i = 0, 1, \ldots, n$.

16.71. Show that for every positive integers n, k and any prime number p, the polynomial $X^n + pX + p^k$ is irreducible over $\mathbb{Q}$.

16.72. (a) Show that a polynomial $X^{2n+1} - a$, $a \in \mathbb{Q}$, $n > 0$, is never normal (see Exercise 7.9 for the definition of a normal polynomial).
(b) Show that a polynomial $X^4 - a$, $a \in \mathbb{Q}$, is normal if and only if $a = b^2$, $b \in \mathbb{Q}$.
(c) Show that if an irreducible over $\mathbb{Q}$ polynomial $X^n - a$, $a \in \mathbb{Q}$ is normal, then $n = 2^k$ or $n = 2^k3^l$, where $k > 0$.

16.73. Show that if $f(X)$ is a monic polynomial with integer coefficients of odd degree n whose reduction modulo one prime p is a product of a first degree polynomial by an irreducible polynomial, and for another prime q its reduction is a product of a quadratic polynomial by an irreducible polynomial, then the Galois group of $f(X)$ over $\mathbb{Q}$ is the symmetric group S_n.

16.74. Find the orders of the Galois groups over $\mathbb{Q}$ for the following polynomials:

(a) $X^6 - 3X^2 + 6$; (b) $X^6 + 5X^2 - 10$; (c) $X^6 + 3X^3 + 3$; (d) $X^6 - 7X^2 + 7$.

16.75. Give an example of a polynomial $f(X) \in \mathbb{Z}[X]$ of degree n with n real zeros whose Galois group over $\mathbb{Q}$ is S_n for $n = 3, 4, 5$.

16.76. Find the order of the Galois group of $X^{10} - 5$ over $\mathbb{Q}$.

16.77. Show that the splitting field of the polynomial $X^4 - 7X^2 + 3X + 1$ is real and its Galois group is A_4.

16.78. Let $f(X)$ be an irreducible polynomial of degree 4 with two real and two nonreal zeros. Show that the Galois group of $f(X)$ is either D_4 or S_4.

16.79. (a) Let K be a field and $f(X) = X^n + pX + q \in K[X]$ a trinomial such that $q \neq 0$. Show that a splitting field of $f(X)$ over K is also a splitting field of the trinomial $g(X) = X^n + tX + t$, where $t = \frac{p^n}{q^{n-1}}$.
(b) Show that the Galois group of an irreducible trinomial $g(X) = X^3 + tX + t$ is cyclic of order 3 if and only if there is an $a \in K$ such that $t = -\frac{a^2+27}{4}$ (notice that the discriminant $\Delta(g(X)) = -4t^3 - 27t^2$).
(c) Let $x_1 = x$ be a zero of an irreducible trinomial $g(X)$ (in (b)) in its splitting field. Show that the two other zeros are:

$$x_2 = \frac{6x^2 - 9x - ax - a^2 - 27}{2a}, \qquad x_3 = -\frac{6x^2 - 9x + ax - a^2 - 27}{2a}.$$

(d) Let $r(X) \in K[X]$ be such that $x_2 = r(x)$. Show that $f(r(X)) = f(X)g(X)$ for some polynomial $g(X) \in K[X]$. Find $g(X)$.

16.80. Let $K \subseteq L$ be an algebraic field extension and let $f(X) \in L[X]$. Show that there exists a nonzero polynomial $g(X) \in L[X]$ such that $f(X)g(X) \in K[X]$ (this exercise is taken from [B], where there is also a solution).

16.81. Let $f(X) \in K[X]$ be a nonzero polynomial. Show that for every positive integer d there exists a $g(X) \in K[X]$ such that $f(X)$ divides $g(X^d)$.

16.82. Let $f(X)$ be a polynomial of degree n over a field K whose Galois group is S_n. Show that any splitting field of $f(X)$ over K cannot be generated by fewer than $n-1$ of its zeros. Is it true that if $f(X)$ (still of degree n) has the Galois group A_n over K, then its splitting field cannot be generated by fewer than $n-2$ of its zeros?

16.83. Let $f(X)$ be an irreducible polynomial of prime degree p over a field K and let M be a subfield of a splitting field of $f(X)$ over K. Show that $f(X)$ is still irreducible over M if and only if p does not divide $[M : K]$. (Suggestion: use Exercises 4.2(a) and 5.3.)

16.84. Let L be a splitting field of an irreducible polynomial $f(X) \in K[X]$ and let $f(\alpha) = 0$ for $\alpha \in L$. If $K \subseteq M$ is a Galois extension, where $M \subseteq L$, then:
(a) the polynomial $f(X)$ is irreducible over M if and only if $M \cap K(\alpha) = K$;
(b) all irreducible factors of $f(X)$ in $M[X]$ have the same degree (see Exercise 7.8) and the number of them equals $[M \cap K[\alpha] : K]$.

16.85. Let $K \subseteq L$ be a Galois extension (of fields of characteristic 0) and let N be a field containing L. Let M be any field such that $K \subseteq M \subseteq N$. Let $f(X) \in K[X]$ be an irreducible polynomial of degree n and let $L = K(\alpha_1, \ldots, \alpha_n), f(\alpha_i) = 0$, be its splitting field (see Exercise 9.22). Let $g(X) \in M[X]$ be a polynomial of degree $m < n$. Define **Tschirnhausen's**[1] **transformation** of $f(X)$ by $g(X)$ as the polynomial

$$\varphi(T) = \prod_{i=1}^{n}(T - g(\alpha_i)).$$

(a) Prove that $\varphi(T) \in M[T]$.
(b) Show that if all $g(\alpha_i)$ are different, then $LM = M(\alpha_1, \ldots, \alpha_n) = M(g(\alpha_1), \ldots, g(\alpha_n))$.
(c) Let $f(X) = X^n + a_{n-1}X^{n-1} + \cdots + a_1X + a_0 \in K[X]$ and let $M^* = K(Y_0, Y_1, Y_2, Y_3)$ be a field of rational functions in the variables Y_1, Y_2, Y_3, Y_4. Choose $g(X) = Y_0X^3 + Y_1X^2 + Y_2X + Y_3$. Show that Tchirnhausen's transformation of $f(X)$ by $g(X)$ is a polynomial in $M^*[T]$

$$\varphi(T) = \sum_{r=0}^{n} f_r(Y_0, Y_1, Y_2, Y_3)T^{n-r}$$

such that $f_r(Y_0, Y_1, Y_2, Y_3) \in K[Y_0, Y_1, Y_2, Y_3]$ are homogeneous polynomials of degree r for $r = 0, 1, \ldots, n$ ($f_0 = 1$). In particular, f_1 is a linear form and f_2 a quadratic form of Y_0, Y_1, Y_2, Y_3.
(d) Prove the following **theorem of Bring**[2] **and Jerrard**[3]: The system of equations $f_1(Y_0, Y_1, Y_2, Y_3) = f_2(Y_0, Y_1, Y_2, Y_3) = f_3(Y_0, Y_1, Y_2, Y_3) = 0$ has a nontrivial solution $(\beta_0, \beta_1, \beta_2, \beta_3)$, where the β_i are in a field M generated over K by zeros of polynomials of degree at most equal to 3.
(e) Show that for the field $M = K(\beta_0, \beta_1, \beta_2, \beta_3)$ whose existence follows from (d), the splitting field ML of $f(X)$ over M is a splitting field of a polynomial $X^5 + aX + b \in M[X]$ (or even $X^5 + X + c$ over $M(\sqrt[4]{a})$).

Hint to (d) The quadratic form $f_2(Y_0, Y_1, Y_2, Y_3)$ can be diagonalized over K and represented as a sum of squares of 4 linear forms in the variables Y_i. Consider a system of 3 homogeneous linear equations in the 4 variables Y_i whose infinitely many solutions depend on 1 parameter (factor two sums of two squares in f_2 over $K(i)$ to get two linear forms besides f_1). Find a value of the parameter which also gives a zero of the form f_3.

16.86. Define $\kappa(n) = m$ as the smallest integer m such that every polynomial $f(X) \in \mathbb{Q}[X]$ of degree n with $[\mathbb{Q}_f : \mathbb{Q}] \geq m$ is irreducible over $\mathbb{Q}$. Find the values $\kappa(n)$ for $n \leq 5$.

[1] Ehrenfried Walther von Tschirnhausen (or Tschirnhaus), 10 April 1651–11 October 1708.
[2] Erland Samuel Bring, 19 August 1736–20 May 1798.
[3] George Birch Jerrard, 1804–23 November 1863.

Finite Fields

16.87. Let $K = \mathbb{F}_2(\alpha)$ be an extension of degree 6. Find all intermediate fields $M = \mathbb{F}_2(\beta) \subseteq K$ giving an expression of β as a sum of powers of α.

16.88. Show that over a finite field every irreducible polynomial is normal (see Exercise 7.9 for the definition of a normal polynomial).

16.89. Let K be an algebraically closed field (see p. 21) containing a finite field $\mathbb{F}_p$. Show that the algebraic closure $\overline{\mathbb{F}}_p$ of $\mathbb{F}_p$ in K (see p. 29) is equal to the union $\bigcup_{n=1}^{\infty} \mathbb{F}_{p^n}$. Show also that $\overline{\mathbb{F}}_p = \bigcup_{n=1}^{\infty} \mathbb{F}_{p^{n!}}$.

16.90. Let $\mathbb{F}$ be a finite field and $\mathbb{F} \subset K \subset \overline{\mathbb{F}}$, where K is a field and $\overline{\mathbb{F}}$ an algebraic closure of $\mathbb{F}$ (see p. 29). Show that if K is not finite, then the order of each nontrivial automorphism σ over K is not finite.

16.91. Let p be a prime number. Show that the polynomial $f(X) = X^p - X + 1$ has no zeros in the finite field $\mathbb{F}_p$ (see Exercise 9.26) and use this in order to prove that $f(X)$ is irreducible as a polynomial over the field of rational numbers.

16.92. Find the factorization of the polynomial $X^p + X + 1$, where p is a prime number, as a product of irreducible polynomials in $\mathbb{F}_p[X]$ (compare Exercise 9.26 but observe the completely different behaviour of this polynomial!).

16.93. Characterize all the prime numbers p such that the reduction modulo p of the polynomial $X^n - a$, $a \in \mathbb{Z}$, is not separable over the field $\mathbb{F}_p$. Do the same for the polynomials $X^n + X + a$.

16.94. Let $K \subset L$ be a field extension of finite fields of odd characteristic p. Assume that L has degree n over K.
(a) Find infinitely many examples when $K \subset L$ is a simple radical extension (see p. 31) and infinitely many when this is not the case.
(b) Show that all quadratic and quartic extensions of finite fields are simple radical and give examples showing that cubic extensions may or may not be simple radical.
(c) Find in terms of n and p conditions which assure that the extension $K \subset L$ is simple radical.

16.95. Show that the multiplicative group K^* of a field K is cyclic if and only if K is a finite field.

Roots of Unity and Cyclotomic Fields

16.96. Let ε be a zero of the polynomial $f(X) = \frac{X^{25}-1}{X^5-1}$. Show that $K = \mathbb{Q}(\varepsilon)$ is a splitting field of $f(X)$ over $\mathbb{Q}$ and find all subfields of K (a primitive element for each such subfield expressed as a polynomial of ε).

16.97. Show that a finite extension over its prime subfield contains only finitely many roots of unity.

16.98. Let K be a field of characteristic different from 2.
(a) Show that if the group of roots of unity in K is finite, then its order is even.
(b) Show that the group of roots of unity in a number field K of finite degree over $\mathbb{Q}$ is finite.
(c) How do the number of roots of unity in the quadratic fields $\mathbb{Q}(\sqrt{d})$ and in the biquadratic fields $\mathbb{Q}(\sqrt{d_1}, \sqrt{d_2})$ (d, d_1, d_2 square-free integers not equal 1 and $d_1 \neq d_2$) depend on d, d_1, d_2?

16.99. Let $n > 2$ be an integer and k an integer such that $\gcd(k, n) = 1$. Then, we have:

(a) $[\mathbb{Q}(\cos \frac{2k\pi}{n}) : \mathbb{Q}] = \frac{\varphi(n)}{2}$;

(b) $$[\mathbb{Q}(\sin \tfrac{2k\pi}{n}) : \mathbb{Q}] = \begin{cases} \varphi(n) & \text{if } 4 \nmid n; \\ \frac{\varphi(n)}{2} & \text{if } n \equiv 0 \pmod 8; \\ \frac{\varphi(n)}{4} & \text{if } n \equiv 4 \pmod 8, n > 4; \end{cases}$$

(c) $$[\mathbb{Q}(\tan \tfrac{2k\pi}{n}) : \mathbb{Q}] = \begin{cases} \varphi(n) & \text{if } 4 \nmid n; \\ \frac{\varphi(n)}{4} & \text{if } n \equiv 0 \pmod 8; \\ \frac{\varphi(n)}{2} & \text{if } n \equiv 4 \pmod 8, n > 4. \end{cases}$$

This exercise is inspired by the results in [Ca].

16.100. (a) Show that $\mathbb{Q}(\varepsilon_m, \varepsilon_n) = \mathbb{Q}(\varepsilon_m + \varepsilon_n)$, where $\varepsilon_m, \varepsilon_n$ are m-th and n-th primitive roots of unity.
(b) Let $K \subseteq L$ be a field extension and $\alpha, \beta \in L$ be elements such that $\alpha^m, \beta^n \in K$ for some positive integers m, n. Find conditions assuring that $K(\alpha, \beta) = K(\alpha\beta)$. What can be said about $K(\alpha, \beta) = K(\alpha + \beta)$?

16.101. Find all roots of unity in the cyclotomic field $\mathbb{Q}(\varepsilon)$, where ε is a primitive n-th root of unity.

16.102. Find the number of quadratic fields over $\mathbb{Q}$ contained in the cyclotomic field $\mathbb{Q}(\varepsilon_n)$ when $n > 2$ is an integer (use **T.10.1**, A.3.11, **T.9.2**; the answer depends on the number of primes dividing $n = 2^r n', 2 \nmid n'$ and on r).

16.103. Let K be a finite field of characteristic p with $q = p^r$ elements. Show that the cyclotomic polynomial $\Phi_{n,K}$, where $p \nmid n$, has a zero in K if and only if $n \mid q-1$.

16.104. (a) Consider the chain of subfields of the cyclotomic field $K = \mathbb{Q}(\varepsilon_5)$ generated by the Gaussian periods θ_d for $d \mid 4$ (see Exercises 10.7, 10.9 and p. 167):

$$\mathbb{Q} = \mathbb{Q}(\theta_4) \subset \mathbb{Q}(\theta_2) \subset \mathbb{Q}(\theta_1) = K.$$

Find quadratic equations for θ_d, $d = 2, 1$, with coefficients in $\mathbb{Q}(\theta_{2d})$ and express θ_2 and $\theta_1 = \varepsilon_5$ by quadratic roots from numbers in $\mathbb{Q}$, respectively, $\mathbb{Q}(\theta_2)$. Using these quadratic equations, find a formula for $\cos \frac{2\pi}{5}$ (which gives the possibility of a straightedge-and-compass construction of a regular pentagon).

(b) Consider the chain of subfields of the cyclotomic field $K = \mathbb{Q}(\varepsilon_{17})$ generated by the Gaussian periods θ_d for $d \mid 16$ (see Exercise 10.7 and p. 167):

$$\mathbb{Q} = \mathbb{Q}(\theta_{16}) \subset \mathbb{Q}(\theta_8) \subset \mathbb{Q}(\theta_4) \subset \mathbb{Q}(\theta_2) \subset \mathbb{Q}(\theta_1) = K$$

Find quadratic equations for each θ_d with coefficients in $\mathbb{Q}(\theta_{2d})$ for $d = 8, 4, 2, 1$ and express each such θ_d by quadratic roots from numbers in these fields. Using these quadratic equations, find a formula for $\cos \frac{2\pi}{17}$ (which gives a possibility of a straightedge-and-compass construction of a regular 17-gon—a regular heptadecagon).

Geometric Constructions

16.105. (a) Show that a given angle α can be trisected by a straightedge-and-compass construction if and only if the polynomial $X^3 - 3X - 2\cos\alpha$ is reducible over the field $\mathbb{Q}(\cos\alpha)$.
(b) Find all positive integers n such that for any given angle α, the angle $\frac{\alpha}{n}$ is constructible (by a straightedge-and-compass construction).

16.106. Let d be a positive integer. Show that an angle of d degrees is constructible (by straightedge-and-compass construction when $(0, 0)$ and $(1, 0)$ are given) if and only if $3 \mid d$.

16.107. Let cubes of volumes 1 and 2 be given (that is, the length of their sides). Is it possible to construct (by straightedge-and-compass construction) cubes of volumes 3 and 4?

16.108. A part of a disk bounded by arcs of two circles is usually called a lune. We may assume that the common chord of the two intersecting circles has length equal to 2. Denote the radii of the two circles by r and R and the central angles defining

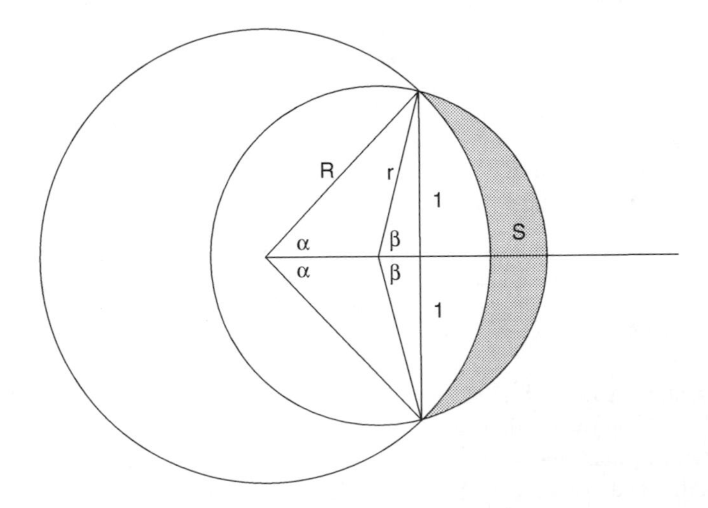

Fig. 16.1 Lune of Hippocrates

the arcs by 2α and 2β, respectively (see Fig. 16.1). Assume that the smaller circle (of radius r) has its center inside the larger one and the lune is a part of the smaller disk outside of the larger one. It is easy to express the area S of the lune:

$$S = \beta r^2 - \alpha R^2 + \sqrt{R^2 - 1} - \sqrt{r^2 - 1}.$$

A general question is whether it is possible to choose r, R, α, β as constructible numbers (over $\mathbb{Q}$). If this is the case, then $\sqrt{S}$ is constructible (by straightedge-and-compass construction), that is, the lune can be squared (which is impossible to do for the whole circle—see Exercise 14.1(c)).
A lune is called a **lune of Hippocrates**[4] when the central angles 2α and 2β are commensurable, that is, $\alpha = m\theta$, $\beta = n\theta$ $(m < n)$ for some θ and positive integers m and n, and the areas of the sectors of the disks corresponding to them are equal, that is, $\alpha R^2 = \beta r^2$. Since $r \sin\beta = R\sin\alpha = 1$, we have

$$\frac{\sin m\theta}{\sin n\theta} = \frac{r}{R} = \sqrt{\frac{m}{n}}$$

and the question is for which relatively prime values m, n the number $\sin\theta$ is constructible, since all numbers r, R, α, β are then constructible.
(a) Let $x = e^{2i\theta} = \cos 2\theta + i\sin 2\theta$. Show that x is a zero of the polynomial $f(X, m, n) = nX^{n-m}(X^m - 1)^2 - m(X^n - 1)^2$.
(b) Consider the polynomials $f(X, m, n)$ for $(m, n) = (1, 2), (1, 3), (2, 3), (1, 5), (3, 5)$ and show (e.g. using **T.14.2** and Maple) that in each of these cases there exists a lune of Hippocrates.

Remark. Hippocrates found the first three examples. The remaining two were found by several mathematicians (probably first by Martin Johan Wallenius in 1766 and somewhat later by Leonhard Euler in 1771)—see [MP]. A proof that these 5 examples are the only existing ones was given by N.G. Chebotarev[5] in 1934 and completed by A.W. Dorodnov in 1947. The natural (simplifying) conditions imposed by Hippocrates in antiquity are necessary in order to get a squarable lune, but they could not have been motivated earlier than the results on transcendental numbers by Gelfond–Schneider and Baker were proved. For an interesting discussion of these conditions, see [G].

---

[4]Hippocrates of Chios, 470–410 BCE.

[5]Nikolai Grigorievich Chebotaryov, 15 June 1894–2 July 1947. His name is (most) often spelled Chebotarev. In his own papers, he used the transcription Tschebotaröw, which we use referring to his book [Tsch].

Chapter 17
Proofs of the Theorems

Theorems of Chap. 1

For a proof of Theorem **T.1.1**, see Exercise 13.14 and its solution on p. 228.

Theorems of Chap. 2

For a proof of Theorem **T.2.1**, see A.5.1.

Theorems of Chap. 3

For the proofs of the Theorems **T.3.1** and **T.3.2**, see A.4.1 and A.4.2, respectively.

T.3.3 Gauss's Lemma. *A nonconstant polynomial with integer coefficients is a product of two nonconstant polynomials in* $\mathbb{Z}[X]$ *if and only if it is reducible in* $\mathbb{Q}[X]$. *More precisely, if* $f \in \mathbb{Z}[X]$ *and* $f = gh$, *where* $g, h \in \mathbb{Q}[X]$ *are nonconstant polynomials, then there are rational numbers* r, s *such that* $rg, sh \in \mathbb{Z}[X]$ *and* $rs = 1$, *so* $f = (rg)(sh)$.

We start with a definition. A polynomial with integer coefficients $f(X)$ is called **primitive** if the greatest divisor of its coefficients is equal to 1.

Lemma 3.1. *The product of two primitive polynomials is also primitive.*

Proof. Let $f(X)$ and $g(X)$ be two integer primitive polynomials. Assume that there is a prime number p dividing all the coefficients of $f(X)g(X)$. We take the reduction of this product modulo p. The result will be a zero polynomial, that is, we have $\bar{f}(X)\bar{g}(X) = 0$ in the ring $\mathbb{F}_p[X]$. But a product of two polynomials with coefficients in a field is zero only if one of the factors is 0. If, say, $\bar{f}(X) = 0$, then all the

J. Brzeziński, *Galois Theory Through Exercises*, Springer Undergraduate Mathematics Series, https://doi.org/10.1007/978-3-319-72326-6_17

coefficients of $f(X)$ are divisible by p, which is impossible ($f(X)$ is primitive). This proves that the product of primitive polynomials is primitive. □

Note now that if $f(X)$ is an arbitrary polynomial with rational coefficients, then $f(X) = c(f)f_0(X)$, where $f_0(X)$ is primitive and $c(f)$ is a rational number. In fact, we can find the least positive integer k such that the coefficients of $kf(X)$ are integers and then divide $kf(X)$ by the greatest common divisor l of its coefficients. This is $f_0(X)$. Thus defining $c(f) = \frac{k}{l}$, we have $f(X) = c(f)f_0(X)$. The rational number $c(f)$ is called the **content** of $f(X)$.

Proof of Gauss's Lemma. Of course, if a polynomial $f \in \mathbb{Z}[X]$ is a product of two nonconstant polynomials in $\mathbb{Z}[X]$, then it is reducible in $\mathbb{Q}[X]$. Conversely, suppose that $f \in \mathbb{Z}[X]$ and $f = gh$, where $g, h \in \mathbb{Q}[X]$ are nonconstant polynomials. We have $g = c(g)g_0$ and $h = c(h)h_0$, where $g_0, h_0 \in \mathbb{Z}[X]$ are primitive polynomials. Hence $f = c(g)c(h)g_0h_0$. Let $c(g)c(h) = \frac{m}{n}$, where m, n are relatively prime integers and $n > 0$. Assume that $n \neq 1$ and choose a prime p dividing n. Then the equality $nf = mg_0h_0$ shows that p divides the right-hand side and since p does not divide m, it must divide all the coefficients of the product g_0h_0. But this is clearly impossible, since the product is a primitive polynomial. The conclusion is that p cannot exist and consequently $n = 1$. Hence, we have a decomposition $f(X) = (mg_0)(h_0)$ into a product of two integer polynomials. Now taking $r = \frac{m}{c(g)}$ and $s = \frac{1}{c(h)}$, we have two rational numbers such that $rg, sh \in \mathbb{Z}[X]$ and $rs = 1$. □

Theorems of Chap. 4

T.4.1. *Let $\alpha \in L \supseteq K$ be algebraic over K.*
(a) Any minimal polynomial of α over K is irreducible and divides every polynomial in $K[X]$ which has α as its zero.
(b) An irreducible polynomial $f \in K[X]$ such that $f(\alpha) = 0$ is a minimal polynomial of α over K.
(c) All minimal polynomials of α over K can be obtained by multiplying one of them by nonzero elements in K.

Proof. (a) Let p be a minimal polynomial of α over K. If $p = p_1p_2$, where $\deg(p_1) < \deg(p)$ and $\deg(p_2) < \deg(p)$, then $p(\alpha) = 0$ implies $p_1(\alpha) = 0$ or $p_2(\alpha) = 0$, which contradicts the choice of p as a polynomial of the smallest possible degree having α as its zero.
(b) As in (a), let p be a minimal polynomial of α over K. We have $f(X) = p(X)q(X) + r(X)$, where $\deg(r) < \deg(p)$ or $r \equiv 0$ is the zero polynomial. The equalities $f(\alpha) = 0$ and $p(\alpha) = 0$ imply that also $r(\alpha) = 0$, so r must be the zero polynomial according to the definition of p, that is, $p \mid f$.
(c) Let both p and p' be minimal polynomials of α over K. According to (a), they divide each other, so $p' = cp$, where c is a nonzero element of K. □

T.4.2 Simple Extension Theorem. *(a) If $\alpha \in L \supseteq K$ is algebraic over K, then each element in $K(\alpha)$ can be uniquely represented as $b_0 + b_1\alpha + \cdots + b_{n-1}\alpha^{n-1}$, where $b_i \in K$ and n is the degree of the minimal polynomial $p(X)$ of α over K. Thus $[K(\alpha) : K] = n$ and $1, \alpha, \ldots, \alpha^{n-1}$ is a basis of $K(\alpha)$ over K.*
(b) If $\alpha \in L \supseteq K$ is transcendental over K, then $K[\alpha]$ is isomorphic to the polynomial ring $K[X]$ by an isomorphism mapping X to α.

Proof. Consider the ring homomorphism (see p. 247 if you want to refresh your knowledge about this notion)

$$\varphi : K[X] \longrightarrow K[\alpha],$$

where $\varphi(f(X)) = f(\alpha)$. We have

$$\text{Ker}(\varphi) = \{f \in K[X] : \varphi(f) = f(\alpha) = 0\} = (p(X)),$$

since every polynomial having α as its zero is a multiple of $p(X)$ by **T.4.1**(a). It is clear that the image of φ is the whole ring $K[\alpha]$. By the fundamental theorem on ring homomorphisms (see A.3.7), we get $K[X]/(p(X)) \cong K[\alpha]$. As we know, each class in $K[X]/(p(X))$ can be uniquely represented by a polynomial (see A.5.8)

$$b_0 + b_1X + \cdots + b_{n-1}X^{n-1}, \; b_i \in K,$$

so that each element in $K[\alpha]$ can be uniquely written as the image

$$b_0 + b_1\alpha + \cdots + b_{n-1}\alpha^{n-1}, b_i \in K,$$

of such a polynomial. Finally, we observe that $K[\alpha]$ is a field, since the polynomial $p(X)$ is irreducible (see A.5.7). In particular, we have $K[\alpha] = K(\alpha)$.

If α is transcendental, then it is clear that the kernel of the homomorphism φ is (0), since by the definition of α, the equality $\varphi(f(X)) = f(\alpha) = 0$ implies that f must be the zero polynomial. Of course, φ is surjective. Thus φ is an isomorphism of the rings $K[X]$ and $K[\alpha]$. □

T.4.3 Tower Law. *Let $K \subseteq L$ and $L \subseteq M$ be finite field extensions. Then $K \subseteq M$ is a finite field extension and $[M : K] = [M : L][L : K]$.*

Proof. Let e_i, $i = 1, \ldots, l$, be a basis of L over K, and f_j, $j = 1, \ldots, m$, a basis of M over L. If $x \in M$, then there is a unique presentation $x = \sum_{j=1}^{m} l_jf_j$, where $l_j \in L$. For each j there is a unique presentation $l_j = \sum_{i=1}^{l} a_{ij}e_i$, where $a_{ij} \in K$. Therefore,

$$x = \sum_{j=1}^{m} l_jf_j = \sum_{j=1}^{m}\sum_{i=1}^{l} a_{ij}e_if_j$$

and the presentation of x as a linear combination of e_if_j with coefficients $a_{ij} \in K$ is unique. This shows that lm products e_if_j form a basis of M over K, so $[M : K] = [M : L][L : K]$. □

T.4.4. *A field extension $L \supseteq K$ is finite if and only if it is algebraic and finitely generated.*

Proof. Assume that $L \supseteq K$ is a finite extension. Then there is a basis $e_1, \dots, e_n$ of L over K, so $L = K(e_1, \dots, e_n)$, that is, L is finitely generated over K. Take $x \in L$ and consider the powers $1, x, x^2, \dots, x^n \in L$. Since the number of these elements is $n+1$, they must be linearly dependent over K, that is, $a_0 + a_1x + a_2x^2 + \cdots + a_nx^n = 0$, where not all coefficients $a_i \in K$ are equal to 0. Thus x is an algebraic element over K, that is, L is an algebraic extension of K.

Assume now that $L \supseteq K$ is algebraic and finitely generated. This means that $L = K(\alpha_1, \alpha_2, \dots, \alpha_r)$ and we have the following tower of fields:

$$K \subseteq K(\alpha_1) \subseteq K(\alpha_1, \alpha_2) \subseteq \cdots \subseteq K(\alpha_1, \dots, \alpha_r).$$

For each $i = 0, \dots, n-1$, α_{i+1} is algebraic over K, so it is algebraic over $K(\alpha_1, \alpha_2, \dots, \alpha_i)$. Hence the extension $K(\alpha_1, \alpha_2, \dots, \alpha_i) \subseteq K(\alpha_1, \alpha_2, \dots, \alpha_i, \alpha_{i+1})$ is finite by **T.4.2**(a). Consequently the extension $K \subseteq L$ is finite as its degree is the product of the degrees $[K(\alpha_1, \alpha_2, \dots, \alpha_i, \alpha_{i+1}) : K(\alpha_1, \alpha_2, \dots, \alpha_i)]$ by **T.4.3**. □

T.4.5. *If $K \subseteq M \subseteq L$ are field extensions such that M is algebraic over K and $\alpha \in L$ is algebraic over M, then it is algebraic over K.*

Proof. Let α be a zero of a polynomial $f(X) = X^n + \alpha_{n-1}X^{n-1} + \cdots + \alpha_1X + \alpha_0$, where $\alpha_i \in M$ for $i = 0, 1, \dots, n-1$. Consider the tower of field extensions:

$$K \subseteq K(\alpha_0, \dots, \alpha_{n-1}) \subseteq K(\alpha_0, \dots, \alpha_{n-1}, \alpha).$$

Since the elements $\alpha_0, \dots, \alpha_{n-1}$ are algebraic over K, the first extension is finite by **T.4.4**. Since α is algebraic over the field $K(\alpha_0, \dots, \alpha_{n-1})$, the second extension is finite by the same theorem. Thus, the extension $K(\alpha_0, \dots, \alpha_{n-1}, \alpha) \supseteq K$ is finite by **T.4.3** and the element α is algebraic over K by **T.4.4**. □

T.4.6. *If $L \supseteq K$ is a field extension, then all elements in L algebraic over K form a field.*

Proof. Let $\alpha, \beta \in L$ be two elements algebraic over K. Then the extensions $K(\alpha, \beta) \supseteq K(\alpha) \supseteq K$ are finite. Hence according to Theorem **T.4.4** they are algebraic. But $\alpha \pm \beta$, $\alpha\beta \in K(\alpha, \beta)$, and $\alpha/\beta \in L$, when $\beta \neq 0$. Hence the set of algebraic elements in L is closed with respect to the four arithmetical operations, that is, the elements of L algebraic over K form a field. □

Theorems of Chap. 5

T.5.1 Kronecker's Theorem. *(a) If f is an irreducible polynomial over K, then there exists a field $L \supseteq K$ such that $L = K(\alpha)$ and $f(\alpha) = 0$.*
(b) If $\tau : K \to K'$ is a field isomorphism, f an irreducible polynomial over $K, L = K(\alpha)$, where $f(\alpha) = 0$ and $L' = K'(\alpha')$, where $\tau(f)(\alpha') = 0$, then there is an isomorphism $\sigma : K(\alpha) \to K'(\alpha')$ such that in the diagram:

$$\begin{array}{ccc} K(\alpha) & \xrightarrow[\sigma]{} & K'(\alpha') \\ \uparrow & & \uparrow \\ K & \xrightarrow[\tau]{} & K' \end{array}$$

we have $\sigma(\alpha) = \alpha'$ and $\sigma|_K = \tau$.
In particular, if $K = K'$ and $\tau = id$, then σ is an isomorphism over K (that is, the isomorphism σ maps each element in K to itself) of the two simple extensions of K by two arbitrary roots of $f(X) = 0$.

Proof. (a) The quotient $L = K[X]/(f(X))$ is a field, since $f(X)$ is irreducible in $K[X]$ (see A.5.7). The class $[X] = \alpha$ is a solution in L of the equation $f(X) = 0$ and $L = K(\alpha)$ (see A.5.8).
(b) The isomorphism $\tau : K \to K'$ can be extended to an isomorphism of the polynomial rings $\tau : K[X] \to K'[X]$ (just applying τ on K to the coefficients of the polynomials). This isomorphisms maps the irreducible polynomial $f(X)$ onto the irreducible polynomial $\tau(f)(X)$ in $K'[X]$. Thus, we have an isomorphism of the quotient rings (see A.3.8):

$$\tau^* : K[X]/(f(X)) \to K'[X]/(\tau(f)(X))$$

such that the class of $[X] = \alpha$ in the first ring maps onto the class of $[X] = \alpha'$ in the second one. Since $K[X]/(f(X)) = K(\alpha)$ and $K'[X]/(\tau(f)(X)) = K'(\alpha')$, $\sigma = \tau^*$ is the required extension of τ. □

T.5.2. *(a) Every polynomial $f \in K[X]$ has a splitting field over K.*
(b) If $\tau : K \to K'$ is an isomorphism of fields, L is a splitting field of a polynomial $f \in K[X]$ and L' is a splitting field of the polynomial $\tau(f) \in K'[X]$, then there exists an isomorphism $\sigma : L \to L'$

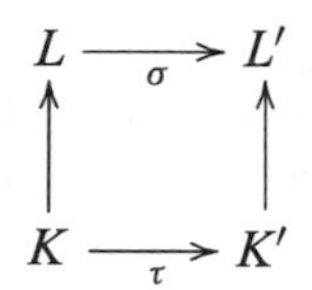

which extends τ (that is, $\sigma|_K = \tau$). In particular, if $K = K'$ and $\tau = id$, then two splitting fields for f over K are K-isomorphic (that is, the isomorphism σ maps each element in K to itself).
Moreover, if f has no multiple zeros, then there are exactly $[L : K]$ different possibilities for σ when τ is given.

Proof. (a) We apply induction with respect to $\deg f = n$ and an arbitrary field K. If $\deg f = 1$, then of course $L = K$. If $\deg f > 1$ and f_1 is an irreducible factor of f, then according to **T.5.1**(a), there is a field $L_1 = K(\alpha_1)$ in which $f_1(X) = 0$ has a solution α_1. Thus $f(X) = (X - \alpha_1)g(X)$, where the $\deg g = \deg f - 1$ and the leading coefficients a of f and g are the same. The polynomial g has a splitting field L over L_1, that is, $g(X) = a(X - \alpha_2)\cdots(X - \alpha_n)$ and $L = L_1(\alpha_2, \ldots, \alpha_n)$, so $L = K(\alpha_1, \alpha_2, \ldots, \alpha_n)$ and $f(X) = a(X - \alpha_1)(X - \alpha_2)\cdots(X - \alpha_n)$.
(b) Let $L = K(\alpha_1, \ldots, \alpha_n)$ and $L' = K(\alpha'_1, \ldots, \alpha'_n)$, where $f(X) = a(X-\alpha_1)\cdots(X-\alpha_n)$, $a \in K$, and $\tau(f)(X) = a'(X - \alpha'_1)\cdots(X - \alpha'_n)$, $a' \in K'$.
We apply induction with respect to the degree $[L : K]$ for arbitrary pairs of fields K and L. If $[L : K] = 1$, then $f(X) = a(X - \alpha_1)(X - \alpha_2)\cdots(X - \alpha_n)$, $\alpha_i \in K$, so $\tau(f)(X) = a(X - \alpha'_1)(X - \alpha'_2)\cdots(X - \alpha'_n)$ and $\alpha'_i \in K$, that is, $L' = K'$. We take $\sigma = \tau$.
Assume now that $[L : K] > 1$. Then there exists an i such that $\alpha_i \notin K$. We may assume that $i = 1$, so α_1 is a solution to the equation $f_1(X) = 0$, where $f_1(X)$ is an irreducible factor of $f(X)$ and $\deg f_1 > 1$. Let α'_1 be a zero of $\tau(f_1)(X)$, which is an irreducible factor of $\tau(f)(X)$. According to **T.5.1**(b), we have an extension σ_1 of τ to an isomorphism of $K(\alpha_1)$ onto $K(\alpha'_1)$. Now L is a splitting field of f over $K(\alpha_1)$ and $[L : K(\alpha_1)] = [L : K]/[K(\alpha_1) : K] < [L : K]$. According to the inductive assumption, there exists an isomorphism $\sigma : L \to L'$ which extends σ_1. Of course, σ extends τ, since σ_1 extends τ.
If $f(X)$ has no multiple zeros and α is a zero of $f_1(X)$, then $\sigma(\alpha)$ is a zero of $\tau(f_1)(X)$, since $0 = \sigma(f_1(\alpha)) = \tau(f_1)(\sigma(\alpha))$. Thus, $\tau(f_1)(X)$ has d different zeros, where $d = \deg f_1$. Hence, we can choose σ_1 in exactly d different ways. Since all zeros of $f(X)$ are different over $K(\alpha_1)$, we can use induction as above to the extension L of $K(\alpha_1)$. Each choice of σ_1 gives $[L : K(\alpha_1)] = [L : K]/[K(\alpha_1) : K] = [L : K]/d$ different choices of σ. Thus the total number of extensions σ of τ equals $d \cdot [L : K]/d = [L : K]$.
When $K = K'$ and $\tau = id$, it is now evident that two splitting fields of $f(X)$ over K are isomorphic. □

T.5.3. *A polynomial $f \in K[X]$ has no multiple zeros in any extension $L \supseteq K$ if and only if* $\gcd(f, f') = 1$.

Proof. Let $\alpha \in L$ be a multiple zero of $f(X)$, that is, $f(X) = (X - \alpha)^2 q(X)$ (the multiplicity of α is at least 2), where $q(X) \in L[X]$. Then $f'(X) = 2(X - \alpha)q(X) + (X - \alpha)^2 q'(X)$, so $f(\alpha) = f'(\alpha) = 0$. Hence $\gcd(f(X), f'(X)) \neq 1$ as it is divisible by $X - \alpha$.

Conversely, let $d = \gcd(f, f')$ be a nonconstant polynomial and let α be a zero of $d(X)$ in some field containing K (here we use **T.5.1**). We have $f(\alpha) = f'(\alpha) = 0$.

Hence $f(X) = (X - \alpha)q(X)$, so $f'(X) = q(X) + (X - a)q'(X)$. The last equality shows that $q(\alpha) = 0$. Hence $q(X) = (X - \alpha)q_1(X)$ and $f(X) = (X - \alpha)^2 q_1(X)$, that is, the polynomial $f(X)$ has multiple zeros (α has at least multiplicity 2). □

T.5.4. *(a) The number of elements in a finite field is a power of a prime number.*
(b) If p is a prime number and $n \geq 1$, then any splitting field of $X^{p^n} - X$ over $\mathbb{F}_p$ is a finite field with p^n elements.
(c) Two finite fields with the same number of elements are isomorphic. More precisely, every finite field with p^n elements is a splitting field of $X^{p^n} - X$ over $\mathbb{F}_p$.

Proof. (a) Let K be a finite field. The prime subfield of K is of course also finite, so let $\mathbb{F}_p$ for a prime number p be the prime subfield contained in K (see **T.2.1**). Let $\alpha_1, \ldots, \alpha_n$ be a basis of K over $\mathbb{F}_p$. Each element of K is uniquely represented as $a_1\alpha_1 + \cdots + a_n\alpha_n$, where $a_i \in \mathbb{F}_p$, so the number of elements of K equals the number of such linear combinations, that is, it is equal to p^n.
(b) Let K be a splitting field of the polynomial $X^q - X$, where $q = p^n$, over $\mathbb{F}_p$. The existence of K follows from **T.5.2**. Let

$$M = \{\alpha \in K \mid \alpha^q = \alpha\}$$

be the set of all solutions of the equation $X^q = X$ in K. If $\alpha, \alpha_1, \alpha_2 \in M$, then (see Exercise 2.12)

$$(\alpha_1 + \alpha_2)^q = \alpha_1^q + \alpha_2^q = \alpha_1 + \alpha_2, \qquad (\alpha_1\alpha_2)^q = \alpha_1^q\alpha_2^q = \alpha_1\alpha_2,$$

and

$$\left(\frac{1}{\alpha}\right)^q = \frac{1}{\alpha^q} = \frac{1}{\alpha} \quad \text{if} \quad \alpha \neq 0,$$

so the elements of M form a subfield of K. The subfield M is of course a splitting field of the polynomial $X^q - X$ over $\mathbb{F}_p$. Thus, by the definition of a splitting field (see p. 27), we have $K = M$. We also see that the field K has $q = p^n$ elements, since all zeros of the polynomial $f(X) = X^q - X$ are different. In fact, we have $f'(X) = qX^{q-1} - 1 = -1$, that is, the polynomials $f(X)$ and $f'(X)$ are relatively prime (see **T.5.3**).
(c) Let K be any field with $q = p^n$ elements. Since all nonzero elements of K form a group with respect to multiplication and the order of this group is $q - 1$, we have $\alpha^{q-1} = 1$ for each nonzero $\alpha \in K$ (see A.2.6). Hence all elements of K including 0 satisfy the equation $X^q = X$. Thus K is a splitting field of the polynomial $X^q - X$ over $\mathbb{F}_p$. □

T.5.5. *For every field K there exists an algebraic closure $\overline{K}$ and two algebraic closures of the same field K are K-isomorphic.*

Proof. As a first step, we prove that for every field K, there exists a field $K_1 \supseteq K$ such that every irreducible polynomial $f(X) \in K[X]$ has a zero in K_1.

Let $\mathcal{S}$ denote the set of all irreducible polynomials $f \in K[X]$. For each $f \in \mathcal{S}$, we take a variable X_f and form the polynomial ring $R = K[\dots X_f \dots]$ generated over K by all the variables X_f. Consider the ideal $\mathcal{I}$ in this polynomial ring generated by all $f(X_f)$, where $f \in \mathcal{S}$. We claim that this ideal is proper, since $1 \notin \mathcal{I}$. If this is not true, then there are polynomials $f_i(X_{f_i})$ such that

$$1 = g_1 f_1(X_{f_1}) + \cdots + g_r f_r(X_{f_r})$$

for some polynomials $g_i \in R$. The polynomials g_i depend on some finite number of variables X_f. Let M be an extension of K in which each polynomial $f(X_{f_i})$ has a zero α_i. Take a ring homomorphism mapping each variable X_{f_i} to α_i and all remaining variables X_f to 0. Applying this to the above equality gives $1 = 0$, which is impossible. Hence 1 cannot belong to the ideal $\mathcal{I}$.

Since $\mathcal{I}$ is a proper ideal, we have a maximal ideal $\mathcal{M}$ containing $\mathcal{I}$ (see A.13.2) and we have the natural surjection of the ring R onto its quotient $R/\mathcal{M}$, which is a field K_1 by A.5.5. Notice now that the field K is mapped into the field K_1, since the kernel of the surjection of R onto K_1 is zero when restricted to the field K (a field has only two ideals (0) and itself, but in this case the kernel is (0), since 1 is not in the kernel—see A.5.4). Thus, we can identify K with a subfield of K_1. Moreover, every polynomial $f(X_f)$ has a zero in K_1, since the image of X_f is a zero of this polynomial ($f(X_f)$ equals zero in K_1—see A.5.8). This proves the first assertion of the theorem.

Now we repeat the process and construct a chain of fields

$$K = K_0 \subseteq K_1 \subseteq K_2 \subseteq \cdots \subseteq K_n \subseteq \cdots$$

such that each polynomial over K_1 has a zero in K_2, each polynomial over K_2 has a zero in K_3 and so on. Define $\overline{K} = \bigcup_{i=0}^{\infty} K_i$. We claim that $\overline{K}$ is an algebraic closure of K.

In fact, if $f(X) \in \overline{K}$ is an irreducible polynomial, then the coefficients of $f(X)$ are already in some field K_i. Hence, this polynomial has a zero in K_{i+1}, that is, it has a zero in $\overline{K}$. As $f(X)$ is irreducible in $\overline{K}[X]$ and has a zero in $\overline{K}$, it has degree 1. Thus the field $\overline{K}$ is algebraically closed, since the only irreducible polynomials over $\overline{K}$ have degree 1 (see p. 21).

Now we prove that two algebraic closures $\overline{K}$ and $\overline{K'}$ of the same field K are isomorphic. Consider the set of all pairs (L, φ) such that L is a subfield of $\overline{K}$ containing K and φ an injection of L into $\overline{K'}$ over K. This set is not empty, since we have the pair consisting of K and the identity as φ. Such pairs are ordered when we declare a pair (L, φ) less than or equal to (L', φ') when $L \subseteq L'$ and φ' restricted to L is equal to φ. If $\mathcal{Y}$ is a set of pairs (L, φ) such that for two pairs (L_1, φ_1) and (L_2, φ_2), we have $L_1 \subseteq L_2$ or $L_2 \subseteq L_1$, then taking $L^0 = \bigcup_{L \in \mathcal{Y}} L$ and defining φ_0 on L_0 so that its restriction to L is φ (for $(L, \varphi) \in \mathcal{Y}$), we get an upper bound of the set $\mathcal{Y}$. Thus, we can apply Zorn's Lemma (see A.13.1), which says that there exist a maximal pair. Let (M, φ) be such a maximal pair and let $M' = \varphi(M)$ be the corresponding subfield of $\overline{K'}$. We want to prove that $M = \overline{K}$ and $M' = \overline{K'}$. If not, then there is an $\alpha \in \overline{K}$ such that $\alpha \notin M$. The element α is algebraic over M (since it is algebraic over

K), so let us take its minimal polynomial $f(X)$ over M. Since $f(X) \in M[X]$, we can take the image of this polynomial applying φ to its coefficients. This polynomial, as a polynomial with coefficients in M', has a zero $\alpha' \in \overline{K'}$. By **T.5.1**, there is an isomorphism $\psi : M(\alpha) \to M'(\alpha')$ which restricted to M is equal to φ. Thus the pair (M, φ) is not maximal if α could be chosen outside of M. This contradiction shows that $M = \overline{K}$. Now the image $\varphi(\overline{K})$ is an algebraically closed subfield of $\overline{K'}$, which is algebraic over it. But this is only possible when $\overline{K'} = \varphi(\overline{K})$, since an algebraically closed field has no nontrivial algebraic extensions. □

Theorems of Chap. 6

T.6.1. *All K-automorphisms of L form a group with respect to the composition of automorphisms.*

Proof. We have to check that if $\sigma, \tau \in G(L/K)$ are automorphisms of L over K, then both $\sigma\tau$ and σ^{-1} also are automorphisms of L over K. It is clear that the composition and inverse of bijective functions are bijective. Thus, we have to check that if $x, y \in L$ and $\sigma(x+y) = \sigma(x) + \sigma(y)$, $\sigma(xy) = \sigma(x)\sigma(y)$ and similarly for τ, then $\sigma\tau(x+y) = \sigma\tau(x) + \sigma\tau(y)$ and $\sigma\tau(xy) = \sigma\tau(x)\sigma\tau(y)$ as well as $\sigma^{-1}(x+y) = \sigma^{-1}(x) + \sigma^{-1}(y)$ and $\sigma^{-1}(xy) = \sigma^{-1}(x)\sigma^{-1}(y)$. To check these equalities requires a few formal and easy steps, so we restrict to only two of them. We have

$$\sigma\tau(xy) = \sigma(\tau(xy)) = \sigma(\tau(x)\tau(y)) = \sigma(\tau(x))\sigma(\tau(y)) = \sigma\tau(x)\sigma\tau(y).$$

Define $x' = \sigma^{-1}(x)$ and $y' = \sigma^{-1}(y)$. Then $\sigma(x'+y') = \sigma(x') + \sigma(y') = x + y$, so $\sigma^{-1}(x+y) = x' + y' = \sigma^{-1}(x) + \sigma^{-1}(y)$. A similar argument gives $\sigma^{-1}(xy) = \sigma^{-1}(x)\sigma^{-1}(y)$.

Of course, we have $\sigma\tau(x) = x$ and $\sigma^{-1}(x) = x$ for each $x \in K$, so $\sigma\tau$ and σ^{-1} are K-automorphisms. □

T.6.2. *If G is a group of automorphisms of L (finite or infinite), then L^G is a subfield of L and $[L : L^G] = |G|$.*

Proof. In order to check that L^G is a field, we have to show that if $x, y \in L^G$, then also $x \pm y, xy, x/y \in L^G$ ($y \neq 0$ in the quotient). This is evident, since $\sigma(x) = x$ and $\sigma(y) = y$ for $\sigma \in G$ clearly implies $\sigma(x \pm y) = x \pm y$, $\sigma(xy) = xy$ and $\sigma(x/y) = x/y$. Of course, we have $0, 1 \in L^G$, so L^G really is a subfield of L.

First we show that $[L : L^G] \geq |G|$. Let $G = \{\sigma_1, \ldots, \sigma_n\}$. Here G need not be a group—simply a set of different automorphisms of L. Assume that L has dimension $m < n$ over L^G and let $e_1, \ldots, e_m$ be a basis of L over L^G. Consider the following system of m homogeneous linear equations:

$$\sigma_1(e_1)x_1 + \cdots + \sigma_n(e_1)x_n = 0$$

$$\ldots$$

$$\sigma_1(e_i)x_1 + \cdots + \sigma_n(e_i)x_n = 0$$

$$\cdots$$

$$\sigma_1(e_m)x_1 + \cdots + \sigma_n(e_m)x_n = 0.$$

Since the number of equations m is less than the number of variables n, the system has a nontrivial solution $(x_1, \ldots, x_n) \in L^n$. Let $x = \sum_{i=1}^{m} a_i e_i$, where $a_i \in L^G$, be an arbitrary element of L. Multiply the i-th equation in the system by a_i for $i = 1, \ldots, m$ and notice that for every $j = 1, \ldots, n$, we have $a_i\sigma_j(e_i) = \sigma_j(a_i e_i)$, since $a_i \in L^G$. Then add all the equations. The result is the following equality:

$$\sigma_1(x)x_1 + \cdots + \sigma_n(x)x_n = 0,$$

which holds for every $x \in L$. However, this contradicts Dedekind's Lemma (see below **T.6.3**), since $(x_1, \ldots, x_n) \neq (0, \ldots, 0)$. Hence $[L : L^G] \geq |G|$. Notice that the proof implies that the degree $[L : L^G]$ is always at least equal to the number of automorphisms in any set (not necessarily a group) G. In particular, if there exists an infinite group of automorphisms of L, then the dimension of L over L^G is not finite.

Now we prove that $[L : L^G] \leq |G|$, when G is a finite group of automorphisms of L. Let $G = \{\sigma_1, \ldots, \sigma_n\}$ and assume that $m = [L : L^G] > |G| = n$. Assume that $e_1, \ldots, e_m$ is a basis of L over L^G and consider the homogeneous linear system:

$$x_1\sigma_1^{-1}(e_1) + \cdots + x_m\sigma_1^{-1}(e_m) = 0$$

$$\cdots$$

$$x_1\sigma_i^{-1}(e_1) + \cdots + x_m\sigma_i^{-1}(e_m) = 0$$

$$\cdots$$

$$x_1\sigma_n^{-1}(e_1) + \cdots + x_m\sigma_n^{-1}(e_m) = 0.$$

Since the number of equations n is less than the number of variables m, the system has a nontrivial solution $(x_1, \ldots, x_m) \in L^m$. Assume that $x_1 \neq 0$. Since the system is homogeneous, any multiple $c(x_1, \ldots, x_m)$, where $c \in L$, is also a solution of the system. Therefore, we can choose x_1 arbitrarily in L. Let's do it in such a way that

$$\sigma_1(x_1) + \cdots + \sigma_n(x_1) \neq 0.$$

Such a choice is possible by Dedekind's Lemma (the sum on the left cannot be 0 for all $x_1 \in L$). With this choice of x_1, we apply σ_i to the i-th equation of the system above and we get:

$$\sigma_1(x_1)e_1 + \cdots + \sigma_1(x_m)e_m = 0$$

$$\cdots$$

$$\sigma_i(x_1)e_1 + \cdots + \sigma_i(x_m)e_m = 0$$

$$\cdots$$

$$\sigma_m(x_1)e_1 + \cdots + \sigma_m(x_m)e_m = 0.$$

Adding all the equations above, we get the following equality:

$$\mathrm{Tr}_G(x_1)e_1 + \cdots + \mathrm{Tr}_G(x_m)e_m = 0,$$

where $\mathrm{Tr}_G(x_1) = \sigma_1(x_1) + \cdots + \sigma_n(x_1) \neq 0$. This gives a contradiction, since $e_1, \ldots, e_m$ are linearly independent over L^G and $\mathrm{Tr}_G(x_i) \in L^G$. Hence $[L : L^G] \leq |G|$. □

T.6.3 Dedekind's Lemma. *If $\sigma_1, \sigma_2, \ldots, \sigma_n$ are different automorphisms of a field L and the equality $a_1\sigma_1(x) + a_2\sigma_2(x) + \cdots + a_n\sigma_n(x) = 0$, where $a_i \in L$, holds for every $x \in L$, then $a_1 = a_2 = \ldots = a_n = 0$.*

Proof. We prove Dedekind's Lemma by induction on the number of automorphisms n. If $n = 1$, then the equality $a_1\sigma_1(x) = 0$, for every $x \in L$, implies that $a_1\sigma_1(1) = a_1 = 0$.

Assume now that the lemma is true when the number of automorphisms is less than $n > 1$ and let

$$a_1\sigma_1(x) + a_2\sigma_2(x) + \cdots + a_n\sigma_n(x) = 0 \qquad (*)$$

for every $x \in L$, where $a_i \in L$. Since the last equality holds for every $x \in L$, we may choose an arbitrary $\alpha \in L$ and replace x by αx. Then, we get:

$$a_1\sigma_1(\alpha)\sigma_1(x) + a_2\sigma_2(\alpha)\sigma_2(x) + \cdots + a_n\sigma_n(\alpha)\sigma_n(x) = 0. \qquad (**)$$

Now, we multiply the equality $(*)$ by $\sigma_n(\alpha)$ and subtract from it the equality $(**)$:

$$a_1(\sigma_n(\alpha)-\sigma_1(\alpha))\sigma_1(x)+a_2(\sigma_n(\alpha)-\sigma_2(\alpha))\sigma_2(x)+\cdots+a_{n-1}(\sigma_n(\alpha)-\sigma_{n-1}(\alpha))\sigma_{n-1}(x) = 0.$$

Using the inductive assumption, we get that all the coefficients in the last equality are equal to 0, that is, $a_i(\sigma_n(\alpha) - \sigma_i(\alpha)) = 0$ for $i = 1, \ldots, n-1$. Since the automorphisms $\sigma_1, \ldots, \sigma_n$ are different, for every $i = 1, \ldots, n-1$ there is an element $\alpha_i \in L$ such that $\sigma_n(\alpha_i) \neq \sigma_i(\alpha_i)$. Choosing $\alpha = \alpha_i$, the equality $a_i(\sigma_n(\alpha) - \sigma_i(\alpha)) = 0$ implies that $a_i = 0$ for $i = 1, \ldots, n-1$. Now we go back to the equality $(*)$, which gives $a_n\sigma_n(x) = 0$ for every $x \in L$. When $x = 1$, we get $a_n = 0$ and the proof is complete. □

Theorems of Chap. 7

T.7.1. *A finite extension $L \supseteq K$ is normal if and only if L is a splitting field of a polynomial with coefficients in K.*

Proof. Let $L = K(\alpha_1, \ldots, \alpha_n)$ be a normal extension of K and let f_i be the minimal polynomial of α_i over K for $i = 1, \ldots, n$. According to the definition of a normal extension, every polynomial f_i splits into linear factors in L. The same is true for the product $f = f_1 \cdots f_n$, so L is a splitting field of f over K, since it contains all its zeros and is generated over K by some of them: $\alpha_1, \ldots, \alpha_n$.

Conversely, assume that L is a splitting field of a polynomial $f \in K[X]$. We want to show that if an irreducible polynomial $g \in K[X]$ has a zero $\alpha \in L$, then g splits in L into linear factors. Let M be a splitting field of the polynomial g over L and let α' be any zero of g in M. We want to show that $\alpha' \in L$, which will show that g already splits in L. Consider the following chain of field extensions:

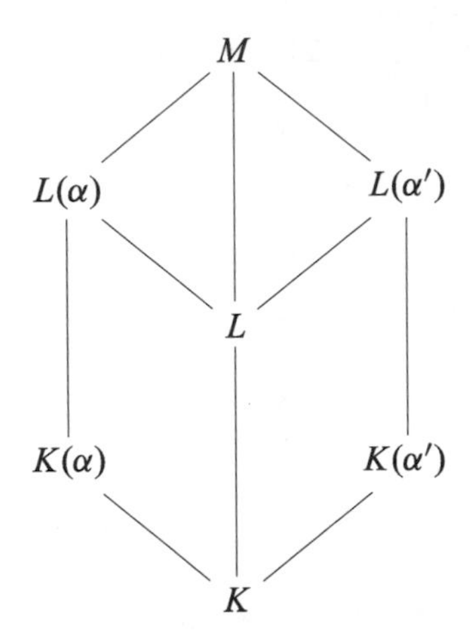

Since L is a splitting field of f over K, both $L(\alpha)$ and $L(\alpha')$ are splitting fields of f over $K(\alpha)$ and $K(\alpha')$, respectively. Since $K(\alpha) \cong K(\alpha')$, there is an isomorphism of the splitting fields $L(\alpha)$ and $L(\alpha')$ of f extending an isomorphism of $K(\alpha)$ and $K(\alpha')$ (see **T.5.2**). Hence $[L(\alpha) : K(\alpha)] = [L(\alpha') : K(\alpha')]$. Of course, $[K(\alpha) : K] = [K(\alpha') : K]$. Thus

$$[L(\alpha) : L] = \frac{[L(\alpha) : K]}{[L : K]} = \frac{[L(\alpha) : K(\alpha)][K(\alpha) : K]}{[L : K]}$$

$$= \frac{[L(\alpha') : K(\alpha')][K(\alpha') : K]}{[L : K]} = \frac{[L(\alpha') : K]}{[L : K]} = [L(\alpha') : L].$$

But $[L(\alpha) : L] = 1$, since $\alpha \in L$, so $[L(\alpha') : L] = 1$, which means that $L(\alpha') = L$, that is, $\alpha' \in L$. $\square$

T.7.2. *Let $L = K(\alpha_1, \ldots, \alpha_n)$ be a finite extension. Then a normal closure of $L \supseteq K$ is unique up to a K-isomorphism. More precisely, every normal closure of $L \supseteq K$ is a splitting field over K of $f = f_1 \cdots f_n$, where f_i is the minimal polynomial of α_i over K.*

Proof. If N is a normal closure of $L \supseteq K$, then every f_i splits in N, so N contains its splitting field. Let $f = f_1 \cdots f_n$. Then N also contains a splitting field of f. But this splitting field of f is normal, so N coincides with it. The uniqueness up to a K-isomorphism is a direct consequence of **T.5.2**. □

Theorems of Chap. 8

T.8.1. (*a*) *All fields of characteristic* 0 *and all finite fields are perfect.*
(*b*) *If char* $(K) = p$, *then an irreducible polynomial* $f \in K[X]$ *is not separable if and only if* $f' \equiv 0$, *which is equivalent to* $f(X) = g(X^p)$, *where* $g \in K[X]$.

Proof. (a) Let K be a field of characteristic 0 and let $f(X) = a_nX^n + \cdots + a_1X + a_0 \in K[X]$ be an irreducible polynomial. Let $\gcd(f, f') = d$. Since $\deg f > 0$ and $\mathrm{char}(K) = 0$, we have $f' \not\equiv 0$ (zero polynomial). We have $d \mid f$ and $\deg d \leq \deg f' < \deg f$, so d must be a constant polynomial as a divisor of an irreducible polynomial f. Thus f is separable by **T.5.3**.
Assume now that K is a finite field of characteristic p. First note that the function $\sigma(x) = x^p$ is an automorphism of the field K (see Exercise 6.5(b)) and since K is finite, we have $\sigma(K) = K$ (the number of elements in $\sigma(K)$ and K is the same). This equality says that each element of K is a p-th power. Let now $f(X)$ be a minimal polynomial of an algebraic element over K in an extension of this field. We want to show that $f(X)$ is separable. If not, then according to (b) below, there is a polynomial $g \in K[X]$ such that $f(X) = g(X^p) = b_mX^{pm} + \cdots + b_1X^p + b_0$ for the same $b_i \in K$. But all b_i are p-powers in K, so we have $b_i = c_i^p$ for some $c_i \in K$. Hence

$$f(X) = b_mX^{pm} + \cdots + b_1X^p + b_0 = c_m^pX^{pm} + \cdots + c_1^pX^p + c_0^p = (c_mX^m + \cdots + c_1X + c_0)^p,$$

which shows that $f(X)$ is not irreducible, contrary to the assumption. Hence $f(X)$ is separable, which proves our claim concerning the finite field K.
(b) Assume that $f \in K[X]$ is irreducible and not separable. Hence by **T.5.3**, we have $\gcd(f, f') = d$, where d is not a constant polynomial. Since $d|f$ is irreducible, we have $d = f$ (up to a constant), so $f|f'$. This is only possible when f' is identically equal to zero. Since $f'(X) = na_nX^{n-1} + \cdots + 2a_2X + a_1$, we get that the only nonzero coefficients a_i in $f(X)$ may be those which correspond to X^i for i divisible by p—if X^i has nonzero coefficient in $f(X)$ and $p \nmid i$, then the coefficient of X^{i-1} in the derivative is also nonzero, so the derivative is nonzero. Hence $f(X)$ is a polynomial in X^p, that is, $f(X) = g(X^p)$ for some $g(X) \in K[X]$. Conversely, it is clear that an irreducible polynomial $f(X) = g(X^p)$ is not separable, since $f'(X) \equiv 0$, so $\gcd(f, f') = f$, which means that $f(X)$ has multiple zeros in an extension of K. □

T.8.2 Primitive Element Theorem. *If $L = K(\alpha_1, \ldots, \alpha_n)$, where $\alpha_1, \ldots, \alpha_n$ are algebraic and all with possibly one exception are separable over K, then there is a primitive element of L over K. In particular, every finite separable extension has a primitive element.*

Proof. First, we assume that K is an infinite field. It suffices if we show that if $L = K(\alpha, \beta)$, then $L = K(\theta)$ for a suitable $\theta \in L$. Let f and g be the minimal polynomials of α and β over K and let

$$\begin{aligned} f(X) &= (X - \alpha_1) \cdots (X - \alpha_n), \\ g(X) &= (X - \beta_1) \cdots (X - \beta_m), \end{aligned}$$

where $\alpha_1 = \alpha$, $\beta_1 = \beta$. According to the assumption all zeros of f or g are different (f or g is separable polynomial). Assume that g is separable. Choose $c \in K$ so that

$$\alpha_i + c\beta_j \neq \alpha_1 + c\beta_1$$

for all $(i, j) \neq (1, 1)$. The existence of c follows from the fact that

$$\alpha_i + x\beta_j = \alpha_1 + x\beta_1$$

holds for finitely many $x \in K$ (fewer than mn "bad" x). Define:

$$\theta = \alpha + c\beta.$$

We have $K(\theta) \subseteq K(\alpha, \beta)$. We want to show that $K(\alpha, \beta) \subseteq K(\theta)$. It suffices if we show that $\beta \in K(\theta)$, since then $\alpha = \theta - c\beta \in K(\theta)$. Consider the polynomials:

$$f(\theta - cX) \quad \text{and} \quad g(X).$$

These polynomials have coefficients in $K(\theta)$ and they have a common zero β, since

$$f(\theta - c\beta) = f(\alpha) = 0 \quad \text{and} \quad g(\beta) = 0.$$

They do not have any other common zeros, since if $f(\theta - c\beta_j) = 0$ for some j, then $\theta - c\beta_j = \alpha_i$ for some i. Thus $\alpha_i + c\beta_j = \theta = \alpha + c\beta$, which occurs only if $i = j = 1$. This shows that

$$\gcd(f(\theta - cX), g(X)) = X - \beta.$$

But the greatest common divisor of two polynomials with coefficients in $K(\theta)$ is a polynomial with coefficients in $K(\theta)$, so $\beta \in K(\theta)$.

If K is a finite field,[1] then L is a finite field as well and its multiplicative group of nonzero elements is cyclic (see Exercise 5.8). Any generator of this group is a (field) primitive element of L over K. □

Theorems of Chap. 9

T.9.1. *Let $L \supseteq K$ be a finite field extension and $G(L/K)$ its Galois group. Then the following conditions are equivalent:*
(a) $[L : K] = |G(L/K)|$;
(b) $L^{G(L/K)} = K$;
(c) There is a group G of K-automorphisms of L such that $K = L^G$ and then, $G = G(L/K)$;
(d) $L \supseteq K$ is normal and separable;
(e) L is a splitting field of a separable polynomial over K.

Proof. (*a*) ⇒ (*b*) We have the following chain of fields:

$$K \subseteq L^{G(L/K)} \subseteq L$$

and according to **T.6.2**, we get $[L : L^{G(L/K)}] = |G(L/K)|$. Since also $[L : K] = |G(L/K)|$, we get $[L^{G(L/K)} : K] = [L : K]/[L : L^{G(L/K)}] = 1$ by **T.4.3**. Thus, we have $L^{G(L/K)} = K$.
(*b*) ⇔ (*c*) Taking $G = G(L/K)$, we get that (b) implies (c). The converse follows if we note that $G \subseteq G(L/K)$ gives $K \subseteq L^{G(L/K)} \subseteq L^G \subseteq L$. Since $L^G = K$, we get $L^{G(L/K)} = K$, which gives (b). Moreover, if $L^G = K$, then Artin's Lemma (see **T.6.2**) and the inclusion $G \subseteq G(L/K)$ imply that $G = G(L/K)$.
(*b*) ⇒ (*d*) We have to show that if $f(X) \in K[X]$ is an irreducible polynomial such that $f(\alpha) = 0$ for some $\alpha \in L$, then $f(X)$ splits in L as a product of linear factors (normality) and all solutions of the equation $f(X) = 0$ in L are different (separability).
Let $G = \{\sigma_1, \ldots, \sigma_n\}$ and let $X = \{\alpha_1, \ldots, \alpha_r\}$ be the set of all different images $\sigma_i(\alpha)$ for $i = 1, \ldots, n$. We assume that $\sigma_1 = id$, so $\alpha_1 = \alpha$. Notice that by the definition of X, if $\alpha_i \in X$, then $\sigma(\alpha_i) \in X$ for $\sigma \in G$, so every $\sigma \in G$ gives a permutation of the elements of X. Consider the polynomial

$$g(X) = (X - \alpha_1) \cdots (X - \alpha_r).$$

We claim that $g(X) \in K[X]$. In fact, the coefficients of this polynomial are $\alpha_1 + \cdots + \alpha_r$, $\alpha_1\alpha_2 + \cdots + \alpha_{r-1}\alpha_r$, …, $\alpha_1 \cdots \alpha_r$, (that is, all fundamental symmetric expressions

[1] There exists a proof of the theorem on primitive elements, which works for both finite and infinite fields K. See [BR].

in $\alpha_1, \dots, \alpha_r$—see section "Symmetric Polynomials" in Appendix). Every $\sigma \in G$ permutes all the α_i, which leaves the coefficients unchanged:

$$\sigma(\alpha_1 + \cdots + \alpha_r) = \sigma(\alpha_1) + \cdots + \sigma(\alpha_r) = \alpha_1 + \cdots + \alpha_r,$$

$$\sigma(\alpha_1\alpha_2 + \cdots + \alpha_{r-1}\alpha_r) = \sigma(\alpha_1\alpha_2) + \cdots + \sigma(\alpha_{r-1}\alpha_r) = \alpha_1\alpha_2 + \cdots + \alpha_{r-1}\alpha_r,$$

$$\dots$$

$$\sigma(\alpha_1 \cdots \alpha_r) = \sigma(\alpha_1) \cdots \sigma(\alpha_r) = \alpha_1 \cdots \alpha_r.$$

Thus by (b), the coefficients of $g(X)$ belong to $L^{G(L/K)} = K$. Since $f(X)$ is irreducible in $K[X]$ and has a zero α, and $g(X)$ is a polynomial in $K[X]$ which also has a zero α, we get that $f(X)$ divides $g(X)$. Hence all zeros of $f(X)$ are in L and are different.
(d) $\Rightarrow$ *(e)* Let $L = K(\alpha_1, \dots, \alpha_n)$ be finite, normal and separable. Let $f_i(X)$ be the minimal polynomial of α_i over K for $i = 1, \dots, n$. Then $f_i(X)$ is irreducible and has a zero in L, so it is separable and has all its zeros in L. Let $f(X)$ be the product of different polynomials among $f_1(X), \dots, f_n(X)$. Then $f(X)$ is separable and L is its splitting field over K.
(e) $\Rightarrow$ *(a)* Let $L = K(\alpha_1, \dots, \alpha_n)$ be a splitting field of a separable polynomial $f(X) = a(X - \alpha_1) \cdots (X - \alpha_n)$, $a \in K$. We use induction with respect to the degree $[L : K]$. If $[L : K] = 1$, then of course, $|G(L/K)| = 1$. Assume that $[L : K] > 1$. Then there exists i such that $\alpha_i \notin K$. We may assume that $i = 1$, so $\alpha = \alpha_1$ is a solution to the equation $f_1(X) = 0$, where $f_1(X)$ is an irreducible factor of $f(X)$ and $\deg f_1 = d > 1$. Since f is separable, $f_1(X) = 0$ has d different solutions in L. Let α' be any of the solutions to $f_1(X) = 0$ in L. As we know (see **T.5.1**(b)), there is an isomorphism $\tau : K(\alpha) \to K(\alpha')$ such that $\tau(\alpha) = \alpha'$ and τ is the identity on K. The number of such τ is exactly d since $\tau(f_1(\alpha)) = f_1(\tau(\alpha)) = 0$, so there are exactly d possibilities for $\tau(\alpha)$ as solutions to the equation $f_1(X) = 0$ in L.
We have the following field extensions:

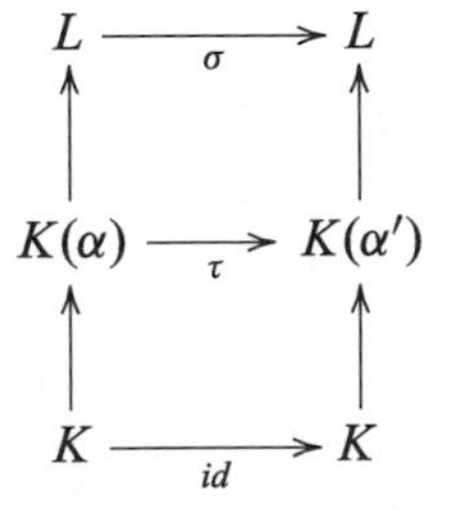

$$\begin{array}{ccc}
L & \xrightarrow{\ \sigma\ } & L \\
\uparrow & & \uparrow \\
K(\alpha) & \xrightarrow{\ \tau\ } & K(\alpha') \\
\uparrow & & \uparrow \\
K & \xrightarrow{\ id\ } & K
\end{array}$$

Now consider $f(X)$ as a separable polynomial with coefficients in $K(\alpha)$. Then L is its splitting field. We have $[L : K(\alpha)] = [L : K]/[K(\alpha) : K] < [L : K]$, so we can use the inductive assumption. Each choice of τ gives $[L : K(\alpha)] = [L : K]/[K(\alpha) : K] = [L : K]/d$ different choices of σ. Thus the total number of extensions σ of $id : K \to K$ equals $d \cdot [L : K]/d = [L : K]$. □

Before we prove the next theorem, recall from Chap. 9 the following notations: If $L \supseteq K$ is a field extension, $\mathcal{F}$ the set of all fields between K and L and $\mathcal{G}$ the set of all subgroups to $G(L/K)$, then we define two functions:

$$f : \mathcal{G} \to \mathcal{F} \quad \text{and} \quad g : \mathcal{F} \to \mathcal{G}$$

in the following way:

$$f(H) = L^H = \{x \in L : \forall_{\sigma \in H} \sigma(x) = x\}$$

and

$$g(M) = G(L/M) = \{\sigma \in G(L/K) : \forall_{x \in M} \sigma(x) = x\}.$$

T.9.2 Fundamental Theorem of Galois Theory. *If $L \supseteq K$ is a finite Galois extension, then f and g are mutually inverse bijections between the set $\mathcal{F}$ of all fields between K and L and the set $\mathcal{G}$ of all subgroups of $G(L/K)$, that is, $f \circ g = id_{\mathcal{F}}$, $g \circ f = id_{\mathcal{G}}$. Moreover, the functions f and g are inclusion-reversing, that is, $f(H_1) \supseteq f(H_2)$ if $H_1 \subseteq H_2$, and $g(M_1) \supseteq g(M_2)$ if $M_1 \subseteq M_2$.*

Proof. By **T.6.2**, we know that $f(H) = L^H$ is a field between K and L for every subgroup H of $G(L/K)$, that is, $f(H) \in \mathcal{F}$. It follows from **T.6.1** that $g(M) = G(L/M)$ is a subgroup of $G(L/K)$ for every field M between K and L, that is, $g(M) \in \mathcal{G}$.

We show $fg = id_{\mathcal{F}}$ and $gf = id_{\mathcal{G}}$, which says that f and g are mutually inverse bijections (see A.9.1). We have

$$fg(M) = f(G(L/M)) = L^{G(L/M)} = M$$

according to **T.9.1**(b), since by **T.9.3**(a) the extension $L \supseteq M$ is Galois. We also have

$$gf(H) = g(L^H) = G(L/L^H) = H,$$

where the last equality is a direct consequence of **T.9.1**(c).

The last statement concerning the inversion of inclusions follows immediately from the definitions of $f(H)$ and $g(M)$. □

T.9.3. *Let $K \subseteq L$ be a Galois extension and M a field between K and L.*
(a) The extension $L \supseteq M$ is a Galois extension.
(b) The extension $M \supseteq K$ is a Galois extension if and only if $G(L/M)$ is normal in $G(L/K)$. If this holds, then $G(M/K) \cong G(L/K)/G(L/M)$.

Proof. (a) Since $K \subseteq L$ is a Galois extension, it is a splitting field of a separable polynomial $f(X) \in K[X]$. L is a splitting field of the same polynomial over M. Thus $M \subseteq L$ is a Galois extension.

(b) Let $K \subseteq M$ be a Galois extension and let $\sigma \in G(L/K)$. We claim that the restriction of σ to M belongs to the Galois group $G(M/K)$. In fact, if $\alpha \in M$ and $g(X)$ is the minimal polynomial of α over K, then $g(\sigma(\alpha)) = \sigma(g(\alpha)) = 0$, so $\sigma(\alpha) \in M$, since all solutions of $g(X) = 0$ belong to M (M is Galois over K, $g(X)$ is irreducible in K and has one zero α in M). Consider the group homomorphism

$$\varphi : G(L/K) \to G(M/K)$$

such that $\varphi(\sigma) = \sigma_{|M}$. The kernel of this homomorphism consists of all automorphisms $\sigma \in G(L/K)$ which map onto the identity $\sigma_{|M} = id_M$, that is, the kernel is $G(L/M)$. Thus $G(L/M)$ is a normal subgroup of $G(L/K)$.
Conversely, assume that $G(L/M)$ is a normal subgroup of $G(L/K)$. Then for every $\sigma \in G(L/K)$ and every $\tau \in G(L/M)$, we have $\sigma^{-1}\tau\sigma \in G(L/M)$. Let $\alpha \in M$. Then $\sigma^{-1}\tau\sigma(\alpha) = \alpha$, that is, $\tau\sigma(\alpha) = \sigma(\alpha)$. Hence $\sigma(\alpha) \in M$ as an element fixed by all automorphisms $\tau \in G(L/M)$ (remember that $M \subseteq L$ is Galois by (a)). Hence, the homomorphism φ is defined—it maps every K-automorphism of L onto its restriction to M. It follows from Theorem **T.5.2**(b) that φ is surjective, since every automorphism $\tau \in G(M/K)$ can be extended to an automorphism $\sigma \in G(L/K)$ (that is, σ restricted to M equals τ). According to the fundamental theorem on group homomorphisms (see A.2.11), we have

$$\frac{G(L/K)}{G(L/M)} \cong \varphi(G(L/K)) = G(M/K).$$

This gives:

$$|G(M/K)| = \left|\frac{G(L/K)}{G(L/M)}\right| = \frac{|G(L/K)|}{|G(L/M)|} = \frac{[L:K]}{[L:M]} = [M:K],$$

that is, $K \subseteq M$ is a Galois extension. □

Theorems of Chap. 10

T.10.1. *(a) The degree $[\mathbb{Q}(\varepsilon) : \mathbb{Q}] = \varphi(n)$, where ε is a primitive n-th root of unity and φ is Euler's function.*
(b) Each automorphism σ in the Galois group $G(\mathbb{Q}(\varepsilon)/\mathbb{Q})$ is given by $\sigma_k(\varepsilon) = \varepsilon^k$, where $k \in \{1, \dots, n\}$ and $\gcd(k, n) = 1$. The mapping $\sigma_k \mapsto k \pmod n$ gives an isomorphism $G(\mathbb{Q}(\varepsilon)/\mathbb{Q}) \cong (\mathbb{Z}/n\mathbb{Z})^*$.

Proof. (a) Let ε be any primitive n-th root of unity. Let $f(X)$ be the minimal polynomial of ε. Since $\varepsilon^n = 1$, the polynomial $f(X)$ divides $X^n - 1$. Let $X^n - 1 = f(X)q(X)$. Hence the field $\mathbb{Q}(\varepsilon)$ is the splitting field of $f(X)$ and of $X^n - 1$, since all zeros of these polynomials are the powers ε^k for $k = 0, 1, \dots, n-1$. The polynomial $f(X)$ is monic, has rational coefficients and divides $X^n - 1$. Hence, it has integer

coefficients by Exercise 3.11. We want to prove that the zeros of $f(X)$ are exactly all $\varphi(n)$ primitive n-th roots of unity ε^k, where $\gcd(k, n) = 1$.
First we show that every zero of $f(X)$ is an n-th primitive root of unity. Let ε^k be a zero of $f(X)$ in $\mathbb{Q}(\varepsilon)$. According to **T.5.1**, there is an automorphism σ of this field such that $\sigma(\varepsilon) = \varepsilon^k$. The automorphism σ maps n-th roots of unity onto n-th roots of unity and since the powers of ε are all such roots, the powers of ε^k must also be all of them. Hence k must be relatively prime to n, so that ε^k is a primitive n-th root of unity (see A.2.2). (Notice that the same argument shows that if a primitive n-th root of unity is a zero of an irreducible polynomial, then all other zeros of this polynomial are also primitive n-th roots of unity.)
Now we prove that any primitive n-th root of unity is among the zeros of $f(X)$. Let p be an arbitrary prime relatively prime to n. We want to show that ε^p is also a zero of $f(X)$. If it is true, then it will follow that all ε^k, for k relatively prime to n, are zeros of $f(X)$. In fact, the exponent $k = p_1 \cdots p_r$ is a product of prime numbers, so starting with ε as a zero of $f(X)$, we get $\varepsilon^{p_1}, \varepsilon^{p_1 p_2}, \ldots$ as zeros of $f(X)$. After r steps, we get ε^k as a zero of $f(X)$.
Assume that ε^p is not a zero of $f(X)$ and let $g(X)$ be its minimal polynomial. Observe that $f(X)$ and $g(X^p)$ have a common zero ε. Since $f(X)$ is irreducible, it divides $g(X^p)$. Let $g(X^p) = f(X)h(X)$, where $h(X)$ is a monic integer polynomial.
Now we want to prove that $f(X)$ and $g(X)$ have a common zero. The zeros of both these polynomials are n-th roots of unity, so if all these zeros are different, then $X^n - 1 = f(X)g(X)q(X)$ for some polynomial $q(X) \in \mathbb{Z}[X]$. Reduce the equalities:

$$X^n - 1 = f(X)g(X)q(X) \qquad \text{and} \qquad g(X^p) = f(X)h(X)$$

modulo p, that is, consider the images of all involved polynomials in $\mathbb{F}_p[X]$. Since $p \nmid n$, the derivative nX^{n-1} of $X^n - 1$ is nonzero, so all zeros of this polynomial are different (see **T.5.3**). Hence the zeros of $f(X)$ and $g(X)$ are also different when considered over $\mathbb{F}_p$. But the second equality says that $g(X)^p = f(X)h(X)$, since $g(X^p) = g(X)^p$ over $\mathbb{F}_p$. Hence the zeros of $f(X)$ are among the zeros of $g(X)$. This contradiction shows that $f(X)$ and $g(X)$ already have common zeros over $\mathbb{Q}$. Since both f and g are monic and irreducible with common zero, they are equal. Hence $f(\varepsilon) = 0$ implies $f(\varepsilon^p) = 0$ for each $p \nmid n$. As we noted above, this implies that all $\varphi(n)$ primitive n-th roots of unity are the zeros of $f(X)$, so $f(X) = \Phi_n(X)$, proving that this polynomial is irreducible.
(b) Let ε be a primitive n-th root of unity. We already know that $\Phi_n(X)$ is the minimal polynomial of ε and its zeros are ε^k for k such that $\gcd(k, n) = 1$ and $k \in \{1, \ldots, n-1\}$. By **T.5.1**, all automorphisms of $\mathbb{Q}(\varepsilon)$ are defined by the images of ε, that is, $\sigma_k(\varepsilon) = \varepsilon^k$ for all k as above. Now notice that $\sigma_k\sigma_l(\varepsilon) = \sigma_k(\sigma_l(\varepsilon)) = \sigma_k(\varepsilon^l) = \varepsilon^{kl} = \sigma_{kl}(\varepsilon)$, where the index kl (in σ_{kl}) is taken modulo n, since $\varepsilon^{kl} = \varepsilon^{kl \pmod n}$ as $\varepsilon^n = 1$. Notice also that $\gcd(kl, n) = 1$, when $\gcd(k, n) = \gcd(l, n) = 1$. Thus the mapping $\varphi : G(\mathbb{Q}(\varepsilon)/\mathbb{Q}) \to (\mathbb{Z}/n\mathbb{Z})^*$, where $\varphi(\sigma_k) = k$ is a bijection and $\varphi(\sigma_k\sigma_l) = \varphi(\sigma_{kl}) \equiv kl \pmod n = \varphi(\sigma_k)\varphi(\sigma_l)$, is a group isomorphism. □

T.10.2. *Let K be a field whose characteristic does not divide n.*
(a) We have:

$$X^n - 1 = \prod_{d|n} \Phi_{d,K}(X) \quad \text{and} \quad \Phi_{n,K}(X) = \prod_{d|n}(X^d - 1)^{\mu(\frac{n}{d})},$$

where μ denotes the Möbius function (see section "Some Arithmetical Functions" in Appendix and Exercise 5.7).
(b) The cyclotomic polynomials $\Phi_{n,K}(x)$ are monic and their coefficients are integer multiples of the identity (that is, the element 1) *of K.*
(c) All irreducible factors of $\Phi_{n,K}(X)$ are of the same degree.
(d) If $K = \mathbb{Q}$, then $\Phi_n(X) = \Phi_{n,\mathbb{Q}}(X)$ is irreducible over $\mathbb{Q}$.

Proof. (a) Since the characteristic of K does not divide n, the polynomial $X^n - 1$ has n different zeros, as its derivative nX^{n-1} is relatively prime to it (see **T.5.3**). Every n-th root of unity is a zero of exactly one polynomial $\Phi_{d,K}(X)$ for $d \mid n$. This proves the first identity in (a).
The second identity follows immediately from the first one and A.11.3 if we choose $G = K(X)^*$, the group of rational functions $p(X)/q(X)$, where $p(X), q(X) \in K[X]$, with multiplication as the group operation and $f(n) = X^n - 1$, $g(n) = \Phi_{n,K}(X)$.
(b) We prove the claim by induction. If $n = 1$, then $\Phi_{1,K}(X) = X - 1$, so (b) is true in this case. Assume that it is true for all positive integers $m < n$. Hence the coefficients of the polynomial $\prod \Phi_{d,K}(X)$, where $d \mid n$ and $d < n$, are all integer multiples of the identity in the field K and it is monic. According to (a), we have

$$\Phi_{n,K}(X) = \frac{X^n - 1}{\prod_{d<n, d|n} \Phi_{d,K}(X)}$$

so the division algorithm also shows that all coefficients of $\Phi_{n,K}(X)$ are integer multiples of the identity in K and it is monic.
(c) Let ε be any primitive n-th root of unity and let $f(X)$ be its minimal polynomial over K. Then $K(\varepsilon)$ is a splitting field of $f(X)$ as well as of the polynomial $X^n - 1$, all of whose zeros are in $K(\varepsilon)$, since they are powers of ε. If η is any other primitive n-th root of unity, then $K(\varepsilon) = K(\eta)$, so the minimal polynomial of η over K has the same degree as the degree of $f(X)$ (see **T.4.2**).
(d) This was proved in **T.10.1**(a). □

Theorems of Chap. 11

T.11.1. *Let $K \subseteq L$ be a Galois extension and $G(L/K) = \{\sigma_1 = 1, \sigma_2, \dots, \sigma_n\}$ its Galois group. Then the following properties of the extension $K \subseteq L$ hold and are equivalent:*
(a) There exists an $\alpha \in L$ such that $\sigma_1(\alpha), \sigma_2(\alpha), \dots, \sigma_n(\alpha)$ is a basis of L over K;
(b) L^+ is a cyclic $K[G]$-module, that is, there is an $\alpha \in L$ such that $L^+ = K[G]\alpha$ for some $\alpha \in L$.

In order to prove the existence of normal bases, we need two auxiliary results:

Lemma 11.2. *Let $K \subseteq L$ be a Galois extension and $G(L/K) = \{\sigma_1 = id, \sigma_2, \ldots, \sigma_n\}$. The elements $\alpha_1, \ldots, \alpha_n \in L$ form a basis of L over K if and only if* $\det[\sigma_i(\alpha_j)] \neq 0$.

Proof. The given elements are linearly dependent if and only if there are elements $a_1, \ldots, a_n \in K$ not all equal to 0 such that $a_1\alpha_1 + \cdots + a_n\alpha_n = 0$. Letting all σ_i act on this equality, we get

$$\begin{aligned} a_1\sigma_1(\alpha_1) + \cdots + a_n\sigma_1(\alpha_n) &= 0 \\ a_1\sigma_2(\alpha_1) + \cdots + a_n\sigma_2(\alpha_n) &= 0 \\ \ldots \\ a_1\sigma_n(\alpha_1) + \cdots + a_n\sigma_n(\alpha_n) &= 0. \end{aligned}$$

The above system of linear equations has a nonzero solution $a_1, \ldots, a_n$ if and only if the determinant of the coefficient matrix (consisting of $\sigma_i(\alpha_j)$) equals 0. □

Lemma 11.3. *Let $f(X_1, \ldots, X_n) \in K[X_1, \ldots, X_n]$ be a nonzero polynomial and let A be an infinite subset of K. Then there exist $a_1, \ldots, a_n \in A$ such that $f(a_1, \ldots, a_n) \neq 0$.*

Proof. We use induction with respect to n. If $n = 1$, then some element of A is not a zero of $f(X)$, since the number of zeros of $f(X)$ is finite. Assume that $n > 1$ and the statement is true for polynomials whose number of variables is less than n. Write $f(X_1, \ldots, X_n)$ as a polynomial with respect to X_n with coefficients depending on $X_1, \ldots, X_{n-1}$, that is,

$$f(X_1, \ldots, X_n) = c_k(X_1, \ldots, X_{n-1})X_n^k + \cdots + c_1(X_1, \ldots, X_{n-1})X_n + c_0(X_1, \ldots, X_{n-1}).$$

Let $c(X_1, \ldots, X_{n-1})$ be the product of those $c_i(X_1, \ldots, X_{n-1})$, for $i = 0, 1, \ldots, k$, which are nonzero. By the inductive assumption, we can find $a_1, \ldots, a_{n-1} \in A$ such that $c(a_1, \ldots, a_{n-1}) \neq 0$. We then consider $f(a_1, \ldots, a_{n-1}, X_n)$, which is a nonzero polynomial in the variable X_n. Hence we can find $a_n \in A$ such that $f(a_1, \ldots, a_{n-1}, a_n) \neq 0$ by the case $n = 1$. Thus we have $a_1, \ldots, a_n \in A$ satisfying our requirement.

Proof of NBT in the Infinite Case. Let $e_1, \ldots, e_n$ be any basis of L over K and consider n linear forms in the polynomial ring $L[X_1, \ldots, X_n]$ given by:

$$\begin{aligned} X_{\sigma_1} &= \sigma_1(e_1)X_1 + \cdots + \sigma_1(e_n)X_n \\ X_{\sigma_2} &= \sigma_2(e_1)X_1 + \cdots + \sigma_2(e_n)X_n \\ &\ldots \\ X_{\sigma_n} &= \sigma_n(e_1)X_1 + \cdots + \sigma_n(e_n)X_n \end{aligned} \qquad (*)$$

Regard the X_{σ_i} as functions of $X_1, \ldots, X_n$. Define $f(X_1, \ldots, X_n) = \det[X_{\sigma_i\sigma_j}]$. We claim that this polynomial is nonzero. In fact, we can choose $X_{\sigma_1} = 1, X_{\sigma_2} = 0, \ldots, X_{\sigma_n} = 0$ and find the suitable values of $X_1, \ldots, X_n$ solving the system $(*)$. Such a solution exists (and is unique), since $\det[\sigma_i(e_j)] \neq 0$ (see Lemma 11.2). For these values of $X_1, \ldots, X_n$ the determinant defining $f(X_1, \ldots, X_n)$ has elements 0 or 1 with exactly one 1 in each row and each column. Thus the value of $f(X_1, \ldots, X_n)$ is ± 1, which implies that this is a nonzero polynomial.

As K is an infinite subset of L, using Lemma 11.3, we choose $a_1, \ldots, a_n \in K$ such that $f(a_1, \ldots, a_n) \neq 0$ and define $\alpha = a_1e_1 + \cdots + a_ne_n$. Then by $(*)$ and the definition of $f(X_1, \ldots, X_n)$, we get

$$f(a_1, \ldots, a_n) = \det[\sigma_i\sigma_j(\alpha)] \neq 0.$$

Thus, $\sigma_1(\alpha), \ldots, \sigma_n(\alpha)$ are linearly independent over K by Lemma 11.2, so these elements form a normal basis of L over K. □

Proof of NBT in the Finite Case. Let $K \subseteq L$ be a field extension of finite fields. Now we want to prove the normal basis theorem in this case. As we know, the field L is a cyclic Galois extension of K. In fact, the argument given below applies to any cyclic Galois extension $K \subseteq L$ (not necessarily finite), so it gives an alternative proof of the theorem in this case.

Let $G = G(L/K) = \langle\sigma\rangle$ be the Galois group, where σ is any of its generators and $[L : K] = |G(L/K)| = n$, so σ has order n. The field L can be considered as a module over the polynomial ring $K[X]$ when we define $X\alpha = \sigma(\alpha)$ for $\alpha \in L$ (see section "Modules Over Rings" in Appendix). As all modules over $K[X]$ having finite dimension over K, the module L is a direct sum of modules of the form $K[X]/(f)$, where f is a nonzero polynomial in $K[X]$ dividing the annihilator of L (see A.6.3). Since $\sigma^n = 1$, we have $(X^n - 1)\alpha = \sigma^n(\alpha) - \alpha = 0$ for all $\alpha \in L$, so $X^n - 1$ belongs to the annihilator of L. Thus the annihilator of L divides $X^n - 1$ (see A.6.3). If the annihilator is generated by a polynomial $a_0 + a_1X + \cdots + a_{n-1}X^{n-1} \in K[X]$ of degree less than n, then

$$(a_0 + a_1X + \cdots + a_{n-1}X^{n-1})\alpha = a_0\alpha + a_1\sigma(\alpha) + \cdots + a_{n-1}\sigma^{n-1}(\alpha) = 0$$

for any $\alpha \in L$. But according to Dedekind's Lemma **T.6.3**, the automorphisms $\sigma^0 = 1, \sigma, \ldots, \sigma^{n-1}$ are linearly independent over L and consequently over K, so all $a_i = 0$. Thus $X^n - 1$ must be the annihilator of L and L is isomorphic to $K[X]/(X^n - 1)$ as a $K[X]$-module. If $\varphi : K[X]/(X^n - 1) \to L$ is an isomorphism and $\varphi(1) = \alpha$, then every element of L is of the form

$$(a_0 + a_1X + \cdots + a_{n-1}X^{n-1})\alpha = a_0\alpha + a_1\sigma(\alpha) + \cdots + a_{n-1}\sigma^{n-1}(\alpha),$$

since every element of $K[X]/(X^n - 1)$ is of the form $(a_0 + a_1X + \cdots + a_{n-1}X^{n-1}) \cdot 1$. Thus $\alpha, \sigma(\alpha), \ldots, \sigma^{n-1}(\alpha)$ form a basis of L over K, since the number of these elements is n and they generate L over K. □

T.11.2 Hilbert's Theorem 90. *Let $L \supseteq K$ be a cyclic extension of degree n and let σ be a generator of the Galois group $G = G(L/K)$. If $\alpha \in L$, then*
(a) $\mathrm{Nr}_G(\alpha) = 1$ if and only there is a $\beta \in L$ such that $\alpha = \frac{\beta}{\sigma(\beta)}$ (multiplicative version);
(b) $\mathrm{Tr}_G(\alpha) = 0$ if and only if there is a $\beta \in L$ such that $\alpha = \beta - \sigma(\beta)$ (additive version).

Proof. (a) First recall that $\mathrm{Nr}_G(\alpha) = \alpha\sigma(\alpha)\cdots\sigma^{n-1}(\alpha)$. Define

$$\alpha_{\sigma^i} = \alpha\sigma(\alpha)\cdots\sigma^{i-1}(\alpha)$$

for $i = 1, \ldots, n$ and observe that $\alpha_{\sigma^n} = 1$ and $\sigma(\alpha_{\sigma^i}) = \frac{1}{\alpha}\alpha_{\sigma^{i+1}}$. Since $\alpha_\sigma = \alpha \neq 0$, by Dedekind's Lemma **T.6.3**, we can find $x \in L$ such that

$$\beta = \alpha_\sigma\sigma(x) + \alpha_{\sigma^2}\sigma^2(x) + \cdots + \alpha_{\sigma^{n-1}}\sigma^{n-1}(x) + \alpha_{\sigma^n}\sigma^n(x) \neq 0.$$

Now observe that $\sigma(\beta) = \frac{1}{\alpha}\beta$, which proves the result as $\sigma(\beta) \neq 0$.
(b) The proof is similar. Using Dedekind's Lemma **T.6.3**, we can find $x \in L$ such that $\mathrm{Tr}_G(x) = x + \sigma(x) + \cdots + \sigma^{n-1}(x) \neq 0$. Define

$$\alpha_{\sigma^i} = \alpha + \sigma(\alpha) + \cdots + \sigma^{i-1}(\alpha)$$

for $i = 1, \ldots, n$ and observe that $\alpha_{\sigma^n} = \mathrm{Tr}_G(\alpha) = 0$ and $\sigma(\alpha_{\sigma^i}) = \alpha_{\sigma^{i+1}} - \alpha$. As in (a), consider

$$\gamma = \alpha_\sigma\sigma(x) + \alpha_{\sigma^2}\sigma^2(x) + \cdots + \alpha_{\sigma^{n-1}}\sigma^{n-1}(x) + \alpha_{\sigma^n}\sigma^n(x)$$

and notice that $\sigma(\gamma) = \gamma - \mathrm{Tr}_G(x)\alpha$. Hence $\alpha = \frac{1}{\mathrm{Tr}_G(x)}(\gamma - \sigma(\gamma))$. Now we choose $\beta = \frac{\gamma}{\mathrm{Tr}_G(x)}$ and obtain $\alpha = \beta - \sigma(\beta)$. □

T.11.3. *Let K be a field containing n different n-th roots of unity. If L is a cyclic extension of K of degree n, then there exists an $\alpha \in L$ such that $L = K(\alpha)$ and $\alpha^n \in K$.*

We give two proofs of this theorem. The first depends on Dedekind's Lemma **T.6.3**. The second is more direct, but needs a little more knowledge of linear algebra.

First Proof. Let σ be a generator of the Galois group $G(L/K)$. According to Dedekind's Lemma **T.6.3**, there exists an $x \in L$ such that

$$\alpha = x + \varepsilon\sigma(x) + \cdots + \varepsilon^{n-1}\sigma^{n-1}(x) \neq 0.$$

Notice now that $\sigma(\alpha) = \varepsilon^{n-1}\alpha$ (this says that α is an eigenvector belonging to the eigenvalue $\varepsilon^{n-1} = \varepsilon^{-1}$ of σ). Hence $\sigma^i(\alpha) = \varepsilon^{i(n-1)}\alpha^n$ for $i = 1, \ldots, n$, which shows that the images of α by all automorphisms of the Galois group are different. Hence, by Exercise 9.24, we have $L = K(\alpha)$. Moreover, we have $\sigma(\alpha^n) = (\sigma(\alpha))^n = (\varepsilon^{n-1}\alpha)^n = \alpha^n$, so $\alpha^n \in L^{G(L/K)} = K$. □

Second Proof. Let σ be a generator of the Galois group of L over K. Since $\sigma^n = id$, the eigenvalues of σ as a linear mapping of L over K satisfy the equation $X^n - 1 = 0$. In fact, if $\sigma(\alpha) = \varepsilon\alpha$ for $\alpha \in L, \alpha \neq 0$, then $\sigma^n(\alpha) = \varepsilon^n\alpha$, that is, $\varepsilon^n = 1$. But $\varepsilon \in K$, so we can choose ε in K as a generator of the cyclic group of all solutions of $X^n - 1 = 0$ and find an eigenvector $\alpha \in L$ belonging to it. We have $\sigma^i(\alpha) = \varepsilon^i\alpha$ for $i = 1, \ldots, n$, so α is fixed only by the identity automorphism of L over K (if $i \neq n$, then $\varepsilon^i\alpha \neq \alpha$). Hence $L = K(\alpha)$. Moreover, $\sigma^i(\alpha^n) = \varepsilon^{in}\alpha^n = \alpha^n$. Thus, α^n is fixed by all the elements of the Galois group of L over K, that is, $\alpha^n = a \in K$. □

T.11.4. *Let K be a field containing m different m-th roots of* 1.
(a) If $K \subseteq L$ is a Kummer extension of exponent m, then every subextension of fields $M \subseteq N$, where $K \subseteq M \subseteq N \subseteq L$, is also a Kummer extension.
(b) All Kummer extensions of K of exponent m are exactly the splitting fields of sets of binomial polynomials $X^m - a$ for some $a \in K$. In particular, all finite Kummer extensions of K are $L = K(\sqrt[m]{a_1}, \ldots, \sqrt[m]{a_r})$ for some elementsa$_1, \ldots, a_r \in K$.
(c) There is a one-to-one correspondence between the isomorphism classes of finite Kummer extensions of K of exponent m and the subgroups A of K^* containing K^{*m} such that the index $[A : K^{*m}]$ is finite. In this correspondence, to a Kummer extension L of K corresponds the subgroup A of K^* consisting of all $a \in K$ such that $a = \alpha^m$ for some $\alpha \in L$, and to a subgroup A of K^* corresponds any splitting field over K of all binomials $X^m - a$ for $a \in A$. Moreover,

$$|G(L/K)| = [L : K] = [A : K^{*m}]. \tag{11.1}$$

Proof. (a) It is clear that the field M contains m different m-th roots of unity. Since the Galois group $G(L/K)$ is abelian, the Galois group $G(N/M)$ is also abelian as the quotient $G(L/M)/G(L/N)$ (see **T.9.3**). Since every element of $G(L/M)$ has order dividing m, the same is true about the elements of the quotient, that is, the elements of $G(M/N)$.
(b) Let L be a splitting field of a set of binomials $X^m - a$ for $a \in K$. We shall prove that L is a Kummer extension of K of exponent m. According to the assumptions, the field K contains m different m-th roots of unity, so we have to prove that L is an abelian Galois extension of K and $\sigma^m = 1$ for each $\sigma \in G(L/K)$.
The extension $K \subseteq L$ is normal as a splitting field of polynomials over K (see **T.7.1**) and separable, since every polynomial $X^m - a$, where $a \in K, a \neq 0$, is separable. In fact, as we noted earlier, such a binomial has m different zeros. Thus L is a Galois extension of K. Since L is generated by the zeros of binomials $X^m - a$ for $a \in K$, every automorphism $\sigma \in G(L/K)$ is uniquely defined by its action on the generators of this form. So let $\sigma, \tau \in G(L/K)$ and let $\alpha, \beta \in L$ be such that $\alpha^m = a, \beta^m = b$, where $a, b \in K$. Then we have $\sigma(\alpha) = \varepsilon\alpha$ and $\sigma(\beta) = \eta\beta$, where ε, η are m-th

roots of unity. Hence $\sigma\tau(\alpha) = \sigma(\eta\alpha) = \varepsilon\eta\alpha = \tau\sigma(\alpha)$, which shows that the group $G(L/K)$ is abelian. Moreover, we have $\sigma^m(\alpha) = \varepsilon^m\alpha = \alpha$, so $\sigma^m = 1$, which implies that the order of each element of $G(L/K)$ divides m.
Now we prove by induction with respect to the degree of L over K that every finite Kummer extension of K of exponent m is a splitting field of a finite number of binomials $X^m - a$, where $a \in K$. The claim is trivial for K (degree over K equal to 1). Assume that the claim is true for fields of degree less than n and let L be a field of degree $n > 1$ over K. Let H be a nontrivial cyclic subgroup of the Galois group $G(L/K)$ such that $G(L/K) = H \times H'$ for a subgroup H' of $G(L/K)$, which may be trivial if $G(L/K)$ is already cyclic. Such subgroups H and H' exist by the fundamental theorem of abelian groups (see A.6.2). Let $M = L^H$ and $M' = L^{H'}$. We have $G(M/K) = H'$ and $G(M'/K) = H$ (see Exercise 9.15). The order of the cyclic group H divides m, since L is a Kummer extension of exponent m. By Hilbert's Theorem 90, we have $M' = K(\alpha)$, where $\alpha^{m'} \in M$ for some divisor m' of m. Hence $\alpha^m = a' \in K$ and M' is a splitting field of $X^m - a'$. The degree of M over K is less than the degree of L over K so using the inductive assumption, we get that M is a splitting field of a finite number of binomials $X^n - a$, where $a \in K$. The whole extension L is a splitting field of these binomials together with the binomial $X^m - a'$ whose zero is α.
(c) Let $\mathcal{F}_m$ denote all Kummer field extensions L of K of exponent m, and $\mathcal{A}_m$ all subgroups of K^* containing K^{*m}. We have two functions:

$$f : \mathcal{A}_m \to \mathcal{F}_m \quad \text{and} \quad g : \mathcal{F}_m \to \mathcal{A}_m$$

such that $f(A) =$ a splitting field of all binomials $X^m - a$ for $a \in A$ and $g(L) =$ the group of all $a \in K^*$ such that $a = \alpha^m$ for some $\alpha \in L^*$. We want to prove that $f \circ g$ and $g \circ f$ are identities on $\mathcal{F}_m$ and $\mathcal{A}_m$, so f and g are bijective functions between these two sets (see A.9.1).
If $L \in \mathcal{F}_m$, then L is a splitting field of some set of binomials $X^m - a$ with $a \in K$. The group $g(L)$ contains all these $a \in K$, so we certainly have $f(g(L)) = L$, since L is defined as a splitting field of some $X^n - a$ with $a \in g(L)$ and for each $a \in g(L)$, the field L contains a splitting field of $X^m - a$. Thus the composition $f \circ g$ is the identity on $\mathcal{F}_m$ (so that g is injective and f surjective—see A.9.1).
Let A be a subgroup of K^* containing K^{*m}. Then, of course, $A \subseteq g(f(A))$, since each element of A is an m-th power of an element of the field $f(A)$ corresponding to A. But A and $g(f(A))$ define the same field, that is, $f(g(f(A))) = f(A)$ by what we have already explained for $L = f(A)$, so

$$[A : K^{*m}] = [g(f(A)) : K^{*m}] = [L : K]$$

by (11.1). Since $A \subseteq g(f(A))$, the last equalities imply $g(f(A) = A$, that is, the composition $g \circ f$ is the identity on $\mathcal{F}_m$ (so that f is injective and g surjective—see A.9.1).

Let L be a Kummer extension of K defined by a subgroup A of K^* containing K^{*m}. We define a bimultiplicative function (see section "Characters and Pairing" in Appendix):

$$\varphi : G(L/K) \times A \to \mathbb{C}^*$$

in the following way. If $(\sigma, a) \in G(L/K) \times A$, then there is an $\alpha \in L$ such that $a = \alpha^m$. Hence $\sigma(\alpha) = \varepsilon\alpha$, where ε is an m-th root of unity. Notice that ε is independent of the choice of a solution of the equation $X^m = a$. In fact, if also $\beta^m = a$, then $\beta = \eta\alpha$, where $\eta \in K$ is an m-th root of unity. Thus, we have $\sigma(\beta) = \sigma(\eta\alpha) = \eta\varepsilon\alpha = \varepsilon\beta$. Now we define $\varphi(\sigma, a) = \varepsilon = \sigma(\alpha)/\alpha$. We check that φ is bilinear. If $\alpha^m = a, \sigma(\alpha) = \varepsilon\alpha, \beta^m = b, \tau(\beta) = \eta\beta$, then

$$\varphi(\sigma, ab) = \frac{\sigma(\alpha\beta)}{\alpha\beta} = \frac{\sigma(\alpha)}{\alpha}\frac{\sigma(\beta)}{\beta} = \varphi(\sigma, a)\varphi(\sigma, b)$$

and

$$\varphi(\sigma\tau, a) = \varepsilon\tau = \frac{\sigma(\alpha)}{\alpha}\frac{\tau(\beta)}{\beta} = \varphi(\sigma, a)\varphi(\tau, a).$$

The left kernel of φ is the set of all $\sigma \in G(L/K)$ such that $\varphi(\sigma, a) = 1$ for each $a \in A$, that is, $\sigma(\alpha) = \alpha$ for each solution of $X^m - a = 0$ when $a \in A$. But such α generate L, so σ is the identity of $G(L/K)$. The right kernel of φ is the set of all $a \in A$ such that $\varphi(\sigma, a) = 1$ for each $\sigma \in G(L/K)$. Hence $\sigma(\alpha) = \alpha$ for each $\sigma \in G(L/K)$, which means that $\alpha \in K$, that is, $a = \alpha^m \in K^{*m}$. Now by A.12.2, we get an isomorphism $A/K^{*m} \cong G(L/K)^*$. In particular, we have $[A : K^{*m}] = |G(L/K)^*| = |G(L/K)| = [L : K]$. □

Theorems of Chap. 12

T.12.1. *(a) If G is solvable and H is a subgroup of G, then H is solvable.*
(b) If N is a normal subgroup of G and G is solvable, then the quotient group G/N is solvable.
(c) If N is a solvable normal subgroup of G such that the quotient group G/N is solvable, then G is solvable.

Proof. (a) If G is solvable, then there exists a chain

$$G = G_0 \supset G_1 \supset \ldots \supset G_n = \{e\}$$

such that G_{i+1} is normal in G_i and the quotient group G_i/G_{i+1} is abelian for $i = 0, 1, \ldots, n-1$. We claim that

$$H = G_0 \cap H \supset G_1 \cap H \supset \ldots \supset G_n \cap H = \{e\} \quad (12.2)$$

is the corresponding chain of subgroups of H. In fact, it is clear that the subgroup $G_{i+1} \cap H$ is normal in $G_i \cap H$ as G_{i+1} is normal in G_i. The quotient $(G_i \cap H)/(G_{i+1} \cap H)$ is abelian, since we have an embedding of pairs $(G_{i+1} \cap H, G_i \cap H) \to (G_{i+1}, G_i)$, for $i = 0, 1, \ldots, n-1$, which induces (see A.2.12) an injective homomorphism $(G_{i+1} \cap H)/(G_i \cap H) \to G_{i+1}/G_i$. Thus $(G_{i+1} \cap H)/(G_i \cap H)$ is abelian, since it is isomorphic to a subgroup of the abelian group G_{i+1}/G_i.
(b) Let (12.2) be a chain giving the solvability of G and consider the following chain of subgroups of G/N:

$$G/N = G_0N/N \supset G_1N/N \supset \ldots \supset G_nN/N = \{e\}.$$

One checks immediately that $G_{i+1}N/N$ is normal in G_iN/N, since G_{i+1} is normal in G_i for $i = 0, 1, \ldots, n-1$. We have a mapping of pairs $(G_{i+1}, G_i) \to (G_{i+1}N/N, G_iN/N)$ such that $(g', g) \mapsto (g'N, gN)$ for $g \in G_{i+1}, g' \in G_i$ (see A.2.12). Since the coset gN, $g \in G_i$, is the image of $g \in G_i$, it is clear that the mapping of the pairs induces a surjective homomorphism $G_i/G_{i+1} \to (G_{i+1}N/N)/(G_iN/N)$. Hence, the quotient $(G_{i+1}N/N)/(G_iN/N)$ is abelian, since G_i/G_{i+1} is an abelian group.
(c) Let

$$N = N_0 \supset N_1 \supset \ldots \supset N_k = \{e\}$$

and

$$G/N = G_0/N \supset G_1/N \supset \ldots \supset G_l/N = \{e\}$$

be chains of subgroups of N and G/N, the existence of which follows from the assumption that these groups are solvable. We use the fact that each subgroup of G/N is the image H/N of a subgroup H of G containing N (see A.2.13), so

$$G = G_0 \supset G_1 \supset \ldots \supset G_l = N = N_0 \supset N_1 \supset \ldots \supset N_k = \{e\}$$

is a chain of subgroups of G such that each quotient G_i/G_{i+1} for $i = 0, 1, \ldots, l-1$ and N_j/N_{j+1} for $j = 0, 1, \ldots, k-1$ is abelian. This shows that the group G is solvable. □

T.12.2. *The symmetric group S_n is not solvable when $n \geq 5$.*

Proof. See Exercise 12.2(c) and its solution on p. 171 (with a solution of Exercise 12.2(b) on p. 220).

Theorems of Chap. 13

In the proof of **T.13.1**, we refer to a number of auxiliary results, which we prove below after the proof of this theorem.

T.13.1. *If char* $(K) = 0$, *then an equation* $f(X) = 0, f \in K[X]$, *is solvable by radicals if and only if the Galois group of* f *over* K *is solvable.*

Proof. First we prove that if K_f is a solvable extension of K, then its Galois group $G(K_f/K)$ is solvable. Let $K_f \subseteq L$, where L is a radical extension of K. According to Lemma 13.1.3, we can assume that L is a Galois extension of K. Thus, we have a chain of fields $K \subseteq K_f \subseteq L$ in which L is a radical Galois extension of K. By Lemma 13.1.4, the group $G(L/K)$ is solvable, and since $K \subseteq K_f$ is a Galois extension, its Galois group $G(K_f/K)$ is a quotient of $G(L/K)$. Hence this group is also solvable by **T.12.1**(c).

Now we prove that if the Galois group $G(K_f/K)$ is solvable, then the extension $K \subseteq K_f$ is solvable. Denote by L any Galois extension of K whose Galois group is solvable (in the theorem, $L = K_f$). Using induction with respect to the degree $[L : K] > 1$, we prove that the extension $L \supset K$ is solvable.

If $[L : K]$ is a prime number, then the Galois group $G(L/K)$ is cyclic, so it is, of course, solvable. Assume that $[L : K] \neq 1$ is not a prime number. Then the solvable group $G(L/K)$ contains a normal subgroup H such that $|G(L/K)| > |H| > 1$ (see Exercise 12.3). We have the corresponding tower of extensions $K \subset L^H \subset L$ in which both extensions $L^H \subset L$ and $K \subset L^H$ are Galois with solvable Galois groups, since the first has a subgroup H of $G(L/K)$ as its Galois group, and the second, the quotient group $G(L/K)/H$ (see **T.12.1**). Both are solvable of orders less than $|G(L/K)| = [L : K]$. By the inductive assumption, there exist radical extensions $L^H \subset L \subseteq L'$ and $K \subset L^H \subseteq L''$. Hence by Lemma 13.1.2(b), $K \subseteq L \subseteq L'L''$ is a radical extension of K (where $L'L''$ is taken in any field containing both L' and L''). □

Lemma 13.1.1. *Let* L *be a splitting field of a polynomial* $X^n - a$ *over a field* K. *Then the Galois group* $G(L/K)$ *is solvable.*

Proof. By Exercise 5.11, we have $L = K(\varepsilon, \alpha)$, where α is a zero of $X^n - a$ and ε generates the group of all zeros of $X^n - 1$. Consider the chain of fields:

$$K \subseteq K(\varepsilon) \subseteq K(\varepsilon, \alpha) = L$$

and the corresponding chain of groups:

$$G(L/K) \supseteq G(L/K(\varepsilon)) \supseteq (id).$$

Of course, the extension $K(\varepsilon) \supseteq K$ is Galois as a splitting field of the polynomial $X^n - 1$ over K. Thus the Galois group $G(L/K(\varepsilon))$ is normal in $G(L/K)$ and $G(L/K)/G(L/K(\varepsilon)) \cong G(K(\varepsilon)/K)$. The last group is abelian. In fact, let σ be an automorphism of $K(\varepsilon)$ over K. Then $\sigma(\varepsilon) = \varepsilon^i$ for some i (notice that $\varepsilon^n = 1$ gives $\sigma(\varepsilon)^n = 1$). If now τ is another automorphism of $K(\varepsilon)$ over K and $\tau(\varepsilon) = \varepsilon^j$, then $\sigma(\iota(\varepsilon)) = \varepsilon^{ij} = \tau(\sigma(\varepsilon))$, that is, $\sigma\tau = \tau\sigma$.

Now notice that the group $G(L/K(\varepsilon))$ is abelian. Let $\sigma, \tau \in G(L/K(\varepsilon))$. Then $\sigma(\alpha) = \varepsilon^i\alpha$ and $\tau(\alpha) = \varepsilon^j\alpha$, for some i, j. Thus $\sigma(\tau(\alpha))) = \varepsilon^{i+j}\alpha = \tau(\sigma(\alpha))$, that is, $\sigma\tau = \tau\sigma$. Hence the group $G(L/K)$ is solvable. □

We say that L is a **radical m-extension** of K if

$$K = K_0 \subseteq K_1 \subseteq \ldots \subseteq K_n = L$$

are such that $K_i = K_{i-1}(\alpha_i)$, $\alpha_i^{r_i} \in K_{i-1}$ and $r_i \leq m$ are positive integers for $i = 1, \ldots, n$. We introduce the notion "radical m-extension" only in order to use it in Chap. 14 in the particular case $m = 2$ in some applications to geometric constructions. The next two results are about radical m-extensions, but the restriction to m is just a technical observation, which can be disregarded in the course of the proofs.

Lemma 13.1.2. *Let $K \subseteq L$ and $K \subseteq L'$ be field extensions.*
(a) If $\sigma : L \to L'$ is a K–isomorphism and $K \subseteq L$ is radical, then $K \subseteq L'$ is also radical.
(b) If L is a radical m-extension of K, and L' is a radical m-extension of K', where $K \subseteq K' \subseteq L$, both L, L' contained in a field M, then the compositum LL' is a radical m-extension of K.

Proof. (a) Let $L = K(\alpha_1, \ldots, \alpha_n)$, where

$$K = K_0 \subseteq K_1 \subseteq \ldots \subseteq K_n = L$$

are such that $K_i = K_{i-1}(\alpha_i)$, $\alpha_i^{r_i} \in K_{i-1}$ and r_i are positive integers for $i = 1, \ldots, n$. Then $L' = \sigma(L) = K(\sigma(\alpha_1), \ldots, \sigma(\alpha_n))$ and

$$K = K_0 \subseteq \sigma(K_1) \subseteq \cdots \subseteq \sigma(K_n) = L',$$

where $\sigma(K_i) = \sigma(K_{i-1})(\sigma(\alpha_i))$, $\sigma(\alpha_i)^{r_i} \in \sigma(K_{i-1})$, so L' is a radical extension of K.
(b) Let L be as above ($r_i \leq m$) and $L' = K'(\alpha'_1, \ldots, \alpha'_n)$, where

$$K' = K'_0 \subseteq K'_1 \subseteq \ldots \subseteq K'_n = L',$$

$K'_i = K'_{i-1}(\alpha'_i)$, $(\alpha'_i)^{s_i} \in K'_{i-1}$ and s_i are positive integers for $i = 1, \ldots, n$ ($s_i \leq m$). Then

$$K = K \subseteq K(\alpha_1) \subseteq \cdots \subseteq K(\alpha_1, \ldots, \alpha_n) =$$

$$L \subseteq K(\alpha_1, \ldots, \alpha_n, \alpha'_1) \subseteq \cdots \subseteq K(\alpha_1, \ldots, \alpha_n, \alpha'_1, \ldots, \alpha'_m) = LL'$$

is a corresponding chain of extensions showing that LL' is a radical extension of K (an m-extension if $r_i, s_i \leq m$). □

Lemma 13.1.3. *Let L be a radical extension of a field K. Then a normal closure N of $L \supseteq K$ is also a radical extension of K. Moreover, if L is a radical m-extension of K, then N is also a radical m-extension.*

Proof. Let $L = K(\alpha_1, \ldots, \alpha_n)$, where

$$K = K_0 \subseteq K_1 \subseteq \ldots \subseteq K_n = L$$

are such that $K_i = K_{i-1}(\alpha_i)$, $\alpha_i^{r_i} \in K_{i-1}$ and $r_i \leq m$ are positive integers for $i = 1, \ldots, n$. Thus $K_i = K(\alpha_1, \ldots, \alpha_i)$ for $i = 1, \ldots, n$. Let f_i be the minimal polynomial of α_i over K. Of course, N is a splitting field of $f = f_1 \cdots f_n$ over K (see **T.7.2**). Let β_i be any zero of f_i in N and let $\tau : K(\alpha_i) \to K(\beta_i)$ be the isomorphism over K such that $\tau(\alpha_i) = \beta_i$. Since N is a splitting field of f over $K(\alpha_i)$ and $K(\beta_i)$, there is an automorphism $\sigma : N \to N$, which extends τ. This automorphism maps L onto $\sigma(L)$, which according to Lemma 13.1.2(a) is also a radical m-extension of K. It is clear that N as a splitting field of f is the compositum of all the fields $\sigma(L)$ obtained for all the choices of τ. Thus N is a radical m-extension of K according to Lemma 13.1.2(b). □

Lemma 13.1.4. *Let L be a radical Galois extension of K. Then the Galois group $G(L/K)$ is solvable.*

Proof. Let $L = K(\alpha_1, \ldots, \alpha_n)$, where

$$K = K_0 \subseteq K_1 \subseteq \ldots \subseteq K_n = L$$

are such that $K_i = K_{i-1}(\alpha_i)$, $\alpha_i^{r_i} \in K_{i-1}$ and r_i are positive integers for $i = 1, \ldots, n$. Thus $K_i = K(\alpha_1, \ldots, \alpha_i)$ for $i = 1, \ldots, n$. We prove the claim by induction on the number n of simple radical extensions between two arbitrary fields K and L.
If $n = 1$, then $L = K(\alpha_1)$, where $\alpha_1^{r_1} = a \in K$. Denote r_1 by r and consider the splitting field $K(\alpha_1, \varepsilon)$ of $X^r - a$ over K, where $\varepsilon^r = 1$. By Lemma **13.1.1**, the Galois group $G(K(\alpha_1, \varepsilon)/K)$ is solvable. Since $K(\alpha_1)$ is Galois over K, its Galois group $G(K(\alpha_1)/K)$ is solvable as a quotient of the solvable Galois group $G(K(\alpha_1, \varepsilon)/K)$. (Observe that $K(\alpha_1)$ need not be a splitting field of $X^r - a$ even if it is Galois and contains a zero of this polynomial—the polynomial may be reducible. Take as an example $X^4 - 4$ over the rational numbers and $L = \mathbb{Q}(\sqrt{2})$.)

Assume now that the claim is true when the number of simple radical extensions between two fields is less than $n > 1$. Consider the following chain of field extensions:

$$\begin{array}{ccccc} K = K_0 \subseteq & K_1 \subseteq \ldots & & \subseteq K_n = L \\ & \downarrow & & \downarrow \\ & K_1(\varepsilon) \subseteq & \ldots & \subseteq K_n(\varepsilon) = L(\varepsilon) \end{array}$$

Then it is clear that $L(\varepsilon)$ is a radical Galois extension of K and of $K_1(\varepsilon)$. The number of simple radical extensions between $K_1(\varepsilon)$ and $L(\varepsilon)$ is $n - 1$. Thus, by the inductive assumption, the Galois group $G(L(\varepsilon)/K_1(\varepsilon))$ is solvable. According to Lemma 13.1.1, the Galois group $G(K_1(\varepsilon)/K)$ is solvable. Hence the Galois group $G(L(\varepsilon)/K)$ is solvable, since its normal subgroup $G(L(\varepsilon)/K_1(\varepsilon))$ and the quotient group $G(L(\varepsilon)/K)/G(L(\varepsilon)/K_1(\varepsilon)) \cong G(K_1(\varepsilon)/K)$ are solvable (see **T.12.1**(c)). Now, we obtain that the Galois group $G(L/K)$ is isomorphic to $G(L(\varepsilon)/K)/G(L(\varepsilon)/L)$ (see **T.9.3**(b)) and as a quotient of a solvable group, is solvable. □

Lemma 13.1.5. *Let L be a Galois extension of K with a cyclic Galois group $G(L/K)$. Then L is a solvable extension of K.*

Proof. Let $[L : K] = n$ and let $K' = K(\varepsilon)$ be a splitting field of $X^n - 1$ over K. Then $L' = L(\varepsilon)$ is a Galois extension of $K' = K(\varepsilon)$ whose Galois group is cyclic as it is isomorphic to a subgroup of the Galois group $G(L/K)$ (of order n) according to Exercise 9.22. Moreover, the field K' contains all roots of unity of degree $[L' : K']$, since this degree divides n and all roots of unity of degree n are in K'. Thus according to **T.11.3**, $K' \subseteq L'$ is a radical extension. $K \subseteq K'$ is also a radical extension. Hence $K \subseteq L'$ is a radical extension, so $K \subseteq L$ is a solvable extension. □

T.13.2. *The Galois group over $K(s_1, s_2, \ldots, s_n)$ of the general equation $f(X) = 0$ of degree n is S_n, so $f(X) = 0$ is not solvable by radicals when $n \geq 5$.*

Proof. It is clear that each elementary symmetric function s_i of $X_1, \ldots, X_n$ is fixed by all elements $\sigma \in S_n$, that is, by all permutations $\sigma(X_i) = X_{\sigma(i)}$. Hence, we have the tower of fields:

$$K(s_1, s_2, \ldots, s_n) \subseteq K(X_1, X_2, \ldots, X_n)^{S_n} \subset K(X_1, X_2, \ldots, X_n).$$

We have $[K(X_1, X_2, \ldots, X_n) : K(X_1, X_2, \ldots, X_n)^{S_n}] = |S_n| = n!$ by **T.6.2** and, at the same time, we have $[K(X_1, X_2, \ldots, X_n) : K(s_1, s_2, \ldots, s_n)] \leq n!$, since the splitting field $K(X_1, X_2, \ldots, X_n)$ of the polynomial of degree n

$$f(X) = X^n - s_1 X^{n-1} + s_2 X^{n-2} + \cdots + (-1)^n s_n = 0$$

over the field $K(s_1, s_2, \ldots, s_n)$ has at most degree $n!$ by Exercise 5.3. Thus

$$K(X_1, X_2, \ldots, X_n)^{S_n} = K(s_1, s_2, \ldots, s_n)$$

and the Galois group of $K(X_1, X_2, \ldots, X_n)$ over $K(s_1, s_2, \ldots, s_n)$ is S_n by **T.9.1**. □

T.13.3 Casus Irreducibilis. *Let $f(X)$ be an irreducible polynomial having three real zeros and coefficients in a real number field K. Then the equation $f(X) = 0$ is not solvable by real radicals over the field K.*

Proof. Assume to the contrary that there exists a radical extension:

$$K_0 = K \subset K_1 \subset \ldots \subset K_n = L \subset \mathbb{R}$$

such that $K_i = K_{i-1}(\alpha_i)$, $\alpha_i^{r_i} \in K_i$, r_i are prime numbers for $i = 1, \ldots, n$ and $K_f = K(x_1, x_2, x_3) \subset L$, where x_1, x_2, x_3 are all (real) zeros of $f(X)$. Choose $i \geq 1$ such that K_{i-1} does not contain x_1, x_2, x_3 while some of these numbers, say x_1, belongs to K_i. Since $K_{i-1} \subset K_{i-1}(x_1) \subseteq K_i$ and $[K_{i-1}(x_1) : K_{i-1}] = 3$ divides $[K_i : K_{i-1}] = r_i$, we have $r_i = 3$ (r_i is a prime divisible by 3). The polynomial $X^3 - a_i$, where $a_i = \alpha_i^3$, is irreducible over K_{i-1} (since it has only one real zero α_i, which does not belong to the real field K_{i-1}), but it is reducible in $K_i = K_{i-1}(\alpha_i) = K_{i-1}(x_1)$ (K_i and $K_{i-1}(x_1)$ have the same degree 3 over K_{i-1}). Thus $X^3 - a_i$ has a zero in $K_i = K_{i-1}(x_1)$ and, as a consequence, it splits in the splitting field $K_{i-1}(x_1, x_2, x_3)$ of f over K_{i-1}. But this gives a contradiction, since the field $K_{i-1}(x_1, x_2, x_3)$ is real, while all the zeros of $X^{r_i} - a_i$ with the exception of α_i are nonreal. □

Theorems of Chap. 14

T.14.1. *Let K be the smallest subfield of $\mathbb{R}$ which contains the coordinates of all points belonging to a given set of points X in the plane. A point $P = (a, b)$ can be constructed from X by a straightedge-and-compass construction if and only if there is a chain of fields:*

$$K = K_0 \subset K_1 \subset \ldots \subset K_n = L \subset \mathbb{R} \qquad (*)$$

such that $a, b \in L$ and $[K_{i+1} : K_i] = 2$ for $i = 0, 1, \ldots, n-1$. In particular, the set of numbers which are constructible from X (like $\mathbb{K}$ when $X = \{(0,0), (1,0)\}$) is a field.

We start with an auxiliary result:

Lemma 14.1.1. *Let K be the smallest number field containing the coordinates of all points belonging to a point set X.*
(a) Every number in K can be constructed from X by a straightedge-and-compass construction.
(b) The coordinates of any point which can be directly constructed from X are in a real field L such that $[L : K] \leq 2$.
(c) If a point has its coordinates in a real field L such that $[L : K] \leq 2$, then it can be constructed from X by a straightedge-and-compass construction.

Proof. (a) A line which goes through two points with coordinates in the field K has an equation $ax + by + c = 0$ with coefficients a, b, c in K. The coordinates of the intersection point of two such lines (if they intersect) is defined by the solution of a linear equation system consisting of two such equations. Thus the intersection point of the lines has its coordinates in the same field K.
A circle whose center (p, q) has coordinates in K and whose radius equals to the distance d between two points with coordinates in K has the equation $(x-p)^2 + (y-q)^2 = d^2$. It is clear that $d^2 \in K$. The intersection points of such a circle with a

line as above are given by the real solutions of the system:

$$ax + by + c = 0,$$
$$(x - p)^2 + (y - q)^2 = d^2.$$

In order to solve such a system, we express y in terms of x from the first equation (or conversely) and substitute it into the second equation. We get a quadratic equation with respect to x (or y) of the form $x^2 + Ax + B = 0$, where the coefficients $A, B \in K$ while the solutions $x = -A/2 \pm \sqrt{A^2/4 - B}$ belong to the field $L = K(\sqrt{A^2/4 - B})$ whose degree over K is at most 2. The field L must be real if an intersection point exists (one or two).
If we have two circles as above and we want to find their intersection points, then we have to solve the system:

$$(x - p)^2 + (y - q)^2 = d^2,$$
$$(x - p')^2 + (y - q')^2 = d'^2$$

for some $p, q, p', q', d^2, d'^2 \in K$. We can do this by subtracting the two equations. Then we get an equivalent system consisting of an equation of a line and an equation of a circle, just as in the previous case.
(b) First we show that the field K is the smallest set of numbers containing all the coordinates of the points in X which is closed with respect to addition, subtraction, multiplication and division. Therefore the elements of K are the numbers obtained successively from the coordinates of the points in X by these four arithmetical operations. Thus, we have to show that if $a, b \in K$ are constructible, then $a \pm b$, ab and a/b (provided $b \neq 0$) are also constructible. We may assume that $a, b > 0$. It is clear how to construct $a \pm b$ when a, b are given. In order to construct ab and a/b ($b \neq 0$), we can draw angles according to Fig. 14.1. The segment $x = ab$ is constructed by drawing a line through the point A parallel to the line BE (a standard straightedge-and-compass construction known from high school). Similarly, we construct $x = a/b$ (See Fig. 14.2).
Now we have to show that any element of a quadratic real field L over K can be obtained by a straightedge-and-compass construction. We know that $L = K(\sqrt{d})$, where $d \in K, d > 0$, so each element of L is of the form $a + b\sqrt{d}$, where $a, b \in K$. Thus, we only need to construct $\sqrt{d}$ when d is given, since we already know how to construct the product $b\sqrt{d}$ given b and $\sqrt{d}$, and the sum $a + b\sqrt{d}$ given its terms $a, b\sqrt{d}$. The construction of $\sqrt{d}$ follows from Fig. 14.2: first we draw a line with line segments of length 1 and d, then we find the mid-point of $1 + d$ and draw a semi-circle with radius $(1 + d)/2$. The line perpendicular to the line segment AB through the point C gives the line segment CD whose length is $\sqrt{d}$. Indeed, by a known result on the altitude of a right triangle, we have $CD^2 = AC \cdot BC = 1 \cdot d$, that is, $CD = \sqrt{d}$. □

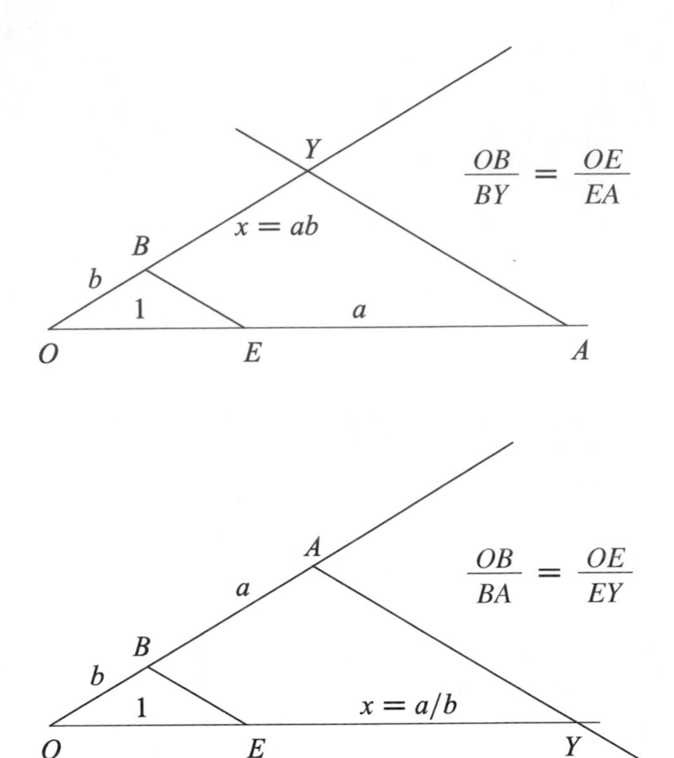

Fig. 14.1 Construction of the product *ab* (*above*) and the quotient a/b (*below*)

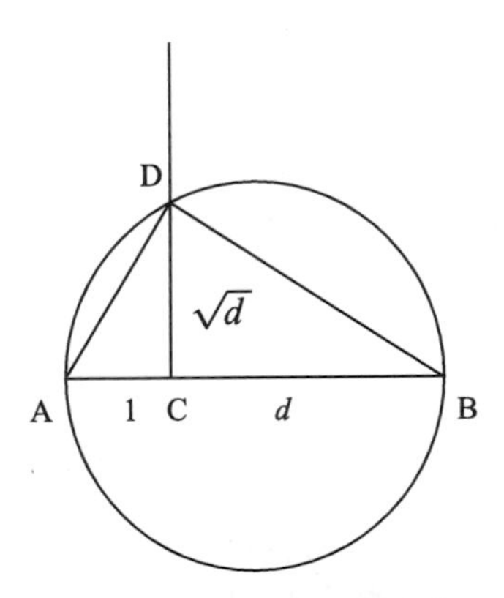

Fig. 14.2 Construction of the square root $\sqrt{d}$

Proof of T.14.1. If a point $P = (a, b)$ can be constructed from X by a straightedge-and-compass construction, then this can be done by constructing a sequence of points $P_0, P_1, P_2, \ldots, P_n = P$ such that each point can be directly constructed from X and the points preceding it. As we know from Lemma 14.1.1(b), the coordinates of every new point are in a real extension of degree at most 2 over the preceding field (starting with K) which contains the coordinates of the points of X and all the points constructed earlier. This shows that the coordinates a, b of P belong to a field L which can be obtained from K by a sequence of extensions each of which has degree at most 2. Thus, we get a sequence $(*)$ if we only take those fields whose degrees are 2.

Conversely, if a point $P = (a, b)$ has its coordinates in an extension L of K, which can be obtained by a chain of quadratic extensions like in $(*)$, then an evident inductive argument using Lemma 14.1.1(c) gives that the elements of each field K_i $(i = 1, \ldots, n)$ can be constructed from X. In particular, the point $P = (a, b)$ can be constructed from X.

Finally, notice that the distance between two points $P = (a, b)$ and $P' = (a', b')$, where $a, b, a', b' \in L$, is $d = \sqrt{(a-a')^2 + (b-b')^2}$, so such a number d belongs to (at most) a quadratic extension of L. □

T.14.2. *Let K be the smallest subfield of $\mathbb{R}$ which contains the coordinates of all points belonging to a point set X in the plane $\mathbb{R}^2$. A point $P = (a, b)$ is constructible from X by a straightedge-and-compass construction if and only if one of the following equivalent conditions hold:*
(a) There exists a Galois extension $L \supseteq K$ such that $a, b \in L$ and $[L : K]$ is a power of 2.
(b) There exists a Galois extension $L \supseteq K$ such that $a + bi \in L$ and $[L : K]$ is a power of 2.
In order to prove this theorem, we need some notations and an auxiliary result. Let K be a subfield of the real numbers and let $r_2(K)$ denote all complex numbers x for which there exists a chain of radical 2-extensions:

$$K = K_0 \subset K_1 \subset \cdots \subset K_n \subset \mathbb{C} \qquad (**)$$

such that $x \in K_n$ and $[K_i : K_{i-1}] = 2$ for $i = 1, \ldots, n$. Let $rr_2(K)$ denote all numbers for which there is such a chain of fields with $K_i \subset \mathbb{R}$. It follows easily from Lemma **13.1.2**(b) that $r_2(K)$ and $rr_2(K)$ are fields. Moreover, if $x \in r_2(K)$, then $\sqrt{x} \in r_2(K)$, while $x \in rr_2(K)$ and $x \geq 0$ imply $\sqrt{x} \in rr_2(K)$.

Lemma 14.2.1. *Let a, b be real numbers. Then the following conditions are equivalent:*

(*a*) $a, b \in rr_2(K)$,
(*b*) $a, b \in r_2(K)$,
(*c*) $a + bi \in r_2(K)$.

Proof. It is evident that (a) implies (b). If $a, b \in K_n$ in a tower $(**)$, then $a + bi \in K_{n+1} = K_n(i)$ and $[K_{n+1} : K_n] \leq 2$, which implies (c). We prove that (c) implies (a) by induction on the number n of fields in a tower $(**)$. Assume that $a + bi \in K_n$. If $n = 1$, then the claim is evident, since K is real and K_1 is a quadratic extension of K. Thus $a, b \in K \subset rr_2(K)$ as the real and imaginary parts of $a + bi$ (K_1 may be real when $b = 0$). Assume that the claim holds when the number of fields in the tower is less than n. Assume that $a + bi \in K_n$. Let $K_n = K_{n-1}(\sqrt{A + Bi})$, where $A + Bi \in K_{n-1}$, $A, B \in \mathbb{R}$ (it may happen that $B = 0$). We have:

$$a + bi = (x + yi) + (z + ti)\sqrt{A + Bi},$$

where $x + yi, z + ti \in K_{n-1}$, $x, y, z, t \in \mathbb{R}$. By the inductive assumption, $x, y, z, t, A, B \in rr_2(K)$. Moreover,

$$\sqrt{A + Bi} = \sqrt{\frac{\sqrt{A^2 + B^2} + A}{2}} + i\sqrt{\frac{\sqrt{A^2 + B^2} - A}{2}},$$

so defining

$$\alpha = \sqrt{\frac{\sqrt{A^2 + B^2} + A}{2}}, \quad \beta = \sqrt{\frac{\sqrt{A^2 + B^2} - A}{2}},$$

we have $\alpha, \beta \in rr_2(K)$, $a = x + z\alpha - t\beta \in rr_2(K)$ and $b = y + t\alpha + z\beta \in rr_2(K)$. □

Proof of T.14.2. First note that (a) and (b) are equivalent. In fact, $L(i)$ is an extension of L such that $[L(i) : L] = 1$ or 2 and $L(i)$ is of course Galois over K (if L is a splitting field of a polynomial with coefficients in K, then $L(i)$ is a splitting field of the same polynomial multiplied by $X^2 + 1$). If L satisfies (a), then we take $L(i)$ as L in (b)—it contains $a+bi$ and its degree over K is a power of 2. If L satisfies (b), then $a-bi \in L$, since the minimal polynomial of $a + bi$ over the real field K has $a - bi$ as its zero, which must belong to the normal extension L of K. Hence $a, b \in L$.
Assume now that a point (a, b) is constructible from a point set X by a straightedge-and-compass construction, that is, $a, b \in rr_2(K)$ and let

$$K = K_0 \subset K_1 \subset \cdots \subset K_n \subset \mathbb{R}$$

be a tower of fields such that $[K_i : K_{i-1}] = 2$ for $i = 1, \dots, n$ and $a, b \in K_n$. Let L be a normal closure of K_n. Then according to Lemma 13.1.2, L is a Galois radical 2-extension of K. Of course, $a, b \in L$ and by **T.4.3**, the degree $[L : K]$ is a power of 2.

Conversely, assume that (a) holds. Then the order of the Galois group $G = G(L/K)$ is a power of 2, that is, $G(L/K)$ is a 2-group. This means that there is a chain

$$G_0 = G(L/K) \supset G_1 \supset \cdots \supset G_{n-1} \supset G_n = \{id\}$$

such that each G_i is a normal subgroup of G_{i-1} and the quotient group G_{i-1}/G_i has 2 elements (see Exercise 12.6 for $p = 2$). The corresponding tower of the fields $L^{G_i} = K_i$ is

$$K_0 = K \subset K_1 \subset \cdots \subset K_n = L$$

and $[K_i : K_{i-1}] = 2$, that is, $a, b \in r_2(K)$. According to Lemma 14.2.1, $a, b \in rr_2(K)$, that is, the point (a, b) is constructible from X by a straightedge-and-compass construction. □

Theorems of Chap. 15

T.15.1. *(a) Let G be a subgroup of S_n. Then for every subgroup H of G there exists a polynomial $F \in K[X_1, \ldots, X_n]$ such that $H = G_F$.*
(b) Let $G = \sigma_1 G_F \cup \cdots \cup \sigma_m G_F$ be the presentation of G as a union of different left cosets with respect to G_F. Then $\sigma_i F(X_1, \ldots, X_n)$, for $i = 1, \ldots, m$, are all different images of F under the permutations belonging to G.

Proof. (a) Consider the monomial $M = X_1X_2^2 \cdots X_n^n$ and notice that for any two permutations $\sigma, \tau \in S_n$, we have $\sigma(M) = \tau(M)$ if and only if $\sigma = \tau$. In fact, since the exponents of all X_i in M are different, two different permutations map the monomial M onto two different monomials (that is, $X_{\sigma(1)}X_{\sigma(2)}^2 \cdots X_{\sigma(n)}^n = X_{\tau(1)}X_{\tau(2)}^2 \cdots X_{\tau(n)}^n$ if and only if $\sigma(i) = \tau(i)$ for every $i = 1, 2, \ldots, n$). Define now $F = \mathrm{Tr}_H(M) = \sum_{\tau \in H} \tau(M)$. The polynomial $F(X_1, \ldots, X_n)$ is invariant with respect to every $\sigma \in H$ and $\sigma(F) = F$ for $\sigma \in G$ if and only if $\sigma \in H$. In fact, for any permutation $\sigma \in H$, the compositions $\sigma\tau$, where $\tau \in H$, give exactly all the elements of H, so $\sigma(F) = F$. On the other hand, if $\sigma \in G \setminus H$, then $\sigma(M)$ is different from all $|H|$ monomials whose sum gives F, that is, we have $\sigma(F) \neq F$. Thus $\sigma(F) = F$, for $\sigma \in G$, if and only if $\sigma \in H = G_F$. Notice that the definition of F only depends on H and not on the group G containing it.
(b) We have

$$\sigma(F) = \sigma'(F) \Leftrightarrow \sigma^{-1}\sigma'(F) = F \Leftrightarrow \sigma^{-1}\sigma' \in H = G_F \Leftrightarrow \sigma G_F = \sigma' G_F.$$

Thus different images $\sigma(F)$, when $\sigma \in G$, correspond to different left cosets of G_F in G. □

T.15.2. *Let $f(X) \in K[X]$, $F(X_1, \ldots, X_n) \in K[X_1, \ldots, X_n]$ and assume that* $\mathrm{Gal}(K_f/K) \subseteq G$, *where G is a subgroup of S_n. Then:*
(a) The resolvent polynomial $r_{G,F}(f)$ has its coefficients in K;
(b) If $K = \mathbb{Q}$, $F(X_1, \ldots, X_n) \in \mathbb{Z}[X_1, \ldots, X_n]$ and $f(X)$ has integer coefficients and the leading coefficient is 1, then the same is true of $r_{G,F}(f)$;
(c) If all the zeros of $r_{G,F}(f)$ are different, then $\mathrm{Gal}(K_f/K)$ *is conjugate in G to a subgroup of G_F if and only if at least one of the zeros of $r_{G,F}(f)$ belongs to K.*

Proof. (a) Let $\sigma \in G$. Then

$$\begin{aligned}\sigma(r_{G,F}(f)(T)) &= \prod_{i=1}^{m}(T - (\sigma\sigma_i F)(\alpha_1, \ldots, \alpha_n)) \\ &= \prod_{i=1}^{m}(T - (\sigma_i F)(\alpha_1, \ldots, \alpha_n)) = r_G(F, f)(T),\end{aligned}$$

since $\sigma\sigma_i$, for $i = 1, \ldots, m$, is also a set of representatives of left cosets of G_F in G. In fact, since $\sigma\sigma_i \in G$, we have $\sigma\sigma_i = \sigma_j\tau$ for some $1 \leq j \leq m$ and $\tau \in G_F$. Thus $\sigma\sigma_i F = \sigma_j\tau F = \sigma_j F$, so $\sigma\sigma_i F(\alpha_1, \ldots, \alpha_n) = \sigma_j F(\alpha_1, \ldots, \alpha_n)$ is also a zero of $r_G(F,f)$. At the same time, it is clear that if $i \neq j$, then $\sigma\sigma_i$ and $\sigma\sigma_j$ represent different cosets of G_F in G. Hence $\sigma\sigma_i F(\alpha_1, \ldots, \alpha_n)$ for $1 \leq i \leq m$ is simply a permutation of the elements $\sigma_i F(\alpha_1, \ldots, \alpha_n)$, which means that $\sigma r_{G,F}(f)$ and $r_{G,F}(f)$ have the same coefficients for every $\sigma \in G$. In particular, this is true for every $\sigma \in \mathrm{Gal}(K_f/K)$, since $\mathrm{Gal}(K_f/K) \subseteq G$. Thus the coefficients of $r_{G,F}(f)$ are fixed by every element in the Galois group $\mathrm{Gal}(K_f/K)$, so they are in K.

(b) The numbers $\alpha_1, \ldots, \alpha_n$ in K_f are zeros of the polynomial $f(X)$ with integer coefficients and the leading coefficient equal to 1. This is the definition of algebraic integers over $\mathbb{Z}$ in a field extension of $\mathbb{Q}$. It is not difficult to prove (but unfortunately outside the scope of this book) that algebraic integers form a ring (for a proof, see [L], Chap. VII, Prop. 1.4). Therefore, the numbers $\sigma(f(\alpha_1, \ldots, \alpha_n))$ are also algebraic integers for each $\sigma \in G(K_f/\mathbb{Q})$ (since each $\sigma(\alpha_i)$ is also a zero of $f(X)$). Thus the coefficients of the resolvent $r_G(F,f)(T)$ are also algebraic integers and, according to (a), are rational numbers. But this means that these coefficients are integers, since algebraic integers in the field of rational numbers are exactly the usual integers (this property of algebraic integers in $\mathbb{Q}$ is easy to check).

(c) If $\sigma \in G$ and $\sigma = \sigma_i\tau$, where $\tau \in G_F$, then $\sigma F = \sigma_i F$ and $\sigma F(\alpha_1, \ldots, \alpha_n) = \sigma_i F(\alpha_1, \ldots, \alpha_n)$ is a zero of $r_G(F,f)$. If all the zeros of $r_G(F,f)$ are different, then $\sigma F(\alpha_1, \ldots, \alpha_n) = F(\alpha_1, \ldots, \alpha_n)$ is equivalent to $\sigma \in G_F$. In fact, $\sigma_i F(\alpha_1, \ldots, \alpha_n) = \sigma_1 F(\alpha_1, \ldots, \alpha_n)$ implies $\sigma_i = \sigma_1 = id$, so $\sigma = \tau \in G_F$.

If $\tau^{-1}\mathrm{Gal}(K_f/K)\tau \subseteq G_F$ for a $\tau \in G$, then $\tau^{-1}\sigma\tau F(\alpha_1, \ldots, \alpha_n) = F(\alpha_1, \ldots, \alpha_n)$ for each $\sigma \in \mathrm{Gal}(K_f/K)$. Hence $\sigma\tau F(\alpha_1, \ldots, \alpha_n) = \tau F(\alpha_1, \ldots, \alpha_n)$, that is, $\tau F(\alpha_1, \ldots, \alpha_n)$, which is a zero of $r_G(F,f)$, belongs to K.

Conversely, assume that $\sigma_i F(\alpha_1, \ldots, \alpha_n) \in K$ for some i. If $\sigma \in \mathrm{Gal}(K_f/K)$, we have $\sigma\sigma_i F(\alpha_1, \ldots, \alpha_n) = \sigma_i F(\alpha_1, \ldots, \alpha_n)$, that is, $\sigma_i^{-1}\sigma\sigma_i F(\alpha_1, \ldots, \alpha_n) = F(\alpha_1, \ldots, \alpha_n)$, so $\sigma_i^{-1}\sigma\sigma_i \in G_F$. Hence $\sigma_i^{-1}\mathrm{Gal}(K_f/K)\sigma_i \subseteq G_F$. □

T.15.3. *Let $K = k(Y_1, \ldots, Y_n)$ be the field of rational functions in the variables Y_i over a field k. Let $f(X) \in k[X] \subset K[X]$ and let $r_{G,F}(f)(T)$ be the resolvent of $f(X)$ with respect to the Galois group* $\mathrm{Gal}(k_f/k) = \mathrm{Gal}(K_f/K) \subseteq S_n$ *and the polynomial*

$$F(X_1, \ldots, X_n) = X_1Y_1 + \cdots + X_nY_n.$$

Let $r_i(T, Y_1, \ldots, Y_n)$, for $i = 1, \ldots, t$, be the irreducible factors of the resolvent $r_{G,F}(f)(T)$ in $k[T, Y_1, \ldots, Y_n]$. Then the Galois group $\mathrm{Gal}(k_f/k)$ *is isomorphic to any group G_{r_i} of those permutations of $Y_1, \ldots, Y_n$ which map $r_i(T, Y_1, \ldots, Y_n)$ onto itself. Moreover, all $r_i(T, Y_1, \ldots, Y_n)$ have the same degree $[k_f : k]$ with respect to T and a common splitting field $K_f = k_f(Y_1, \ldots, Y_n)$ over K.*

Proof. Define $\theta_\varrho = \alpha_1 Y_{\varrho(1)} + \cdots + \alpha_n Y_{\varrho(n)} \in k_f(Y_1, \ldots, Y_n) = K_f$ for $\varrho \in S_n$. The group S_n acts on the set of all θ_ϱ if we define $\sigma(\theta_\varrho) = \theta_{\sigma\varrho}$. Defining $\theta = \theta_{id} =$

$\alpha_1 Y_1 + \cdots + \alpha_n Y_n$, we have $\varrho(\theta) = \theta_\varrho$, so the action of S_n on the set of all θ_ϱ is transitive. Of course,

$$\alpha_1 Y_{\varrho(1)} + \cdots + \alpha_n Y_{\varrho(n)} = \alpha_{\varrho^{-1}(1)} Y_1 + \cdots + \alpha_{\varrho^{-1}(n)} Y_n.$$

Let $\theta_i = \sigma_i(\theta) = \theta_{\sigma_i}$ denote a zero of the polynomial $r_i(X, Y_1, \ldots, Y_n)$ and assume that $\sigma_1 = id$, that is, $\theta_1 = \theta$.

Notice now that each $\theta_i = \alpha_1 Y_{\sigma_i(1)} + \cdots + \alpha_n Y_{\sigma_i(n)} \in k_f(Y_1, \ldots, Y_n)$ is a primitive element of the extension $K = k(Y_1, \ldots, Y_n) \subseteq K_f = k_f(Y_1, \ldots, Y_n)$. In fact, according to Exercise 15.9(a), the Galois group of this extension is isomorphic to $\mathrm{Gal}(k_f/k)$ when $\tau \in \mathrm{Gal}(k_f/k)$ corresponds to an automorphism of $K_f = k_f(Y_1, \ldots, Y_n)$, which acts on the coefficients of the rational functions in K_f as τ and as the identity on Y_i. Thus, for the element $\theta_i \in K_f$, we have

$$\tau(\theta_i) = \tau(\alpha_1) Y_{\sigma_i(1)} + \cdots + \tau(\alpha_n) Y_{\sigma_i(n)},$$

which shows that the number of different images of θ_i is equal to the order of the Galois group $\mathrm{Gal}(K_f/K)$ (diffrent automorphisms give different sets of coefficients of Y_i). Hence, according to Exercise 9.24, θ_i is a primitive element of the extension $K \subseteq K_f$ and every automorphism of K_f over K is uniquely defined by its action on θ_i. Since $r_i(T, Y_1, \ldots, Y_n)$ is the minimal polynomial of θ_i over K, the remaining zeros of this polynomial are the images $\tau(\theta_i)$ under the automorphisms τ of the Galois group $\mathrm{Gal}(K_f/K)$. Thus, all polynomials $r_i(X, Y_1, \ldots, Y_n)$ have the same degree $|\mathrm{Gal}(K_f/K)| = [K_f : K] = [k_f : k]$ and they have a common splitting field K_f over K.

The Galois group $\mathrm{Gal}(k_f/k)$ acts on the set of θ_ϱ in two different ways. First of all each $\tau \in \mathrm{Gal}(k_f/k)$ acts as the permutation $\pi(\tau) \in S_n$ of $\{1, \ldots, n\}$ defined by the relation:

$$\pi(\tau)(i) = j \Leftrightarrow \tau(\alpha_i) = \alpha_j,$$

that is, $\tau(\alpha_i) = \alpha_{\pi(\tau)(i)}$. This gives the usual representation of the Galois group $\mathrm{Gal}(k_f/k)$ as a permutation group. Notice that π is an injective homomorphism of $\mathrm{Gal}(k_f/k)$ into S_n. In particular, $\pi(\tau^{-1}) = \pi(\tau)^{-1}$.

The second action is defined according to Exercise 15.9(a)—the Galois group of the extension $K = k(Y_1, \ldots, Y_n) \subseteq K_f = k_f(Y_1, \ldots, Y_n)$ is isomorphic to $\mathrm{Gal}(k_f/k)$ when $\tau \in \mathrm{Gal}(k_f/k)$ corresponds to an automorphism of $K_f = k_f(Y_1, \ldots, Y_n)$ which acts on the coefficients of the rational functions in K_f as τ and as the identity on Y_i. Thus for the element $\theta_\varrho \in K_f$, we have

$$\tau(\theta_\varrho) = \tau(\alpha_1) Y_{\varrho(1)} + \cdots + \tau(\alpha_n) Y_{\varrho(n)}.$$

Let G_{r_i} denote the group of permutations $\varrho \in S_n$ such that

$$\varrho r_i(T, Y_1, \ldots, Y_n) = r_i(T, Y_{\varrho(1)}, \ldots, Y_{\varrho(n)}) = r_i(T, Y_1, \ldots, Y_n)$$

for $i = 1, \ldots, t$. We will show that π is a group isomorphism of $\mathrm{Gal}(k_f/k)$ onto G_{r_i}.

We have:

$$r_i(T, Y_1, \dots, Y_n) = \prod_{\tau \in G(K_f/K)} (T - \tau(\theta_i)) = \prod_{\tau \in G(K_f/K)} (T - \tau(\alpha_1 Y_{\sigma_i(1)} + \cdots + \alpha_n Y_{\sigma_i(n)})).$$

If $\pi(\tau)$ is the permutation corresponding to $\tau \in \mathrm{Gal}(K_f/K)$, then

$$\pi(\tau) r_i(T, Y_1, \dots, Y_n) = r_i(T, Y_{\pi(\tau)(1)}, \dots, Y_{\pi(\tau)(n)})$$

$$= \prod_{\tau' \in G(K_f/K)} (T - \tau'(\alpha_1 Y_{\pi(\tau)(\sigma_i(1))} + \cdots + \alpha_n Y_{\pi(\tau)(\sigma_i(n))}))$$

$$= \prod_{\tau' \in G(K_f/K)} (T - \tau'(\alpha_{\pi(\tau)^{-1}(1)} Y_{\sigma_i(1)} + \cdots + \alpha_{\pi(\tau)^{-1}(n)} Y_{\sigma_i(n)}))$$

$$= \prod_{\tau' \in G(K_f/K)} (T - \tau'\tau^{-1}(\alpha_1 Y_{\sigma_i(1)} + \cdots + \alpha_n Y_{\sigma_i(n)})) = r_i(T, Y_1, \dots, Y_n)$$

taking into account that $\pi(\tau)^{-1} = \pi(\tau^{-1})$, $\alpha_{\pi(\tau^{-1})(i)} = \tau^{-1}(\alpha_i)$ and that $\tau'\tau^{-1}$ permutes all the elements of $\mathrm{Gal}(K_f/K)$ when $\tau \in \mathrm{Gal}(K_f/K)$ (τ arbitrary but fixed). Hence, for every $\tau \in \mathrm{Gal}(k_f/k)$, we have $\pi(\tau) \in G_{r_i}$.
Conversely, let $\varrho \in G_{r_i}$, that is,

$$\varrho r_i(T, Y_1, \dots, Y_n) = r_i(T, Y_{\varrho(1)}, \dots, Y_{\varrho(n)}) = r_i(T, Y_1, \dots, Y_n).$$

Then on the one hand,

$$r_i(T, Y_1, \dots, Y_n) = \prod_{\tau \in G(K_f/K)} (T - \tau(\theta_i)) = \prod_{\tau \in G(K_f/K)} (T - \tau(\alpha_1 Y_{\sigma_i(1)} + \cdots + \alpha_n Y_{\sigma_i(n)}))$$

$$= \prod_{\tau \in G(K_f/K)} (T - \tau(\alpha_{\sigma_i^{-1}(1)} Y_1 + \cdots + \alpha_{\sigma_i^{-1}(n)} Y_n)),$$

and on the other,

$$r_i(T, Y_{\varrho(1)}, \dots, Y_{\varrho(n)}) = \prod_{\tau \in G(K_f/K)} (T - \tau(\alpha_{\sigma_i^{-1}(1)} Y_{\varrho(1)} + \cdots + \alpha_{\sigma_i^{-1}(n)} Y_{\varrho(n)}))$$

$$= \prod_{\tau \in G(K_f/K)} (T - \tau(\alpha_{\varrho^{-1}\sigma_i^{-1}(1)} Y_1 + \cdots + \alpha_{\varrho^{-1}\sigma_i^{-1}(n)} Y_n)).$$

Hence $\alpha_{\sigma_i^{-1}(1)} Y_1 + \cdots + \alpha_{\sigma_i^{-1}(n)} Y_n$ is among the elements $\tau(\alpha_{\varrho^{-1}\sigma_i^{-1}(1)} Y_1 + \cdots + \alpha_{\varrho^{-1}\sigma_i^{-1}(n)} Y_n)$, that is, there is a $\tau \in \mathrm{Gal}(k_f/k)$ such that $\tau(\alpha_{\varrho^{-1}\sigma_i^{-1}(k)}) = \alpha_{\sigma_i^{-1}(k)}$ for every $k = 1, \ldots, n$. Hence $\pi(\tau)\varrho^{-1}\sigma_i^{-1}(k) = \sigma_i^{-1}(k)$ for $k = 1, \ldots, n$. This means that $\varrho = \pi(\tau)$, so π is bijective, and consequently an isomorphism of $\mathrm{Gal}(k_f/k)$ and G_{r_i}. □

T.15.4 Dedekind's Theorem. *(a) Let $f \in R[X]$ be a separable monic polynomial and assume that its image $f^* \in R^*[X]$ is also separable and* $\deg(f) = \deg(f^*) = n$. *Then the Galois group of f^* over K^* as a permutation group of its zeros is a subgroup of the Galois group of f over K as a permutation group of its zeros.*
(b) In (a), let $R = \mathbb{Z}$, $R^* = \mathbb{F}_p$ and let φ be the reduction modulo a prime number p. If $f \in \mathbb{Z}[X]$ and

$$f^* = f_1^* \cdots f_k^*,$$

where f_i^* are irreducible over $\mathbb{F}_p$, then $Gal(\mathbb{Q}_f/\mathbb{Q})$, considered as a permutation subgroup of S_n, contains a permutation which is a product of cycles of length $\deg(f_i^*)$ *for $i = 1, \ldots, k$.*

Proof. (a) Let $L = K(\alpha_1, \ldots, \alpha_n)$ and $L^* = K^*(\alpha_1^*, \ldots, \alpha_n^*)$ be splitting fields of $f(X)$ over $K[X]$ and $f^*(X)$ over K^*. Consider the polynomials:

$$r(f)(T, Y_1, \ldots, Y_n) = \prod_{\sigma \in S_n} (T - (\alpha_{\sigma(1)}Y_1 + \cdots + \alpha_{\sigma(n)}Y_n)),$$

and

$$r(f^*)(T, Y_1, \ldots, Y_n) = \prod_{\sigma \in S_n} (T - (\alpha^*_{\sigma(1)}Y_1 + \cdots + \alpha^*_{\sigma(n)}Y_n)).$$

The coefficients of $r(f)(T, Y_1, \ldots, Y_n)$ (of the monomials in variables $T, Y_1, \ldots, Y_n$) are symmetric functions of $\alpha_1, \ldots, \alpha_n$ and as such can be expressed as integer polynomials of the coefficients of f (see section "Symmetric Polynomials" in Appendix). In exactly the same way, the coefficients of $r(f^*)(T, Y_1, \ldots, Y_n)$ are symmetric functions of $\alpha_1^*, \ldots, \alpha_n^*$ and can be expressed as integer polynomials of the coefficients of f^* (see section "Symmetric Polynomials" in Appendix). Since the formulae expressing the coefficients of $r(f)$ and $r(f^*)$ by the coefficients of f and f^* are the same, we see that the polynomial $r(f^*)(T, Y_1, \ldots, Y_n)$ is the polynomial $r(f)(T, Y_1, \ldots, Y_n)$ when the homomorphism φ is applied to its coefficients in R. Notice also that all the $n!$ zeros $\alpha_{\sigma(1)}Y_1 + \cdots + \alpha_{\sigma(n)}Y_n$ of $r(f)$ and $\alpha^*_{\sigma(1)}Y_1 + \cdots + \alpha^*_{\sigma(n)}Y_n$ of $r(f^*)$ as polynomials of T are different when $\sigma \in S_n$. In fact, if $\alpha_{\sigma(1)}Y_1 + \cdots + \alpha_{\sigma(n)}Y_n = \alpha_{\tau(1)}Y_1 + \cdots + \alpha_{\tau(n)}Y_n$ for $\sigma, \tau \in S_n$, then $\alpha_{\sigma(i)} = \alpha_{\tau(i)}$ for $i = 1, \ldots, n$, so $\sigma(i) = \tau(i)$ for each i, since all α_i are different. Hence, we have $\sigma = \tau$. Similarly, we get that all $\alpha^*_{\sigma(1)}Y_1 + \cdots + \alpha^*_{\sigma(n)}Y_n$ are different.

Consider now the factorization of $r(f)(T, Y_1, \ldots, Y_n)$ into irreducible factors, the Galois resolvents, as in (15.3):

$$r(f)(T) = r_1(T, Y_1, \ldots, Y_n) \cdots r_t(T, Y_1, \ldots, Y_n).$$

The homomorphism φ applied to the coefficients in R of the polynomial $r(f)$ and its factors r_i gives the factorization:

$$r^*(f)(T) = r_1^*(T, Y_1, \dots, Y_n) \cdots r_t^*(T, Y_1, \dots, Y_n),$$

but the polynomials $r_i^*(T, Y_1, \dots, Y_n)$ need no longer be irreducible. Consider the resolvent $r_1(T, Y_1, \dots, Y_n)$ of $f(X)$ and let

$$r_1^* = r_{11}^* \cdots r_{1u}^*$$

be the factorization of $r_1^*(T, Y_1, \dots, Y_n)$ into irreducible factors r_{1j}^*. These factors are Galois resolvents of $f^*(X)$.
Consider all permutations of $Y_1, \dots, Y_n$ which map the resolvent r_{11}^* into itself. By **T.15.3**, such permutations form the group $\mathrm{Gal}(K_f^*/K^*)$. Applying the same permutation to the Galois resolvent r_1 of $f(X)$, we map it onto one of the resolvents r_i of $f(X)$ for some $i = 1, \dots, t$. We claim that this must be the same resolvent r_1. In fact, if the permutation gives r_i, with $i \neq 1$, then the same permutation transforms r_1^* to r_i^* and since the factor r_{11}^* of r_1^* maps onto itself, it must also be a factor of r_i^*. But if $i \neq 1$, then the polynomials r_1^* and r_i^* are relatively prime, since they are products of different linear factors $T - (\alpha_{\sigma(1)}^* Y_1 + \cdots + \alpha_{\sigma(n)}^* Y_n)$. Thus, every permutation of $Y_1, \dots, Y_n$ which maps r_{11}^* into itself also maps r_1 into itself. Hence, applying **T.15.3**, we get that every element of $\mathrm{Gal}(K_f^*/K^*)$ belongs to the group $\mathrm{Gal}(K_f/K)$.
(b) The Galois group of the polynomial f^* over $\mathbb{F}_p$ is cyclic (see **T.5.4**). Let σ be a generator of this group. By Exercise 9.4, the automorphism σ can be represented as a product of k cycles of length equal to the degrees of the irreducible factors of $f^*(X)$ over $\mathbb{F}_p$. Using (a), we get a permutation of this shape in the Galois group $\mathrm{Gal}(\mathbb{Q}_f/\mathbb{Q})$. □

Chapter 18
Hints and Answers

Problems of Chap. 1

1.1 (a) $-3, \frac{3}{2} \pm \frac{1}{2} i\sqrt{3}$;
(b) $-7, -1 - i\sqrt{3}, -1 + i\sqrt{3}$;
(c) $-\sqrt[3]{9} - \sqrt[3]{3} - 1, \frac{1}{2}(\sqrt[3]{9} + \sqrt[3]{3}) - 1 + \frac{1}{2} i\sqrt{3}\left(\sqrt[3]{3} - \sqrt[3]{9}\right), \frac{1}{2}(\sqrt[3]{9} + \sqrt[3]{3}) - 1 - \frac{1}{2} i\sqrt{3}\left(\sqrt[3]{3} - \sqrt[3]{9}\right)$;
(d) $\sqrt[3]{2} - \sqrt[3]{4}, \frac{1}{2}(\sqrt[3]{4} - \sqrt[3]{2}) - \frac{1}{2}i\sqrt{3}(\sqrt[3]{4} + \sqrt[3]{2}), \frac{1}{2}(\sqrt[3]{4} - \sqrt[3]{2}) + \frac{1}{2}i\sqrt{3}(\sqrt[3]{4} + \sqrt[3]{2})$;
(e) $1 - i\sqrt{3}, 1 + i\sqrt{3}, \sqrt{2}, -\sqrt{2}$;
(f) $1 - i, 1 + i, \frac{1}{2}(1 + \sqrt{13}), \frac{1}{2}(1 - \sqrt{13})$;
(g) $\frac{1\pm\sqrt{2}}{2} \pm \frac{1}{2}\sqrt{-1 + 2\sqrt{2}}$;
(h) $1 \pm \sqrt{7} \pm \sqrt{6 + 2\sqrt{7}}$.

1.3 Notice that $(x_1 - x_2)^2 = x_3^2 + \frac{4q}{x_3}$ and similarly for $(x_2 - x_3)^2$ and $(x_3 - x_1)^2$. Express $x_1^3 + x_2^3 + x_3^3$ and $x_1^3x_2^3 + x_2^3x_3^3 + x_3^3x_1^3$ in terms of p, q using Vieta's formulae: $x_1 + x_2 + x_3 = 0, x_1x_2 + x_2x_3 + x_3x_1 = p, x_1x_2x_3 = -q$.

1.4 The equation has solutions $2, -1 \pm \sqrt{3}$. Write down the trigonometric form of the numbers $-1 \pm i$. Show that for a suitable choice of the values of the third roots, we have

$$\sqrt[3]{-2 + 2i} + \sqrt[3]{-2 - 2i} = 2.$$

1.5 Represent the complex numbers $-\frac{q}{2} \pm \sqrt{\frac{q^2}{4} + \frac{p^3}{27}}$ in the formula (1.5) in trigonometric form.

J. Brzeziński, *Galois Theory Through Exercises*, Springer Undergraduate Mathematics Series, https://doi.org/10.1007/978-3-319-72326-6_18

1.6 (a) The equation satisfied by α is $X^3 - 3cX - 2a = 0$.
(b) Use the binomial formula in the following form:

$$(X+Y)^5 = X^5 + Y^5 + 5XY[(X+Y)^3 - 3XY(X+Y)] + 10X^2Y^2(X+Y)$$

with $X = \sqrt[5]{a+\sqrt{b}}$, $Y = \sqrt[5]{a-\sqrt{b}}$. The equation satisfied by α is $X^5 - 5cX^3 + 5c^2X - 2a = 0$. If $a = 1, c = 0$ the polynomial is irreducible (see Chap. 3 concerning irreducibility). Of course, there are equations of degree 5 which cannot be represented in this form, e.g. those with coefficient of X^2 different from 0.

1.7 If $a = 0$, then the equation $X^n - a = 0$ has only one solution $x = 0$ (with multiplicity n). In general, we have $a = |a|e^{i\varphi}$, so the equality $x^n = |a|e^{i\varphi}$ is equivalent to $x = \sqrt[n]{|a|}e^{\frac{\varphi+2\pi ik}{n}}$ for n different values of $k = 0, 1, \dots, n-1$.

1.8 Notice that the discriminant of $f(X) = aX^2 + bX + c = a(X - x_1)(X - x_2)$ is $\Delta(f) = (x_1 - x_2)^2 = (b^2 - 4ac)/a^2$ (see p. 3).

Problems of Chap. 2

2.1 Only (d) is a field.

2.2 Let K be a subfield of $\mathbb{Q}$. Starting with $1 \in K$ show that all integers, and then, all rational numbers must belong to K.

2.3 For example, $\mathbb{F}_p(X)$, where p is a prime number and X is a variable.

2.4 Use **T. 2.1** and consider two cases depending on whether the prime subfield of K is finite or infinite.

2.5 (a) Show that the sum, difference, product and the quotient of two elements of $K(\alpha)$ belong to $K(\alpha)$.
(b) If ab is a square in K, then it is easy to check the equality of the fields. In the opposite direction, consider two cases: $K(\sqrt{a}) = K(\sqrt{b}) = K$, $K(\sqrt{a}) = K(\sqrt{b}) \neq K$ and in the second case use (a).

2.6 Use Exercise 2.5. (a) $a + b\sqrt{2}$, $a, b \in \mathbb{Q}$;
(b) $a + bi$, $a, b \in \mathbb{Q}$;
(c) $a + b\sqrt{2} + ci + di\sqrt{2}$, $a, b, c, d \in \mathbb{Q}$;
(d) $a + b\sqrt{2} + c\sqrt{3} + d\sqrt{6}$, $a, b, c, d \in \mathbb{Q}$.

2.7 If you need help, see the examples on p. 178.

2.8 (a) $\mathbb{Q}(\sqrt{2}) = \{a + b\sqrt{2} : a, b \in \mathbb{Q}\}$;
(b) $\mathbb{Q}(i, \sqrt{2}) = \{a + bi + c\sqrt{2} + di\sqrt{2} : a, b, c, d \in \mathbb{Q}\}$;
(c) $\mathbb{Q}(i, \sqrt{5}) = \{a + bi + c\sqrt{5} + di\sqrt{5} : a, b, c, d \in \mathbb{Q}\}$;
(d) $\mathbb{Q}(i) = \{a + bi : a, b \in \mathbb{Q}\}$.

2.9 (a) Check that all conditions in the definition of the field are satisfied if and only if every nonzero matrix in the given set has nonzero determinant. The condition that $X^2 + 1 = 0$ has no solutions in K is needed in order to exclude the possibility that a nonzero matrix has determinant equal to 0.

(b) The subfield of L isomorphic to K consists of the diagonal matrices $\begin{bmatrix} a & 0 \\ 0 & a \end{bmatrix}$.

(c) Choose $K = \mathbb{F}_3$ in (a).

2.10 (a) Similarly as in Exercise 2.9 check that addition and multiplication of matrices in the given set satisfy all the conditions in the definition of a field. This time the suitable condition is that the polynomial $X^2 - X + 1$ has no zeros in K, which guarantees that nonzero matrices of given form have nonzero determinants.
(b) Choose $K = \mathbb{F}_2$ in (a). Then the field consists of the following matrices:

$$0 = \begin{bmatrix} 0 & 0 \\ 0 & 0 \end{bmatrix}, \quad 1 = \begin{bmatrix} 1 & 0 \\ 0 & 1 \end{bmatrix}, \quad \alpha = \begin{bmatrix} 0 & 1 \\ 1 & 1 \end{bmatrix}, \quad \beta = 1 + \alpha = \begin{bmatrix} 1 & 1 \\ 1 & 0 \end{bmatrix}.$$

The addition and multiplication tables are as follows:

$+$	0	1	α	β
0	0	1	α	β
1	1	0	β	α
α	α	β	0	1
β	β	α	1	0

$\cdot$	0	1	α	β
0	0	0	0	0
1	0	1	α	β
α	0	α	β	1
β	0	β	1	α

2.11 Show that $2a = 0$ for each $a \in K$. The answer is 2.

2.12 (a) For $m = 1$ use the binomial theorem and notice that the binomial coefficients $\binom{p}{i}$ are divisible by p if $0 < i < p$. Then use induction.
(b) The same argument as in (a) for $m = 1$.

Problems of Chap. 3

3.2 All polynomials of degree 1 are $X, X + 1$. Construct all reducible polynomials of degree 2 using these two. Which are not on the list? Answer: $X^2 + X + 1$. This is the only irreducible polynomial of degree 2. Construct all reducible polynomials of degree 3 using these three: $X, X + 1, X^2 + X + 1$. Which are not on the list? Answer: $X^3 + X + 1$, $X^3 + X^2 + 1$. Continue the process! Answer: $X^4 + X + 1$, $X^4 + X^3 + X^2 + X + 1$, $X^4 + X^3 + 1$, $X^5 + X^2 + 1$, $X^5 + X^3 + 1$.

3.4 (a) $X^4 + 64 = (X^2 + 4X + 8)(X^2 - 4X + 8)$ (see above Exercise 3.1(b));
(b) $X^4 + 1 = (X^2 - \sqrt{2}X + 1)(X^2 + \sqrt{2}X + 1)$ (see above Exercise 3.1(b));
(c) $X^7 + 1 = (X + 1)(X^3 + X + 1)(X^3 + X^2 + 1)$ in $\mathbb{F}_2[X]$;

(d) $X^4 + 2$ is irreducible in $\mathbb{F}_{13}[X]$ (if you need help, see the example on p. 179);
(e) $X^3 - 2$ is irreducible in $\mathbb{Q}[X]$;
(f) $X^6 + 27 = \left(X^2 + 3\right)\left(X^2 + 3X + 3\right)\left(X^2 - 3X + 3\right)$ in $\mathbb{R}[X]$;
(g) $X^3 + 2 = (X + 2)^3$ in $\mathbb{F}_3[X]$;
(h) $X^4 + 2 = (X + 1)(X + 2)(X^2 + 1)$ in $\mathbb{F}_3[X]$.
In (a), (b), (c), (e), (f) and (h) use Exercise 3.1(a) in order to show that the factors are irreducible.

3.7 (a) Assume $f(X) = g(X)h(X)$, where $g(X), h(X)$ are polynomials in $\mathbb{Z}[X]$, $\deg g(X) = k \geq 1$ and $\deg h(X) = l \geq 1$ and look at the relation between the images of $f(X), g(X), h(X)$ after reduction modulo p.
(c) For $n > 1$ consider $pf(X)$ and use Eisenstein's criterion.
(d) Consider the polynomial $f(X + 1)$ and use Eisenstein's criterion.

3.8 (c) Use prime numbers $p = 2$ in (c_1) and (c_2), $p = 5$ in (c_3), $p = 2, 3$ in (c_4) and (c_6), $p = 2, 5$ in (c_5).

3.10 Let $f_1, f_2, \ldots, f_r$ be irreducible polynomials over K and consider the polynomial $f_1 f_2 \ldots f_n + 1$. Use the fact that $K[X]$ has unique factorization (see **T. 3.2**). Observe that the exercise is trivial over infinite fields.

3.11 Use Gauss's Lemma **T. 3.3**.

Problems of Chap. 4

4.1 (a), (b), (c) are algebraic. Use **T. 4.6**. (d), (e), (f) are transcendental. Use **T. 4.6** and the transcendence of e and π.

4.2 Use **T. 4.2** and **T. 4.3**. You may need the result which says that the equation $f(X) = 0$ has a solution in some extension of the field L (this is proved in introductory courses in algebra, see section "Fields" in Appendix or **T. 5.1**(a)).

4.3 The degree of each number is equal to the degree of its minimal polynomial. These polynomials are:

(a) $X^6 - 2X^3 - 2$;
(b) $X^6 - 6X^4 - 4X^3 + 12X^2 - 24X - 4$;
(c) $X^4 + X^3 + X^2 + X + 1$;
(d) $X^2 - 2$;
(e) $X^3 - 2$;
(f) $X^{p-1} + X^{p-2} + \cdots + X + 1$.

In each case, we have to show that the polynomial is irreducible. In order to prove this, one can use the different methods discussed in Chap. 3. In case (b), it may be useful to apply **T. 4.2**, and in case (e), Exercise 4.11(b).

4.4 Use **T. 4.2** and **T. 4.3** in order to construct suitable bases.

(a) 4; a basis: $1, i, \sqrt{2}, i\sqrt{2}$;
(b) 4; a basis $1, \sqrt{2}, \sqrt{3}, \sqrt{6}$;
(c) 6; a basis $1, i, \sqrt[3]{2}, \sqrt[3]{4}, i\sqrt[3]{2}, i\sqrt[3]{4}$;
(d) 3; a basis: $1, \sqrt[3]{2}, \sqrt[3]{4}$;
(e) 2; a basis: $1, X$;
(f) 8; a basis: $1, \sqrt{2}, \sqrt{3}, \sqrt{5}, \sqrt{6}, \sqrt{10}, \sqrt{15}, \sqrt{30}$;
(g) 3; a basis: $1, \alpha, \alpha^2$, where $\alpha = \sqrt[3]{1+\sqrt{3}}$;
(h) 4; a basis: $1, \alpha, \alpha^2, \alpha^3$;
(i) 3; a basis: $1, \alpha, \alpha^2$;
(j) 4; a basis: $1, X, X^2, X^3$.

4.5 Take any $\alpha \in K \setminus \mathbb{Q}$ and modify it so that $\alpha^2 \in \mathbb{Z}$. Use Exercise 2.5(b) in order to show uniqueness.

4.6 Notice that if $z = a + bi$ is algebraic, then so is $\bar{z} = a - bi$ (give a suitable argument!). Use Exercise 3.5 and **T. 4.6** in both directions.

4.7 Consider $z = \cos r\pi + i \sin r\pi$ and use Exercise 4.6.

4.8 (a) $x = \frac{1}{2}\sqrt[3]{4}$; (b) $x = \frac{1}{3}(1 - \sqrt[3]{2} + \sqrt[3]{4})$; (c) $x = -1 + \sqrt[3]{4}$.

4.9 (a) $x = -\sqrt{2} + \sqrt{3}$; (b) $x = \frac{1}{2} + \frac{1}{4}\sqrt{2} - \frac{1}{4}\sqrt{6}$; (c) $x = \frac{1}{2}(-5 + 4\sqrt{2} + 3\sqrt{3} - 2\sqrt{6})$.

4.10 (a) $x = 1 + \alpha^3$; (b) $x = \alpha + \alpha^2$; (c) $x = 1$; (d) $x = \alpha + \alpha^2$.

4.11 (a) Notice that $p(Y) - \frac{p(X)}{q(X)}q(Y)$ is a nonzero polynomial in Y of degree n over $K(\alpha)$, where $\alpha = \frac{p(X)}{q(X)}$, having X as its zero.
(b) Use (a) and **T. 4.3**.
(c) Show that the polynomial $p(Y) - \alpha q(Y)$ is irreducible over the field $K(\alpha)$, for example, using a version of Gauss's Lemma **T. 3.3** and the remark after it for $R = K[\alpha]$. Notice that the degree of this polynomial with respect to α is one.

4.12 In (b) and (c) use **T. 4.3**.
(d) In general, we have $[M_1M_2 : K] \leq [M_1 : K][M_2 : K]$ and for trivial reasons, the inequality may be strict (e.g. $M_1 = M_2$ and bigger than K). If $[M_1M_2 : K] = [M_1 : K][M_2 : K]$, then $M_1 \cap M_2 = K$ and $[M_1M_2 : M_1] \leq [M_2 : K]$ (as a consequence of the inequality above).

4.13 Use Exercise 4.2.

4.14 Consider $[K(\alpha) : K(\alpha^2)]$ and use **T. 4.3**. The answer to the second part of the exercise is no (give a counterexample!).

4.15 (a) Consider a minimal polynomial of an element $\alpha \in L \setminus \mathbb{C}$ (if such exists) and use the fundamental theorem of algebra (see **T. 1.1**).
(b) Consider the minimal polynomial of an element $\alpha \in L \setminus \mathbb{R}$ (if such exists) and use Exercise 3.5(b).

4.16 No. Consider $K = \mathbb{Q}$ and $L = \mathbb{Q}(\alpha) \supset \mathbb{Q}$, where $\alpha^4 + \alpha + 1 = 0$. Prove that there is no quadratic extension of $\mathbb{Q}$ which is contained in L.

4.17 Study the tower of extensions $\mathbb{Q}(e, \pi) \supset \mathbb{Q}(e + \pi, e\pi) \supset \mathbb{Q}$.

4.18 (a) Let $f(X) = \alpha_n X^n + \cdots + \alpha_1 X + \alpha_0$, where α_i for $i = 0, 1, \ldots, n$ are algebraic over K. Consider the tower of fields $K \subseteq K(\alpha_0, \alpha_1, \ldots, \alpha_n) \subseteq K(\alpha_0, \alpha_1, \ldots, \alpha_n, \alpha)$ and use **T. 4.2**, **T. 4.3** and **T. 4.4** (or **T. 4.6** and **T. 4.5**).
(b) Use **T. 4.6** and (a).

4.19 Use **T. 4.6** and **T. 4.5** or Exercise 4.18.

Problems of Chap. 5

5.1 (a) 4; $1, \sqrt{2}, \sqrt{5}, \sqrt{10}$;
(b) 6; $1, \sqrt[3]{2}, \sqrt[3]{4}, \varepsilon, \varepsilon\sqrt[3]{2}, \varepsilon\sqrt[3]{4}, \varepsilon^3 = 1, \varepsilon \neq 1$;
(c) 8; $1, i, \sqrt{2}, i\sqrt{2}, \sqrt[4]{2}, i\sqrt[4]{2}, \sqrt[4]{8}, i\sqrt[4]{8}$;
(d) 8; $1, \alpha, \alpha^2, \alpha^3, i, i\alpha, i\alpha^2, i\alpha^3$, where $\alpha = \sqrt{\frac{1}{2}(-1+\sqrt{5})}$;
(e) 4; $1, \alpha, \alpha^2, \alpha^3$, where $\alpha = e^{\frac{\pi i}{4}}$;
(f) 4; $1, \alpha, \alpha^2, \alpha^3$, where $\alpha = \sqrt[4]{2}$;
(g) 4; $1, \sqrt{2}, \sqrt{3}, \sqrt{6}$;
(h) $p-1$; $1, \alpha, \ldots, \alpha^{p-2}$, where $\alpha = e^{\frac{2\pi i}{p}}$.

5.2 (a) yes; (b) yes; (c) no.

5.3 Let $\alpha_1, \alpha_2, \ldots, \alpha_n$ be zeros of $f(X)$ in $L = K(\alpha_1, \alpha_2, \ldots, \alpha_n)$. Consider the chain of fields

$$K \subseteq K(\alpha_1) \subseteq K(\alpha_1, \alpha_2) \subseteq \cdots \subseteq K(\alpha_1, \alpha_2, \ldots, \alpha_n).$$

5.4 (a) The elements of K are $a + b\alpha$, $a, b \in \mathbb{F}_2$ and $\alpha^2 = \alpha + 1$, which defines the value of the product of any two elements of K (write down a multiplication table using this information); (b) The elements of K are $a + b\alpha + c\alpha^2$, $a, b, c \in \mathbb{F}_2$ and $\alpha^3 = \alpha + 1$, which defines the value of the product of any two elements of K; (c) The elements of K are $a + b\alpha$, $a, b \in \mathbb{F}_3$ and $\alpha^3 = -1$, which defines the value of the product of any two elements of K.

5.5 Use **T. 5.4** and **T. 4.3**.

5.6 (a) Compute the number of elements in the field $\mathbb{F}_p[X]/(f(X))$ and use **T. 5.4** (see A.5.7 and A.5.8).
(b) Consider the field $\mathbb{F}_p[X]/(f(X))$ and use **T. 5.4** and Exercise 5.5.

5.7 (a) Prove the first formula using Exercise 5.6 (this is easy). The second formula is a special case of the following general theorem: If $f : \mathbb{N} \to \mathbb{C}$ and $g : \mathbb{N} \to \mathbb{C}$ are two arbitrary functions and

$$f(n) = \sum_{d|n} g(d),$$

then $g(n) = \sum_{d|n} f(d)\mu\left(\frac{n}{d}\right)$. The last implication is called the Möbius inversion formula and can be easily proved using the equality $\sum_{d|n} \mu(d) = 0$ when $n \neq 1$. Use the inversion formula when $f(n) = p^n$ and $g(n) = nv_p(n)$. For a detailed proof of the Möbius inversion formula see A.11.2.
(b) $v_2(1) = v_2(2) = 1$, $v_2(3) = 2$, $v_2(4) = 3$, $v_2(5) = 6$, $v_2(6) = 9$, $v_2(7) = 18$, $v_2(8) = 30$.
(c) The answer: $(p^q - p)/q$. That this number is an integer also follows from Fermat's Little Theorem (see Exercise 6.3(c)).

5.8 (a) There are many proofs of this result. For one of them see A.5.2.
(b) According to (a), the group L^* is cyclic. Take one of its generators as γ. See also Exercise 8.15.

5.9 (a) Show that the zeros of the primitive polynomials are exactly the generators of the group of nonzero elements of the field. In such a field with p^n elements, compute the number of elements, which generate the group of its nonzero elements. Then find the number of irreducible polynomials which have these elements as their zeros. Use Exercise 5.6.
(b) Using for example **T. 5.4**, show that $\mathbb{F}$ has an extension of degree n. Use Exercise 5.8.

5.10 Use the evident formula for $(f+g)'$ and start the proof with $f = aX^m$, $g = bX^n$.

5.11 (a) Use Exercise 5.8(a).
(b) In one direction, use **T. 5.3**.

5.12 (b) Consider $X^4 + 4$, for example, over $\mathbb{Q}$.

5.13 Show that the polynomial $X^4 + 1$ has a zero in the field $\mathbb{F}_{p^2}$ by studying its multiplicative group of order $p^2 - 1$, which is divisible by 8 if only $p > 2$.

Problems of Chap. 6

6.1 (a) Let $f(X) = a_nX^n + \cdots + a_1X + a_0$ and $f(\alpha) = 0$. Look at $\sigma(f(\alpha))$.
(b) How to express the elements of the field $K(\alpha_1, \ldots, \alpha_n)$ by means of the elements $\alpha_1, \ldots, \alpha_n$ (see Chap. 2)?

6.2 (a) $G = \{\sigma_0, \sigma_1\}$, $\sigma_0 = id.$, $\sigma_1(\sqrt{2}) = -\sqrt{2}$;
(b) $G = \{\sigma_0\}$, $\sigma_0 = id.$;
(c) $G = \{\sigma_0, \sigma_1\}$, $\sigma_0 = id.$, $\sigma_1(\sqrt[4]{2}) = -\sqrt[4]{2}$;
(d)

G	$\sqrt{2}$	$\sqrt{3}$
σ_0	$\sqrt{2}$	$\sqrt{3}$
σ_1	$-\sqrt{2}$	$\sqrt{3}$
σ_2	$\sqrt{2}$	$-\sqrt{3}$
σ_3	$-\sqrt{2}$	$-\sqrt{3}$

(e) $G = \{\sigma_0\}$, $\sigma_0 = id.$;
(f) $G = \{\sigma_0, \sigma_1, \sigma_2, \sigma_3\}$. $\sigma_i(X) = iX$.

6.3 (a) Let σ be an automorphism of a field. Look at $\sigma(1), \sigma(2) = \sigma(1+1)$ and so on.

6.4 Show that $\sigma(x) > 0$ if $x > 0$ (use $x = (\sqrt{x})^2$). Assume that, e.g., $\sigma(x) > x$ and choose $r \in \mathbb{Q}$ such that $\sigma(x) > r > x$. Then use Exercise 6.3(a).

6.5 Consider $L^{G(L/K)}$. Use **T. 6.2** (and **T. 4.3**).

6.6 In (a) and (b) use Exercise 4.11(a).

(c) Check that the mapping sending a matrix $\begin{pmatrix} a & b \\ c & d \end{pmatrix} \in GL_2(K)$ onto a function $\varphi(X) = \frac{aX+b}{cX+d}$ is a group homomorphism. Find its kernel. If necessary, see A.2.11.

6.7 $|G| = 6$, $L^G = \mathbb{Z}_2(\alpha)$, where $\alpha = \frac{(X^3+X+1)(X^3+X^2+1)}{(X^2+X)^2}$.

6.8 $|G| = 6$, $L^G = \mathbb{Z}_3(X^6 + X^4 + X^2)$.

6.9 (a) $L^G = \mathbb{R}(X^2, Y)$.

6.10 (a) $L^G = \mathbb{R}(X^2, XY)$.

6.11 $L^G = \mathbb{Q}(X^2, Y^2)$.

6.12 $L^G = \mathbb{Q}$. Use **T. 6.2** and Exercise 4.11.

6.13 (a) $L^G = K(X + Y, XY)$.
(b) Use **T. 6.2**.

6.14 Check that $\sigma N(x) = \sigma(x)$ ($x \in L^H$, $\sigma \in G$) is well defined, that is, $\sigma\tau(x) = \sigma(x)$ for any $\tau \in N$ when $x \in L^H$ (evident). Consider the chain $L \supseteq L^H \supseteq (L^H)^{G/H} \supseteq L^G$. Use the Tower Law (**T. 4.3**) and $|G| = |H||G/H|$ (see A.2.11).

6.16 No. Construct a counterexample using $L = K(X)$.

6.17 (a) 2; $\sigma_0(a) = a$, $\sigma_1(a) = 2a^2 - a^5$;
(b) 4; $\sigma_0(a) = a$, $\sigma_1(a) = -a$, $\sigma_2(a) = \frac{1}{2}a^5$, $\sigma_3(a) = -\frac{1}{2}a^5$;
(c) 4; $\sigma_0(a) = a$, $\sigma_1(a) = -a$, $\sigma_2(a) = 10a^5 - a^{11}$, $\sigma_3(a) = a^{11} - 10a^5$.

Problems of Chap. 7

7.1 (a), (b), (e) and (h) are not normal, all the remaining are.

7.2 (a) $\mathbb{Q}(\sqrt[4]{2}, i)$;
(b) $\mathbb{Q}(\sqrt{2}, \sqrt[3]{2}, \varepsilon)$, $\varepsilon^3 = 1$, $\varepsilon \neq 1$;
(c) L;
(d) $\mathbb{Q}(X, \varepsilon)$, $\varepsilon^3 = 1$, $\varepsilon \neq 1$;
(e) $\mathbb{Q}(X, i)$, $i^2 = -1$;
(f) $L(\alpha), \alpha^2 + 1 = 0$.

7.3 (a) and (c)—not necessarily! Give examples! (b) Yes.

7.4 Use **T. 5.1**(b) and **T. 5.2**.

7.5 Use **T. 7.1**, Exercise 6.1 and in (b), Exercise 7.4.

7.6 In (a) and (b) use **T. 7.1**.

7.7 Show that the minimal polynomial of β over K is irreducible over $K(\alpha)$.

7.8 Let α_i and α_j be zeros of two irreducible factors of $f(X)$ over L. Consider the field extensions $K(\alpha_i) \subseteq L(\alpha_i)$ and $K(\alpha_j) \subseteq L(\alpha_j)$. Use **T. 5.1** and **T. 5.2**.

7.9 (a) Use **T. 4.2**.
(b) Consider one concrete example of a splitting field of $X^n - 2$ with $n > 2$ (e.g. $n = 4$)—the general argument will be similar.

7.10 (a) 12,2; Two quadratic factors of $f(X)$ over K have the same splitting field: Find the discriminats of these polynomials and show that their product is a square (in fact, 9). The command `galois(f(X))` also gives a solution.
(b) 6,1; (the polynomial is normal).
(c) 36,3; The polynomial $f(X)$ has a quadratic and a cubic irreducible factor over K. The splitting field of $f(X)$ over K has degree 6 as the quadratic factor remains irreducible over the cubic extension of K (see Exercise 4.2). You can use the command `galois(f(X))`.
(d) 48,3; The polynomial $f(X)$ has an irreducible factor of degree 4 over K but it is not normal over K. Use the command `galois(f(X))`, which gives the degree of the normal closure. You can see that it is not sufficient to adjoin one zero of the quadric polynomial to K in order to get a normal closure.
(e) 5040,6; Use `galois(f(X))` and Exercise 5.3 (see also Exercise 16.82)).
(f) 14,2; There are three irreducible quadratic factors of $f(X)$ over K and all have the same splitting field over K. Either use `galois(f(X))` or check that for each two of these factors, the product of their discriminants is a square (so they have the same splitting field).

7.11 Only (b) is not normal.

Problems of Chap. 8

8.1 (a) Find the minimal polynomial of X over the field $\mathbb{F}_p(X^2)$. Use the definition of separability or **T. 8.1**.
(b) For example, $L \supset K$, where $L = \mathbb{F}_p(X)$, $K = \mathbb{F}_p(X^p)$.
(c) $\mathbb{F}_2(X)$ is a splitting field of the polynomial $T^2 - X^2$ over the field $\mathbb{F}_2(X^2)$.

8.3 (a) Consider $K \subseteq K(\alpha^p) \subseteq K(\alpha)$. Use **T. 8.1**(b).
(c) If K is perfect show that every irreducible factor of $X^p - a$, where $a \in K$, must be of degree 1. Use **T. 8.1**(b). Conversely, when $K = K^p$ use the same argument as in the proof of **T. 8.1**(a) in the case of a finite field K (see p. 121).
(d) Show that L is separable over K using (b) and (c).

8.4 (a) Use Exercise 8.3.
(b) Use (a) and **T. 7.2**.

8.5 Use Exercise 8.4.

8.6 Use **T. 8.1**(b).

8.7 As a first step, assume $L = K(\gamma)$.

8.8 Use Exercise 8.4(a).

8.9 Use Exercise 8.7.

8.10 (a) For example $\sqrt{2} + \sqrt{3}$; (b) For example $\sqrt{2} + \sqrt{3} + i$; (c) For example $\sqrt{2} + \sqrt{3} + i$; (d) For example $X + Y$; (e) For example X; (f) For example $\sqrt{2} + \sqrt[3]{2}$ or $\sqrt[6]{2}$.

8.11 Show that if γ exists, then $\gamma^2 \in K$. Find $[L : K]$.

8.12 (a) Compare the degrees of L over M and L over the subfield of M generated over K by the coefficients of the minimal polynomial of γ over M.
(b) If $L = K(\gamma)$, then use the fact that the minimal polynomial of γ over M divides the minimal polynomial of γ over K. In the opposite direction, show that L is finitely generated and algebraic over K. Use Exercise 4.11. Thereafter, consider separately K finite and K infinite. If K is finite use Exercise 5.8. If K is infinite, use the fact that L as a vector space over K is not a union of a finite number of proper subspaces (we prove this auxiliary fact on p. 196). Choose any γ which is not in the union of all proper subfields of L containing K.

8.13 Consider $K \subset L$ from Exercise 8.11. (e.g. $\mathbb{F}_p(X^2, Y^2, X + cY)$, where $c \in K$).

8.14 Use **T. 8.2**.

8.15 (b) No. (c) 4, 6 for $n = 2$; 13, 24 for $n = 3$; 32, 54 for $n = 4$.

8.16 Use Exercises 2.13 and 8.4.

8.17 For example: (a) $c = \sqrt[6]{2}$; (b) $c = \sqrt[15]{3^5 5^3}$; (c) $c = \sqrt[3]{3} + \sqrt[3]{5}$; (d) $c = \sqrt[15]{2}$.

Problems of Chap. 9

9.1 (c), (d), (e), (f) if $p \neq 2$ and (g) are Galois extensions, the remaining are not Galois.

9.2 (a) $L = \mathbb{Q}(\sqrt{2}, \sqrt{5})$; $[L : \mathbb{Q}] = |G(L : \mathbb{Q})| = 4$, $G = G(L : \mathbb{Q}) = \{\sigma_0, \sigma_1, \sigma_2, \sigma_3\}$, where

G	$\sqrt{2}$	$\sqrt{5}$	o
σ_0	$\sqrt{2}$	$\sqrt{5}$	1
σ_1	$-\sqrt{2}$	$\sqrt{5}$	2
σ_2	$\sqrt{2}$	$-\sqrt{5}$	2
σ_3	$-\sqrt{2}$	$-\sqrt{5}$	2

The subgroups of G: $I = \{\sigma_0\}$, $H_1 = \{\sigma_0, \sigma_1\}$, $H_2 = \{\sigma_0, \sigma_2\}$, $H_3 = \{\sigma_0, \sigma_3\}$.

The fields between $\mathbb{Q}$ and L: $L^G = \mathbb{Q}$, $L^I = L$,
$L^{H_1} = \mathbb{Q}(\sqrt{5})$, $L^{H_2} = \mathbb{Q}(\sqrt{2})$, $L^{H_3} = \mathbb{Q}(\sqrt{10})$.

(b) $L = \mathbb{Q}(i, \sqrt{5})$. Compare (a).

(c) $L = \mathbb{Q}(\varepsilon)$, $\varepsilon = e^{\frac{2\pi i}{5}}$, $[L : \mathbb{Q}] = |G(L : \mathbb{Q})| = 4$, $G = \{\sigma_0, \sigma_1, \sigma_2, \sigma_3\}$, where

G	ε	o
σ_0	ε	1
σ_1	ε^2	4
σ_2	ε^3	4
σ_3	ε^4	2

The subgroups of G: $G, I = \{\sigma_0\}$, $H = \{\sigma_0, \sigma_3\}$.
The fields between $\mathbb{Q}$ and L: $L^G = \mathbb{Q}$, $L^I = L$,
$L^H = \mathbb{Q}(\varepsilon + \varepsilon^4) = \mathbb{Q}(\cos\frac{2\pi}{5}) = \mathbb{Q}(\sqrt{5})$.

(d) $L = \mathbb{Q}(i, \sqrt{2})$. Compare (a).

(e) $L = K(\sqrt[4]{2})$; $[L : K] = |G(L/K)| = 4$, $G = \{\sigma_0, \sigma_1, \sigma_2, \sigma_3\}$, where

G	$\sqrt[4]{2}$	o
σ_0	$\sqrt[4]{2}$	1
σ_1	$-\sqrt[4]{2}$	2
σ_2	$i\sqrt[4]{2}$	4
σ_3	$-i\sqrt[4]{2}$	4

The subgroups of G: $G, I = \{\sigma_0\}$, $H = \{\sigma_0, \sigma_1\}$.
The fields between K and L: $L^G = K : L^I = L$,
$L^H = K(\sqrt{2}) = \mathbb{Q}(i, \sqrt{2})$.

(f) $L = \mathbb{Q}(\sqrt[3]{5}, \varepsilon), \varepsilon^3 = 1, \varepsilon \neq 1; [L : \mathbb{Q}] = |G(L : \mathbb{Q})| = 6$, $G = \{\sigma_0, \sigma_1, \sigma_2, \sigma_3, \sigma_4, \sigma_5\}$, where

G	$\sqrt[3]{5}$	ε	o
σ_0	$\sqrt[3]{5}$	ε	1
σ_1	$\sqrt[3]{5}$	ε^2	2
σ_2	$\varepsilon\sqrt[3]{5}$	ε	3
σ_3	$\varepsilon\sqrt[3]{5}$	ε^2	2
σ_4	$\varepsilon^2\sqrt[3]{5}$	ε	3
σ_5	$\varepsilon^2\sqrt[3]{5}$	ε^2	2

The subgroups of G: $G, I = \{\sigma_0\}$, $H_1 = \{\sigma_0, \sigma_1\}$,
$H_2 = \{\sigma_0, \sigma_3\}$, $H_3 = \{\sigma_0, \sigma_5\}$, $H = \{\sigma_0, \sigma_2, \sigma_4\}$.
The fields between $\mathbb{Q}$ and L: $L^G = \mathbb{Q}$, $L^I = L$,
$L^{H_1} = \mathbb{Q}(\sqrt[3]{5})$, $L^{H_2} = \mathbb{Q}(\varepsilon^2\sqrt[3]{5})$, $L^{H_3} = \mathbb{Q}(\varepsilon\sqrt[3]{5})$, $L^H = \mathbb{Q}(\varepsilon)$.

(g) $L = \mathbb{Q}(\alpha, i)$, $\alpha = \sqrt{\frac{1}{2}(\sqrt{5} - 1)}$; $[L : \mathbb{Q}] = |G(L : \mathbb{Q})| = 8$. Let $\beta = \sqrt{\frac{1}{2}(\sqrt{5} + 1)}$ $(\alpha\beta = 1)$. $G = \{\sigma_0, \sigma_1, \sigma_2, \sigma_3, \sigma_4, \sigma_5, \sigma_6, \sigma_7\}$, where

G	α	i	o
σ_0	α	i	1
σ_1	$-\alpha$	i	2
σ_2	$i\beta$	i	2
σ_3	$-i\beta$	i	2
σ_4	α	$-i$	2
σ_5	$-\alpha$	$-i$	2
σ_6	$i\beta$	$-i$	4
σ_7	$-i\beta$	$-i$	4

The subgroups of G: $G, I = \{\sigma_0\}$, $H_1 = \{\sigma_0, \sigma_1\}$,
$H_2 = \{\sigma_0, \sigma_2\}$, $H_3 = \{\sigma_0, \sigma_3\}$, $H_4 = \{\sigma_0, \sigma_4\}$, $H_5 = \{\sigma_0, \sigma_5\}$,
$H_{23} = \{\sigma_0, \sigma_1, \sigma_2, \sigma_3\}$, $H_{45} = \{\sigma_0, \sigma_1, \sigma_4, \sigma_5\}$,
$H_{67} = \{\sigma_0, \sigma_1, \sigma_6, \sigma_7\}$.
The fields between $\mathbb{Q}$ and L: $L^G = \mathbb{Q}$, $L^I = L$,
$L^{H_1} = \mathbb{Q}(\sqrt{5}, i)$, $L^{H_2} = \mathbb{Q}(\alpha + i\beta)$, $L^{H_3} = \mathbb{Q}(\alpha - i\beta)$,
$L^{H_4} = \mathbb{Q}(\alpha)$, $L^{H_5} = \mathbb{Q}(i\alpha)$, $L^{H_{23}} = \mathbb{Q}(i)$, $L^{H_{45}} = \mathbb{Q}(\sqrt{5})$,
$L^{H_{67}} = \mathbb{Q}(i\sqrt{5})$.

(h) $L = K(\varepsilon)$, $\varepsilon^3 = 1$, $\varepsilon \neq 1$; $[L : K] = |G(L/K)| = 2$, $G = \{\sigma_0, \sigma_1\}$, where $\sigma_0 = id.$, $\sigma_1(\varepsilon) = \varepsilon^2$. Only the trivial subgroups and subfields.

9.3 (b) We take in Exercise 9.2(a)–(h) only the zeros of $f(X)$ which are not in the ground field K.

(a) The zeros: $\sqrt{2}, -\sqrt{2}, \sqrt{5}, -\sqrt{5}$ numbered $1, 2, 3, 4$, respectively. Then $\sigma_0, \sigma_1, \sigma_2, \sigma_3$ are $(1), (1, 2), (3, 4), (1, 2)(3, 4)$, respectively.
(b) The zeros: $i, -i, \sqrt{5}, -\sqrt{5}$ numbered $1, 2, 3, 4$, respectively. As in (a).
(c) The zeros: $\varepsilon, \varepsilon^2, \varepsilon^3, \varepsilon^4$ ($\varepsilon = e^{\frac{2\pi i}{5}}$) numbered $1, 2, 3, 4$, respectively. Then $\sigma_0, \sigma_1, \sigma_2, \sigma_3$ are $(1), (1, 2, 4, 3), (1, 3, 4, 2), (1, 4)(2, 3)$, respectively.
(d) The zeros: $\alpha = 1/2\sqrt{2} + 1/2\,i\sqrt{2}, -\alpha, \bar{\alpha}, -\bar{\alpha}$ numbered $1, 2, 3, 4$, respectively. The four automorphisms defined by $\sigma(\sqrt{2}) = \pm\sqrt{2}, \sigma(i) = \pm i$ are $\{(1), (1, 2)(3, 4), (1, 3)(2, 4), (1, 4)(2, 3)\}$.
(e) The zeros: $\sqrt[4]{2}, -\sqrt[4]{2}, i\sqrt[4]{2}, -i\sqrt[4]{2}$ numbered $1, 2, 3, 4$, respectively. Then $\sigma_0, \sigma_1, \sigma_2, \sigma_3$ are $(1), (1, 2)(3, 4), (1, 3, 2, 4), (1, 4, 2, 3)$, respectively.
(f) The zeros: $\sqrt[3]{5}, \varepsilon\sqrt[3]{5}, \varepsilon^2\sqrt[3]{5}$, numbered $1, 2, 3$, respectively. Then $\sigma_0, \sigma_1, \sigma_2, \sigma_3, \sigma_4, \sigma_5$ are $(1), (2, 3), (1, 2, 3), (1, 2), (1, 3, 2), (1, 3)$, respectively.
(g) In the notations of Exercise 9.2(g), the zeros are $\alpha, -\alpha, i\beta, -i\beta$ numbered $1, 2, 3, 4$ in this order. Then the automorphisms $\sigma_0, \sigma_1, \sigma_2, \sigma_3, \sigma_4, \sigma_5, \sigma_6, \sigma_7$ are $(1), (1, 2)(3, 4), (1, 3)(2, 4), (1, 4)(2, 3), (3, 4), (1, 2), (1, 3, 2, 4), (1, 4, 2, 3)$, respectively.
(h) The zeros are $\varepsilon, \varepsilon^2$, where $\varepsilon = e^{\frac{2\pi i}{3}}$, numbered $1, 2$ in the given order. The automorphisms σ_0, σ_1 are $(1), (1, 2)$.

9.4 (a) Use Exercise 6.1.
(b) Use **T. 5.1**(b), **T. 5.2**(b) and Exercise 6.1.
(c) Use **T. 9.1**(b).

9.5 (a) Modify Exercise 9.2(a).
(b) Modify Exercise 9.2(f).
(c) $[L : \mathbb{Q}] = |G(L : \mathbb{Q})| = 8$, since L is a splitting field of $X^4 - 2$ over $\mathbb{Q}$. $G = G(L : \mathbb{Q}) = \{\sigma_0, \sigma_1, \sigma_2, \sigma_3, \sigma_4, \sigma_5, \sigma_6, \sigma_7\}$, where

G	$\sqrt[4]{2}$	i	o
σ_0	$\sqrt[4]{2}$	i	1
σ_1	$-\sqrt[4]{2}$	i	2
σ_2	$i\sqrt[4]{2}$	i	4
σ_3	$-i\sqrt[4]{2}$	i	4
σ_4	$\sqrt[4]{2}$	$-i$	2
σ_5	$-\sqrt[4]{2}$	$-i$	2
σ_6	$i\sqrt[4]{2}$	$-i$	2
σ_7	$-i\sqrt[4]{2}$	$-i$	2

The subgroups of G**:** $G, I = \{\sigma_0\}, H_1 = \{\sigma_0, \sigma_1\}$, $H_2 = \{\sigma_0, \sigma_4\}, H_3 = \{\sigma_0, \sigma_5\}, H_4 = \{\sigma_0, \sigma_6\}$, $H_5 = \{\sigma_0, \sigma_7\}, G_1 = \{\sigma_0, \sigma_1, \sigma_2, \sigma_3\}$, $G_2 = \{\sigma_0, \sigma_1, \sigma_4, \sigma_5\}, G_3 = \{\sigma_0, \sigma_1, \sigma_6, \sigma_7\}$.
The fields between $\mathbb{Q}$ **and** L**:** $L^G = \mathbb{Q}, L^I = L$;
$L^{H_1} = \mathbb{Q}(\sqrt{2}, i), L^{H_2} = \mathbb{Q}(\sqrt[4]{2}), L^{H_3} = \mathbb{Q}(i\sqrt[4]{2})$,
$L^{H_4} = \mathbb{Q}((1 + i)\sqrt[4]{2}), L^{H_5} = \mathbb{Q}((1 - i)\sqrt[4]{2})$,
$L^{G_1} = \mathbb{Q}(i), L^{G_2} = \mathbb{Q}(\sqrt{2}), L^{G_3} = \mathbb{Q}(i\sqrt{2})$.

(d) $[L : K)| = 4$, since K is the fixed field of the group $G = \{\sigma_0, \sigma_1, \sigma_2, \sigma_3\}$, which is the group of K-automorphisms of L, where:

G	X	o
σ_0	X	1
σ_1	$-X$	2
σ_2	$1/X$	2
σ_3	$-1/X$	2

The subgroups of G**:** $G, I = \{\sigma_1\}, H_1 = \{\sigma_0, \sigma_1\}$
$H_2 = \{\sigma_0, \sigma_2\}$, $H_3 = \{\sigma_0, \sigma_3\}$.
The fields between K **and** L**:** $L^G = K, L^I = L$,
$L^{H_1} = \mathbb{R}(X^2)$,
$L^{H_2} = \mathbb{R}(X + \frac{1}{X}), L^{H_3} = \mathbb{R}(X - \frac{1}{X})$.

(e) $[L:K] = |G(L/K)| = 4$, since K is the fixed field of the group $G = G(L/K) = \{\sigma_0, \sigma_1, \sigma_2, \sigma_3\}$ of K-automorphisms of L, where

G	X	Y	o
σ_0	X	Y	1
σ_1	$-X$	Y	2
σ_2	X	$-Y$	2
σ_3	$-X$	$-Y$	2

The subgroups of G**:** $G, I = \{\sigma_0\}$, $H_1 = \{\sigma_0, \sigma_1\}$, $H_2 = \{\sigma_0, \sigma_2\}$, $H_3 = \{\sigma_0, \sigma_3\}$.
The fields between K **and** L**:** $L^G = K, L^I = L$, $L^{H_1} = \mathbb{R}(X^2, Y)$, $L^{H_2} = \mathbb{R}(X, Y^2)$, $L^{H_3} = \mathbb{R}(X^2, XY)$.

(f) $[L:K] = |G(L/K)| = 4$, since K is the fixed field of the group $G = G(L/K) = \{\sigma_0, \sigma_1, \sigma_2, \sigma_3\}$ of K-automorphisms of L, where

G	X	Y	o
σ_0	X	Y	1
σ_1	$-X$	$-Y$	2
σ_2	Y	X	2
σ_3	$-Y$	$-X$	2

The subgroups of G**:** $G, I = \{\sigma_0\}$, $H_1 = \{\sigma_0, \sigma_1\}$, $H_2 = \{\sigma_0, \sigma_2\}$, $H_3 = \{\sigma_0, \sigma_3\}$.
The fields between K **and** L**:** $L^G = K$, $L^I = L$, $L^{H_1} = \mathbb{R}(X^2, XY)$, $L^{H_2} = \mathbb{R}(X+Y, XY)$, $L^{H_3} = \mathbb{R}(X-Y, XY)$.

9.6 Take two different subgroups $H_1, H_2 \subset V$ both of order 2. Use Exercise 4.5 in order to choose d_1, d_2 for the different fields K^{H_1}, K^{H_2}.

9.7 No.

9.8 $L = \mathbb{Z}_3(X), L^G = \mathbb{Z}_3(X^6 + X^4 + X^2)$, $[L : L^G] = 6$.

G	X	o
σ_0	X	1
σ_1	$X+1$	3
σ_2	$X+2$	3
σ_3	$2X$	2
σ_4	$2X+1$	2
σ_5	$2X+2$	2

The subgroups of G**:** $G, I = \{\sigma_0\}, H_1 = \{\sigma_0, \sigma_3\}$, $H_2 = \{\sigma_0, \sigma_4\}, H_3 = \{\sigma_0, \sigma_5\}, H = \{\sigma_0, \sigma_1, \sigma_2\}$.
The fields between L^G **and** L**:** $L^G, L^I = L, L^{H_1} = \mathbb{Z}_3(X^2)$, $L^{H_2} = \mathbb{Z}_3(X^2 - X), L^{H_3} = \mathbb{Z}_3(X^2 + X), L^H = \mathbb{Z}_3(X^3 - X)$.

9.9 $L \supseteq M$ and $L \supseteq K$ are Galois extensions. Use **T. 9.1**(a).

9.10 (b) Consider the index of $H(L/M)$ in $G(L/K)$ and use the homomorphism $\varphi : H(L/M) \to G(M/K)$, where $\varphi(\sigma) = \sigma|_M$. Use **T. 9.1**.
(c) Use (b) and **T. 9.1**(a).
(d) Check directly that $\mathcal{N}(G(L/M)) \subseteq H(L/M)$. Use (b) and the characterization of the normaliser as the biggest subgroup of $G(L/K)$ in which $G(L/M)$ is normal.

9.11 (a) Use the definition of the normal extension together with **T. 9.1**(a) and **T. 4.3**.
(b) How to describe all the zeros of $f_\alpha(X)$ in terms of α and $\sigma \in G(L/K)$? Use (a).
(c) Count how many times a given zero $\beta = \sigma(\alpha)$ of $f_\alpha(X)$ (for a fixed $\sigma \in G(L/K)$) appears among the images $\sigma_1(\alpha), \ldots, \sigma_n(\alpha)$.

9.12 (a) Consider L as a vector space over L^H and use **T. 6.3** in order to show that the image of Tr_H has dimension 1.
(b) Use (a) and the properties of Tr_H as a linear transformation.

9.13 (a) Use **T. 5.4, T. 8.2** and **T. 9.1**(d).
(b) Use Exercise 2.12.
(c) $\mathrm{Nr}_{L/K}(x) = x^{\frac{q^n-1}{q-1}}$, where $n = [L : K]$. The mapping $\mathrm{Nr}_{L/K} : L^* \to K^*$ is a group homomorphism. How many elements belong to its kernel?

9.14 Use Exercise 9.13(b) and **T. 9.2**. For example, $\beta = \alpha^3+\alpha$ (or $\beta = \alpha^3+\alpha+1$).

9.15 (a) Use Exercises 7.6, 8.16 and **T. 9.1**(d).
(b) Consider a natural homomorphism mapping an automorphism $\sigma \in G(M_1M_2/M_2)$ onto its restriction to M_1. What is the image and the kernel of this homomorphism?
(c) Consider the homomorphism $\varphi : G(M_1M_2/K) \to G(M_1/K) \times G(M_2/K)$, where $\varphi(\sigma) = (\sigma|_{M_1}, \sigma|_{M_2})$. Show that $\mathrm{Ker}(\varphi) = \{e\}$. Then compute the orders of $G(M_1M_2/K)$ and $G(M_1/K) \times G(M_2/K)$ using **T. 9.1**(a) and Exercise 7.7.
(d) Use Exercise 7.4(b).

9.16 In each case below, we give an example of a field (chosen among plenty of other possibilities):
(a) $\mathbb{Q}(i)$; (b) $\mathbb{Q}(\alpha)$, $\alpha = \varepsilon + \frac{1}{\varepsilon}$, $\varepsilon = e^{\frac{2\pi i}{7}}$; (c) Exercise 9.2(c);
(d) $\mathbb{Q}(\alpha)$, $\alpha = \varepsilon + \frac{1}{\varepsilon}$, $\varepsilon = e^{\frac{2\pi i}{11}}$; (e) Exercise 9.2(a); (f) $\mathbb{Q}(i, \varepsilon)$, $\varepsilon = e^{\frac{2\pi i}{5}}$;
(g) $\mathbb{Q}(\sqrt{2}, \sqrt{3}, \sqrt{5})$; (h) $\mathbb{Q}(\alpha, \beta)$, $\alpha^3 = 3\alpha + 1$, $\beta^3 = 7\beta + 7$;
(i) Exercise 9.5(c); (j) Exercise 9.2(f);
(k) $L = K_f, f(X) = X^4 - 2X - 2$; (l) $L = K_f, f(X) = X^3 + 20X + 16$.

9.17 Use **T. 9.1**(c).

9.18 (a) In order to show that $[\mathbb{Q}(\varepsilon) : \mathbb{Q}] = p - 1$ use Exercise 3.7(d). Use Exercise 6.1 (and A.2.2) in order to show that σ_k are all automorphisms. Notice that by Exercise 5.8, the group $(\mathbb{Z}/p\mathbb{Z})^*$ is the multiplicative group of the field $\mathbb{F}_p = \mathbb{Z}/p\mathbb{Z}$ and as such is cyclic of order $p - 1$.
(b) Use a special case of Dirichlet's Theorem: For every natural number n, there exists a natural number k such that $nk + 1$ is a prime number (for a proof of this case, see Exercise 10.12).

9.19 No. A counterexample is in the next chapter, but try to find your own.

9.20 $G(L/M_1M_2) = H_1 \cap H_2$ (show that each $\sigma \in G(L/M_1M_2)$ must be in both $H_1 = G(L/M_1)$ and $H_2 = G(L/M_2)$, and conversely); $G(L/M_1 \cap M_2) =$ the smallest subgroup of $G(L/K)$ containing both H_1 and H_2 (that is, the group of all finite products in which every factor belongs either to H_1 or to H_2).
This answer can be also expressed in the following way: $M_1 = L^{H_1}, M_2 = L^{H_2}$ gives $L^{H_1 \cap H_2} = L^{H_1}L^{H_2}$ and $L^{\langle H_1, H_2\rangle} = L^{H_1} \cap L^{H_2}$, where $\langle H_1, H_2\rangle$ denotes the smallest subgroup of $G = G(L/K)$ containing both H_1 and H_2.

9.21 Show that $K(\alpha) \supseteq K$ is a Galois extension.

9.22 In (a) and (b) follow a similar argument as in the solution of Exercise 9.15(b). In (c) use (b).

9.23 What happens when you take the complex conjugate of a number in L?

9.24 Use **T. 9.1**(a) and Exercise 9.11(a).

9.25 (a) Every permutation of the zeros is a composition of transpositions (of only two zeros). What happens to the sign of $\alpha_i - \alpha_j$ when two zeros are shifted? An even permutation is a composition of an even number of transpositions, and an odd permutation needs an odd number of transpositions (see section "Permutations" in Appendix if you need a reminder concerning permutations).
(b) Show that the index of $\mathrm{Gal}_0(L/K)$ in $\mathrm{Gal}(L/K)$ is 1 or 2 and the second possibility occurs if and only if $\sqrt{\Delta(f)} \notin K$.

9.26 (a) Let α be a zero of $f(X)$. Check that all $\alpha + i$ for $i = 1, \ldots, p-1$ are also zeros of this polynomial.
(b) Let $\alpha_i = \alpha + i$ for $i = 0, 1, \ldots, i-1$. Let $f_i(X)$ be the minimal polynomial of α_i over K. Notice that $K(\alpha_i) = K(\alpha)$ for each $i = 0, 1, \ldots, p-1$.
(c) Apply Fermat's Little Theorem (Exercise 6.3(c)).

9.27 Exercise 7.11(a): If $a_1 = a$ is a zero of $f(X)$, then the remaining are $a_2 = 2-a^2$, $a_3 = a^3 - 3a$, $a_4 = a^4 - a^3 - 3a^2 + 2a$, $a_5 = -(a^4 - 4a^2 + 2)$. The Galois group is cyclic of order 5 and is generated by $\sigma(a) = 2 - a^2$.
Exercise 7.11(c): If a is a zero of $f(X)$, then the remaining zeros can be obtained from the factorization of $f(X)$ over the field $\mathbb{Q}(a)$ (using `factor(f(X),a)`). Since the Galois group has order 7, it is cyclic and generated by any σ mapping a onto another zero (different from a) of $f(X)$.
Exercise 7.11(d): If $a_1 = a$ is a zero of $f(X)$, then the remaining zeros are $a_2 = -a$, $a_{3,4} = \pm\frac{1}{6}(-252a+468a^3-215a^5+3a^7)$, $a_{5,6} = \pm\frac{1}{24}(-552a+1566a^3-1064a^5+15a^7)$, $a_{7,8} = \pm\frac{1}{24}(-1212a + 3294a^3 - 1922a^5 + 27a^7)$. The automorphisms are functions fi defined by mapping a onto one of the eight zeros (for example, `f3:= a -> a_3`, where a_3 is given as a function of a). Then we can check the orders of fi in the Galois group computing `simplify(fi(fi(a)))`. We get that $fi(fi(a)) = -a$ for all $i \neq 1, 2$, so all elements in the Galois group with the exception of the identity and the automorphism mapping a onto $-a$ have order 4 (the involution mapping a onto $-a$ has order 2). This shows that the Galois group is the quaternion group (see Exercise 12.1(e)).

9.28 (a) The zeros of the polynomial are $a, a^2 - 2, 2 - a - a^2$, so $K = \mathbb{Q}(a)$. The Galois group is cyclic of order 3 and $G(K/\mathbb{Q}) = \langle\sigma\rangle$, where $\sigma(a) = a^2 - 2$.
(b) The zeros of the polynomial are $a, -a^2, -a^4, a^5, a - a^4, a^2 - a^5$, so $K = \mathbb{Q}(a)$. The Galois group is cyclic of order 6 and $G(K/\mathbb{Q}) = \langle\sigma\rangle$, where $\sigma(a) = a^5$ (define `σ:=a->a^5` and check the order of σ in the group using `simplify(σ(σ(a)))` and `simplify(σ(σ(σ(a))))`. Observe that the order is not 2 or 3, so it must be 6).
(c) The zeros of the polynomial are $a, -\frac{1}{9}(9 - 12a^2 + a^5), \frac{1}{3}(6 + 4a^3 + 2a^4 + a^5)$, $\frac{1}{9}(27 + 9a - 12a^2 + 12a^3 + 6a^4 + 4a^5, -\frac{1}{9}(45 + 9a - 6a^2 + 18a^3 + 9a^4 + 5a^5)$, $-\frac{1}{9}(18+9a+6a^2+6a^3+3a^4+a^5)$. The Galois group of order 6 is S_3 and $G(K/\mathbb{Q}) = \langle\sigma, \tau\rangle$, where $\sigma(a) = \frac{1}{3}(6+4a^3+2a^4+a^5)$ is of order 2, $\tau(a) = -\frac{1}{9}(9-12a^2+a^5)$ is of order 3 and $\sigma\tau = \tau^2\sigma$ (check this by defining the corresponding functions σ, τ in Maple as in (b)).

9.29 Use the command `galois(f(X))`.
(a) The Galois group D_5 (dihedral of order 10—see A.2.4) and the minimal number of zeros generating the splitting field is 2.
(b) The Galois group A_5 (alternating of order 60) and the minimal number of zeros generating the splitting field is 3.
(c) The Galois group S_5 (symmetric of order 60) and the minimal number of zeros generating the splitting field is 4.
(d) The Galois group G_5 (also denoted F_{20} or $G(1,5)$ and called general affine group of order 20—see p. 277). The minimal number of zeros generating the splitting field is 2 (the polynomial $f(X)$ has only one zero in the field $\mathbb{Q}(a)$, where `a:=RootOf(X^5+15X+12)` and has a factor $X^4 + aX^3 + a^2X^2 + a^3X + 15 + a^4$ over this field. The field $\mathbb{Q}(a,b)$, where `b:=RootOf(X^4+aX^3+a^2X^2+a^3X+15+a^4,a)` is the splitting field of $f(X)$ over $\mathbb{Q}$).
You may use Maple in order to adjoin zeros and check how the given polynomials split in the subsequent extensions but it is much more elegant to deduce the minimal number of zeros generating the splitting fields using the knowledge of the order of the Galois groups.

Problems of Chap. 10

10.1 $\Phi_1(X) = X - 1$; $\Phi_2(X) = X + 1$; $\Phi_3(X) = X^2 + X + 1$; $\Phi_4(X) = X^2 + 1$; $\Phi_5(X) = X^4 + X^3 + X^2 + X + 1$; $\Phi_6(X) = X^2 - X + 1$; $\Phi_7(X) = X^6 + X^5 + X^4 + X^3 + X^2 + X + 1$; $\Phi_8(X) = X^4 + 1$; $\Phi_9(X) = X^6 + X^3 + 1$; $\Phi_{10}(X) = X^4 - X^3 + X^2 - X + 1$.

10.2 (a) $\Phi_p = (X^p - 1)/(X - 1)$.
(b) This is a special case of Exercise 10.3(a) for $k = r(n)$.
(c) This is a special case of Exercise 10.3(b) for $r = p, s = n$.
$\Phi_{20}(X) = 1 - X^2 + X^4 - X^6 + X^8$,
$\Phi_{105}(X) = 1 + X + X^2 - X^6 - X^5 - X^8 - 2\,X^{41} - X^{42} - X^{43} + X^{46} + X^{47} + X^{48} - 2\,X^7 - X^9 + X^{12} + X^{13} + X^{14} + X^{15} + X^{16} + X^{17} - X^{20} - X^{22} - X^{24} - X^{26} - X^{28} + X^{31} + X^{32} + X^{33} + X^{34} + X^{35} + X^{36} - X^{39} - X^{40}$.

10.3 (a) Let ε_r for $r = 1, \ldots, \varphi(k)$ be all primitive k-th roots of unity and let $d = n/k$. Show that all d-th roots of ε_r give all primitive n-th roots of unity (notice that $\varphi(n) = d\varphi(k)$, depending on the assumption on primes dividing n and k).
(b) Use the formula expressing the cyclotomic polynomials by binomials $X^d - 1$ in **T. 10.2**(a).

10.4 Notice that if $m > 1$ is odd and ε is a primitive m-th root of unity, then $-\varepsilon$ is a primitive $2m$-th root of unity.

10.5 (a) Use Exercise 9.15(b).

(b) Look at the degree of $\mathbb{Q}(\varepsilon_m)\mathbb{Q}(\varepsilon_n)$ over $\mathbb{Q}(\varepsilon_m)$ using (a), Exercise 9.15(b), **T. 9.1** and **T. 4.1**.
(c) $\mathbb{Q}(\varepsilon_m)\mathbb{Q}(\varepsilon_n) = \mathbb{Q}(\varepsilon_{[m,n]})$ and $\mathbb{Q}(\varepsilon_m) \cap \mathbb{Q}(\varepsilon_n) = \mathbb{Q}(\varepsilon_{(m,n)})$, where $[m,n]$ is the least common multiple and (m,n) the greatest common divisor of m and n.

10.6 Use Exercise 9.23. One can choose $\eta = \varepsilon_n + \overline{\varepsilon_n}$ (notice that $\overline{\varepsilon_n} = 1/\varepsilon_n$).

10.7 (a) Use **T. 9.2** taking into account that the Galois group is cyclic.
(c) Find the number of different images of θ_d under all automorphisms of $G(K/\mathbb{Q})$. Use Exercise 9.24.

10.8 (a) $\mathbb{Q}(i)$, $\mathbb{Q}(\sqrt{2})$, $\mathbb{Q}(\sqrt{-2})$; (b) $\mathbb{Q}(\sqrt{5})$; (c) $\mathbb{Q}(\sqrt{-7})$.

10.9 (a) $\sigma(\varepsilon) = \varepsilon^2$ has order 4, $G_2 = \langle\sigma^2\rangle$, $K^{G_2} = \mathbb{Q}(\theta_2) = \mathbb{Q}(\sqrt{5})$, $\theta_2 = \varepsilon_5 + \sigma^2(\varepsilon_5) = \varepsilon_5 + \varepsilon_5^4$ has minimal polynomial $X^2 + X - 2$;
(b) $\sigma(\varepsilon_7) = \varepsilon_7^3$ has order 6,
$G_2 = \langle\sigma^3\rangle$, $K^{G_2} = \mathbb{Q}(\theta_3)$,
$\theta_3 = \varepsilon_7 + \sigma^3(\varepsilon_7) = \varepsilon_7 + \varepsilon_7^6$ has the minimal polynomial $X^3 + X^2 - 2X - 1$;
$G_3 = \langle\sigma^2\rangle$, $K^{G_3} = \mathbb{Q}(\theta_2) = \mathbb{Q}(\sqrt{-7})$, $\theta_2 = \varepsilon_7 + \sigma^2(\varepsilon_7) + \sigma^4(\varepsilon_7) = \varepsilon_7 + \varepsilon_7^2 + \varepsilon_7^4$ has the minimal polynomial $X^2 + X + 2$.
(c) $\sigma(\varepsilon_{17}) = \varepsilon_{17}^3$ has order 16,
$G_2 = \langle\sigma^8\rangle$, $K^{G_2} = \mathbb{Q}(\theta_8)$, $\theta_8 = \varepsilon_{17} + \sigma^8(\varepsilon_{17}) = \varepsilon_{17} + \varepsilon_{17}^{16}$ has the minimal polynomial $X^8 + X^7 - 7X^6 - 6X^5 + 15X^4 + 10X^3 - 10X^2 - 4X + 1$.
$G_4 = \langle\sigma^4\rangle$, $K^{G_6} = \mathbb{Q}(\theta_4)$, $\theta_4 = \sum_0^3 \sigma^{4i}(\varepsilon_{17}) = \varepsilon_{17} + \varepsilon_{17}^{3^4} + \varepsilon_{17}^{3^8} + \varepsilon_{17}^{3^{12}}$ has the minimal polynomial $X^4 + X^3 - 6X^2 - X + 1$,
$G_8 = \langle\sigma^2\rangle$, $K^{G_8} = \mathbb{Q}(\theta_2)$, $\theta_2 = \sum_0^7 \sigma^{2i}(\varepsilon_{17}) = \varepsilon_{17} + \varepsilon_{17}^{3^2} + \varepsilon_{17}^{3^4} + \varepsilon_{17}^{3^6} + \varepsilon_{17}^{3^8} + \varepsilon_{17}^{3^{10}} + \varepsilon_{17}^{3^{12}} + \varepsilon_{17}^{3^{14}}$ has the minimal polynomial $X^2 + X - 4$.

10.10 Use Exercise 10.7(c) for $\theta_2 = \sum_{i=1}^{\frac{p-1}{2}} \sigma^{2i}(\varepsilon) = \sum_{i=1}^{\frac{p-1}{2}} \varepsilon^{g^{2i}}$, where σ is a generator of the Galois group $G(\mathbb{Q}(\varepsilon)/\mathbb{Q})$ and ε a p-th root of unity ($\varepsilon \neq 1$). Define $\theta_2' = \sum_{j=0}^{\frac{p-3}{2}} \sigma^{2j+1}(\varepsilon) = \sum_{j=0}^{\frac{p-3}{2}} \varepsilon^{g^{2j+1}}$. Compute $\theta_2\theta_2'$ and consider in the product all the terms such that $g^{2i} + g^{2j-1} \pmod{p}$ has a fixed value. These values are given by the values of the quadratic form $X^2 + gY^2$ over the field $\mathbb{F}_p$. Show that for $p \equiv 1 \pmod{4}$ this quadratic form does not represent 0 (that is, the only solution of the equation $X^2 + gY^2 = 0$ is $X \equiv Y \equiv 0 \pmod{4}$) and each nonzero value in $\mathbb{F}_p$ is represented in $p+1$ different ways. Count the number of possibilities for X^2 and $Y \neq 0$ ($X = g^i, Y = g^j$). Show that for $p \equiv 3 \pmod{4}$, the quadratic form represents 0 in $(p-1)/2$ different ways with $XY \neq 0$, and every nonzero element in $\mathbb{F}_p$ in $(p-3)/2$ different ways. Show that θ_2 has the minimal polynomial $X^2 + X - \frac{p-1}{4}$, when $p \equiv 1 \pmod{4}$ and $X^2 + X + \frac{p+1}{4}$, when $p \equiv 3 \pmod{4}$.

10.11 (a1) $X^6 + X^5 + X^4 + X^3 + X^2 + X + 1 = (X^3 + X + 1)(X^3 + X^2 + 1)$;
(a2) $X^6 + 6X^3 + 8 = (X^3 + 2)(X^3 + 4)$;
(a3) $X^{12} + 2X^{11} + X^{10} + 2X^9 + X^8 + 2X^7 + X^6 + 2X^5 + X^4 + 2X^3 + X^2 + 2X + 1 = (X^3 + 2X + 1)(X^3 + 2X^2 + 1)(X^3 + X^2 + 2X + 1)(X^3 + 2X^2 + X + 1)$.
(b) Let ε be a primitive n-th root of unity over $\mathbb{F}_p$ and let k be the degree of the field $\mathbb{F}_p(\varepsilon)$ over $\mathbb{F}_p$. What is the relation between n and k?

10.12 (a) Use **T. 10.2**(a) and show that $X^n - 1$ has a double zero x.
(b) What can be said about the order k of x in the group $\mathbb{Z}_p^*$? What is its relation to n? Consider the cases $k = n$ and $k < n$ and use (a).
(c) Use induction starting from Exercise 10.2(a), and for the inductive step Exercise 10.3(a).
(d) Assume that the number of primes congruent to 1 modulo n is finite and consider their product N with n. Take a prime p which divides $\Phi_n(N^k)$ for a sufficiently large k and show that it must be congruent to 1 modulo n using (b).

10.13 (a) G is a product of cyclic groups of some orders $n_1, \ldots, n_r$. For every order n_i, we can find a prime number $p_i = n_i k_i + 1$, where k_i is an integer (this is a special case of Dirichlet's Theorem on primes in arithmetical progressions—see Exercise 10.12) and we can choose k_i so that all p_i are different. We have a surjection of $\mathbb{Z}_{p_i}^*$ onto $\mathbb{Z}_{n_i}$ and hence, a surjection of $\mathbb{Z}_{p_1}^* \times \cdots \times \mathbb{Z}_{p_r}^* = \mathbb{Z}_{p_1 \cdots p_r}^*$ (see A.3.11) onto $G = \mathbb{Z}_{n_1} \times \cdots \times \mathbb{Z}_{n_r}$.
(b) Use (a) and **T. 9.3**(b).

10.14 (a) Consider a tower $K \subseteq K(\varepsilon) \subseteq K(\varepsilon, \alpha)$, where $\varepsilon^n = 1$, $K(\varepsilon)$ is a splitting field of $X^n - 1$ and $\alpha^n = a$. The maximal order of $G(L/K)$ is $n\varphi(n)$, where φ is Euler's function (for the definition of Euler's function, see p. 279).
(b) Let $\sigma \in G(L/K)$ and let $\sigma(\varepsilon) = \varepsilon^a$, $a \in \mathbb{Z}_n^*$ and $\sigma(\alpha) = \varepsilon^b \alpha$, $b \in \mathbb{Z}_n$. Map σ onto the matrix $\begin{pmatrix} a & b \\ 0 & 1 \end{pmatrix}$.
(c) If $X^p - a$, where $a \in \mathbb{Q}$, is irreducible over $\mathbb{Q}$ then the Galois group is isomorphic to the group of all matrices as in (b), where $a \in \mathbb{Z}_p^*$ and $b \in \mathbb{Z}_p$, so its order is $p(p-1)$. For $p = 2$, we have a cyclic group of order 2, for $p = 3$, the group of order 6 isomorphic to S_3 and for $p = 5$, the group of order 20 (usually denoted by $GA(1, 5)$ as a special case of the **general affine groups** $GA(1, p)$ given by the matrices in (b)).
(d) The Galois group is a cyclic group of order n.

10.15 (a) What is the m-th power of a primitive $2m$-th root of unity?
(b) Take a primitive m-th root of unity and find a zero of the second polynomial.

10.16 (a) 4; (b) 8; (c) 6; (d) 12;
(e) 12; (f) 8; (g) 16; (h) 16.

10.17 The first coefficient with absolute value bigger than 1 appears for $n = 105 = 3 \cdot 5 \cdot 7$. The number 105 is the smallest one having three different odd prime factors. The formulae for $\Phi_{n,\mathbb{Q}}(X)$ say that the size of the coefficients depends on the number of different primes dividing n (see Exercise 10.2(b)). But it is not true that three different factors of n imply that there are coefficients with absolute value bigger than 1. For example, the coefficients of $\Phi_{n,\mathbb{Q}}(X)$ for $n = 231 = 3 \cdot 7 \cdot 11$ are all of absolute value at most 1. However, it is not a bad guess, since it is possible to prove that n with only two odd prime factors give polynomials with coefficients of absolute value at most 1 (see Exercises 10.2(b) and 10.4). It is known that for sufficiently large n, the coefficients can be arbitrarily large.

10.18 (a) Show that the splitting field of $X^6 - a$ has at most degree 12 over $\mathbb{Q}$. When is it less than 12? Answer: $a = -3b^2$ for an integer $b \neq 0$. Show that this happens if and only if $\mathbb{Q}(\sqrt{a}) = \mathbb{Q}(\sqrt{-3})$, which is (the quadratic subfield of) the cyclotomic field $\mathbb{Q}(\varepsilon_6)$.
(b) The Galois group of an irreducible binomial $X^7 - a$ always has order 42. For an irreducible binomial $X^8 - a$, the order of the Galois group is at most 32. It is always divisible by 8 (why?) and, in fact, may be 8 or 16, but usually is 32. The exceptional cases appear when $a = -b^2$ (8) or $a = \pm 2b^2$ (16) for an integer $b \neq 0$. This may be obtained experimentally using Maple (and supplemented by a proof) or deduced from the properties of cyclotomic extensions (see Exercise 10.14). For irreducible binomials $X^9 - a$, the order of the Galois group is always 54. In fact, if the polynomial $X^9 - a$ is reducible over the field $M = \mathbb{Q}(\varepsilon_9)$, where ε_9 is a primitive 9-th root of unity, then $a = b^3$, where $b \in M$ by Capelli's theorem (see Exercise 5.12). Then $X^3 - a$, which is irreducible over $\mathbb{Q}$ (since $X^9 - a$ is irreducible), has a zero b in M. So the splitting field of $X^3 - a$ is contained in M, which is a contradiction, since $G(M/\mathbb{Q})$ is an abelian group, while the Galois group of $X^3 - a$ is S_3. Thus $\mathbb{Q}(\varepsilon_6, \sqrt[9]{a})$ has degree 54 over $\mathbb{Q}$.

Problems of Chap. 11

11.1 (a) $G(\mathbb{Q}(i)/\mathbb{Q}) = \{\sigma_0, \sigma_1\}$, where $\sigma_0(i) = i, \sigma_1(i) = -i$. A normal basis: $\sigma_0(\alpha) = 1 + i, \sigma_1(\alpha) = 1 - i$, where $\alpha = 1 + i$.
(b) $G(\mathbb{Q}(\sqrt{2}, \sqrt{3})/\mathbb{Q}) = \{\sigma_0, \sigma_1, \sigma_2, \sigma_3\}$, where

G	$\sqrt{2}$	$\sqrt{3}$
σ_0	$\sqrt{2}$	$\sqrt{3}$
σ_1	$-\sqrt{2}$	$\sqrt{3}$
σ_2	$\sqrt{2}$	$-\sqrt{3}$
σ_3	$-\sqrt{2}$	$-\sqrt{3}$

A normal basis: $\sigma_0(\alpha) = 1 + \sqrt{2} + \sqrt{3} + \sqrt{6}, \sigma_1(\alpha) = 1 - \sqrt{2} + \sqrt{3} - \sqrt{6}, \sigma_2(\alpha) = 1 + \sqrt{2} - \sqrt{3} - \sqrt{6}, \sigma_3(\alpha) = 1 - \sqrt{2} - \sqrt{3} + \sqrt{6}$, where $\alpha = 1 + \sqrt{2} + \sqrt{3} + \sqrt{6}$.
(c) $G(\mathbb{Q}(\varepsilon)/\mathbb{Q}) = \langle \sigma \rangle$, where $\sigma(\varepsilon) = \varepsilon^2$. A normal basis: $\sigma^0(\alpha) = \varepsilon, \sigma(\alpha) = \varepsilon^2, \sigma^2(\alpha) = \varepsilon^4, \sigma^3(\alpha) = \varepsilon^3$, where $\alpha = \varepsilon$.
(d) $G(\mathbb{F}_2(\gamma)/\mathbb{F}_2) = \langle \sigma \rangle$, where $\sigma(\gamma) = \gamma^2$. A normal basis: $\sigma^0(\alpha) = \gamma, \sigma(\alpha) = \gamma^2, \sigma^2(\alpha) = \gamma^4 = \gamma^2 + \gamma + 1$, where $\alpha = \gamma$

11.2 (a) Let σ be the nontrivial automorphism of L over K. If $\alpha \notin K$, then $L = K(\alpha)$. Let $\beta = \sigma(\alpha) = p - \alpha$, where $\alpha^2 = p\alpha + q$. When do α, β form a basis of L over K?
(b) Let $L = K(\alpha)$ and let σ be a generator of the Galois group $G(L/K)$. Define $\beta = \sigma(\alpha)$ and $\gamma = \sigma(\beta)$. Notice that $\sigma^3 = 1$ so $\sigma(\gamma) = \alpha$. When do α, β, γ form a basis of L over K? You may use Lemma 11.2 on p. 129.

11.3 (a) Use Exercise 9.24. The converse is not true—find a suitable example.

(b) In both cases the answer is "no". Try to construct suitable examples using finite fields.

11.6 (a) Use **T. 11.3** for $K = E = \mathbb{Q}(\varepsilon)$ and $n = 3$. Notice that $E = \mathbb{Q}(\sqrt{-3})$ is the quotient field of $\mathbb{Z}[\varepsilon]$ and any α can be replaced by $r^2\alpha$, where $r \in \mathbb{Z}[\varepsilon]$.
(b) Use Exercise 9.15, **T. 9.2** and the fact that the cyclic group $\mathbb{Z}_2 \times \mathbb{Z}_3 = \mathbb{Z}_6$ has exactly one subgroup of order 2.
(c) Use **T. 11.4** for $n = 3$.
(d) Use (19.1). Examples: the splitting fields of $X^3 - 3X + 1$ ($f = 1 + \varepsilon$), $X^3 - 7X + 7$ ($f = 2 + 2\varepsilon$), $X^3 - 13X + 13$ ($f = 4 + 3\varepsilon$).

11.7 (a) Use **T. 11.3** for $K = E = \mathbb{Q}(i)$ and $n = 4$. Notice that $E = \mathbb{Q}(i)$ is the quotient field of $\mathbb{Z}[i]$ and any α can be replaced by $r^2\alpha$, where $r \in \mathbb{Z}[i]$.
(b) Use Exercises 9.15, 9.23, **T. 9.2** and the fact that the group $G = \mathbb{Z}_2 \times \mathbb{Z}_4$ has exactly two subgroups H of order 2 such that G/H is cyclic (of course, of order 4). Consider the fixed subfields corresponding to these two subgroups.
(c) Use **T. 11.4** for $n = 4$.
(d) Use (c) and check that the automorphism which maps $\sqrt[4]{\alpha}$ onto $\sqrt[4]{\bar{\alpha}}$ has order 2. Take its fixed field. It is possible to show directly that the Galois group of this field is cyclic of order 4. The minimal polynomial of γ is $X^4 - 4\mathrm{Nr}(f)gX^2 + \mathrm{Nr}(f)g^2[4\mathrm{Nr}(f) - \mathrm{Tr}(f)^2]$.
(e) Use (d) and **T. 9.2** (the group C_4 has only one subgroup of order 2). Examples: the splitting fields of $X^4 + 8X^2 + 8$ ($f = -(1 + i), m = -1$), $X^4 - 10X^2 + 20$ ($f = 1 + 2i, m = 2$), $X^4 - 52X^2 + 468$ ($f = 2 + 3i, m = 1$).

11.8 Use **T. 11.2**(b) and Exercise 9.24.

11.9 By Exercise 8.3(d), we know that L is a Galois extension of K. Let q be a prime dividing the order of the Galois group $G(L/K)$. Assume that $[L : K] = q$ and consider two cases: $\mathrm{char}(K) \neq q$ and $\mathrm{char}(K) = q$ (in the first case the characteristic of K may be 0). In the first case show that all q-th roots of unity are in K. Thereafter use **T. 11.3** and show that $L = K(\alpha)$, where $\alpha^q = a \in K$. Take $\beta \in K$ such that $\beta^q = \alpha$ and take the norms from L to K of the two sides (see (6.1)). This gives a contradictions unless $q = 2$.
In the second case, use Exercise 11.8 to show that $L = K(\alpha)$, where $\alpha^q - \alpha = a \in K$. Take $\beta \in K$ such that $\beta^q - \beta = a\alpha^{q-1}$ in order to get a contradiction to the fact that $\alpha \notin K$.
Finally, consider the case when the order of $G(L/K)$ is a power of 2 and consider what happens if i (a solution of $X^2 + 1 = 0$ in L) belongs to K.

Problems of Chap. 12

12.1 (b) Consider the chain $G = G_0 = S_3 \supset G_1 = \langle\sigma\rangle \supset G_2 = \langle I\rangle$, where σ is an element of order 3 (any nontrivial rotation) and I the identity.

(c) Consider the chain $G = G_0 = S_4 \supset G_1 = A_4 \supset G_2 = V_4 \supset G_4 = \langle I \rangle$, where A_4 is the group of even permutations of $1, 2, 3, 4$ (the group of rotations of a regular tetrahedron), V_4 is the subgroup of permutations of order at most 2 in A_4 (it consists of $(1), (1,2)(3,4), (1,3)(2,4), (1,4)(2,3)$) and I is the identity.
(d) Consider the chain $G_0 = D_n \supset G_1 = C_n \supset \langle I \rangle$, where C_n is the group of all rotations of the regular polygon with n sides and I is the identity.
(e) Consider the chain $G_0 = \mathbb{H}^*(\mathbb{Z}) \supset G_1 = \langle -1 \rangle \supset G_2 = \langle 1 \rangle$.

12.2 (c) The group A_n is normal in S_n as a subgroup of index 2 (see A.2.8). Use (b) and **T. 12.1**(a).

12.3 Among all possible chains $G = G_0 \supset G_1 \supset \ldots \supset G_n = \{e\}$ consider a chain with maximal possible value of n. Show that then all the quotients G_i/G_{i+1} are cyclic of prime order. Use A.2.15 and A.2.13.

12.4 (a) Show that the functions $\varphi_{a,b}(x)$ are bijections on the set $\{0, 1, \ldots, n-1\}$ and as such can be considered as permutations belonging to the group S_n of all permutations of this set (with n elements).
Notice that different $x \in \{0, 1, \ldots, n-1\}$ have different images $\varphi_{a,b}(x)$, so $\varphi_{a,b}(x)$ is bijective on the set $\{0, 1, \ldots, n-1\}$.
(b) The number of functions $\varphi_{a,b}(x)$ is $n\varphi(n)$, where φ is Euler's function (for the definition of $\varphi(n)$, see p. 279). Show that $\varphi_{a,b} \circ \varphi_{c,d} = \varphi_{ac,ad+b}(x)$ and $\gcd(ac, n) = 1$. Use A.2.1.
(c) Check that Φ is a group homomorphism (use (b)) and find its kernel $\mathcal{T}_n$. Notice that $\mathcal{T}_n$ is abelian. Use **T. 12.1**(c) for $G = \mathcal{G}_n$ and $N = \mathcal{T}_n$.
(d) Notice that the subgroup $\mathcal{T}_n$ of all translations $\varphi_{1,b}(x) = x + b$ already acts transitively on $\{0, 1, \ldots, p-1\}$. How many solutions does the equation $\varphi_{a,b}(x) = x$ have when a, b are fixed?
(e) Let $\varphi_{a,b}(x)$ be a function of order p. Show that $a = 1$, so $\varphi_{a,b}(x) \in \mathcal{T}_p$. Use (b) above.
(f) Denote by $\mathcal{M}$ the group of matrices in (d) and show that $\Psi : \mathcal{G}_n \to \mathcal{M}$ such that $\Psi(\varphi_{a,b}) = \begin{pmatrix} a & b \\ 0 & 1 \end{pmatrix}$ is a group isomorphism.

12.5 (d) Look at the fixed group G_x of any element $x \in X$ and use the formula $|X||G_x| = |G|$ (see A.9.2(a)).
(e) Show that if $x, x' \in X$ and $gx = x'$ for some $g \in G$, then $H_{x'} = gH_xg^{-1}$. Conclude that two arbitrary orbits Hx and Hx' for $x, x' \in X$ have the same number of elements. Use A.9.2 ($|Hx| = |H|/|H_x|$).

12.6 Show that the center of a p-group, is always nontrivial. Consider the action of G on the set $X = G$ by conjugation, that is, for $g \in G$ and $x \in X$ the action of g takes x to gxg^{-1}. Notice that the fixed elements for this action of G are x such that $gxg^{-1} = x$ for each $g \in G$, so $gx = xg$, that is, such x form the center $C(G)$ of G (see p. 245). Use the formula in A.9.2(a) in this case.
Use the fact that G has nontrivial center together with **T. 12.1** and induction with respect to the order of the group. Take into account that $C(G)$ is a normal subgroup of G.

Problems of Chap. 13

13.3 Consider a longest possible chain of fields

$$K = K_0 \subset K_1 \subset \ldots \subset K_n = L$$

such that $K_i = K_{i-1}(\alpha_i)$, where $\alpha_i^{r_i} \in K_{i-1}$ and r_i are positive integers for $i = 1, \ldots, n-1$. Show that for such a chain all r_i are prime numbers (observe that the next field is strictly bigger than the preceding one).

13.4 (a) Denote by L a splitting field of $f(X)$ over K and consider the Galois group $G(L/K)$ as a subgroup of S_p. Notice that the order of $G(L/K)$ is divisible by p and use Cauchy's theorem (see A.9.3) to prove the existence of a cycle of length p in $G(L/K)$. Show that the complex conjugation is a transposition of two zeros and use A.8.4 to show that $G(L/K) = S_p$.
(b) Study the derivative of the polynomial $f(X)$ and plot its graph in order to show that exactly three zeros are real. Use (a).
(c) In order to show the irreducibility, use Eisenstein's criterion for $p = 2$. Study the sign changes of $f(X)$, in particular, the signs at the integer points from $X = 2$ to $X = 2(p-2)+1$, in order to show that there are exactly $p-2$ real zeros (in fact, in the intervall $(2, 2(p-2)+1)$).

13.5 For example, $X^{n-5}f(X)$, where $n \geq 5$ and $f(X) = X^5 - 4X - 2$ according to Exercise 13.4(b) for $p = 2$.

13.6 (b) Assume that the polynomial $f(X)$ is reducible over K_i. Show that it cannot happen by considering two possibilities: $f(X) = g(X)h(X)$ over K_i, where $\deg(g) = 2$, $\deg(h) = 3$ or $\deg(g) = 1$, $\deg(h) = 4$. Compare the degree of a splitting field M of $f(X)$ over K_{i-1} (this is 120 by Exercise 13.4(a)) with its degree resulting from the two possible factorizations.

13.7 Discuss all possible Galois groups and use **T. 13.1**. Alternatively, show how to solve such equations.

13.8 Compare the splitting fields of $f(X)$ and $f(X^n)$ over K. Alternatively: How can we express the solutions of $f(X^n) = 0$ when the solutions of $f(X) = 0$ are known, and conversely?

13.9 (a) If two of the zeros of a polynomial do not generate its splitting field, then there is an automorphism of this field having these two zeros as fixed points but moving another zero by **T. 9.1** and **T. 9.3**.
(b) Notice that the Galois group $G(K_f/K)$ is transitive on the set of zeros of $f(X)$ by Exercise 7.4 and use the description of solvable groups with this property given in Exercise 12.5. Use (a).

13.10 Let K_f be a splitting field of $f(X)$ over K. What can be said about the field K_f if $f(X)$ has at least two real zeros?

13.11 (a) The discriminant of $f(X)$ is $\Delta(f) = \prod_{1\le i<j\le 5}(\alpha_i - \alpha_j)^2$, where α_i, $i = 1, 2, 3, 4, 5$, denote its zeros. Study the sign of $\Delta(f)$ depending on the cases: all α_i real, exactly three real, only one real.
(b) An example of a solvable equation with $\Delta(f) > 0$ is $f(X) = X^5 - 2 = 0$ (only one real zero). If $f(X) = X^5 + X^2 + 1$, then $f(X)$ also has only one real zero (thus $\Delta(f) > 0$), but $G(\mathbb{Q}_f/\mathbb{Q}) = S_5$ (for a proof, see Exercise 15.10(a)), so the equation $f(X) = 0$ is not solvable by radicals.

13.13 Use the fact that the polynomial $X^p - a$, where $p > 2$ is a prime number and a a real number, has only one real zero $\sqrt[p]{a}$. Use Exercise 5.12. The arguments are similar to those given in the proof of **T. 13.3** on p. 140.

13.14 (a) Let $K = \mathbb{R}(\alpha)$, where α is a zero of an irreducible polynomial $f(X) \in \mathbb{R}[X]$ of odd degree. Use the fact that $f(X)$ has a zero in $\mathbb{R}$.
(b) Use the formula for solving quadratic equations.
(c) Consider a Sylow 2-subgroup H of $G = G(K/\mathbb{R})$ (see A.9.4) and its fixed field K^H over $\mathbb{R}$. Use (a), and then use Exercise 12.3 in combination with (b).

Problems of Chap. 14

14.1 (a) The side of the new cube has length $\sqrt[3]{2}$. Use **T. 14.1** and $[\mathbb{Q}(\sqrt[3]{2}) : \mathbb{Q}] = 3$.
(b) We have to construct the point $(\cos 20°, \sin 20°)$. From $\cos 3\alpha = 4\cos^3\alpha - 3\cos\alpha$, we get that $\cos 20°$ satisfies the equation $4x^3 - 3x = \frac{1}{2}$. Show that $[\mathbb{Q}(\cos 20°) : \mathbb{Q}] = 3$ and use **T. 14.1**.
(c) If the construction is possible, the point $(0, \sqrt{\pi})$ can be constructed from $(0, 0), (1, 0)$. Show that $[\mathbb{Q}(\sqrt{\pi}) : \mathbb{Q}] = \infty$ using the fact that π is a transcendental number (see Chap. 4, p. 19 and Exercise 4.18). Use **T. 14.1**.

14.3 (a) Yes. The radius of the new circle is $\sqrt{r_1^2 + r_2^2}$, where r_1 and r_2 are the radii of the two given circles.
(b) Sometimes. The radius of the new sphere satisfies the equation $X^3 - r_1^3 - r_2^3 = 0$, where r_1 and r_2 are the radii of the two given spheres. For example, if $r_1 = r_2 = 1$, then a construction is impossible, but if $r_1 = 1, r_2 = \sqrt[3]{7}$, then it is possible (use **T. 14.1**).

14.4 Yes. The side of the square is $x = \sqrt{\frac{1}{2}ah}$, where a is the length of a side of the triangle and h the corresponding height—both a and h are given.

14.5 No. The volume of a regular tetrahedron with side a is equal to $\frac{a^3\sqrt{2}}{12}$. Thus the side of the cube is $x = \frac{1}{\sqrt[6]{72}}$. Use **T. 14.1**.

14.6 A regular polygon with n sides is constructible if and only if the angle $\frac{2\pi}{n}$ is constructible, which means that the point $(\cos\frac{2\pi}{n}, \sin\frac{2\pi}{n})$ is constructible. Study the fields $L = \mathbb{Q}(\cos\frac{2\pi}{n}, \sin\frac{2\pi}{n})$ and $L(i) \supseteq \mathbb{Q}(\varepsilon)$, where ε is a primitive n-th root of unity. Use **T. 14.1** and **T. 14.2** as well as the fact that $[\mathbb{Q}(\varepsilon) : \mathbb{Q}] = \varphi(n)$ where φ is Euler's function (see **T. 10.1**).

14.7 (a) $\cos 72° = \frac{\sqrt{5}-1}{4}$; (b) and (c): Use (a) and Exercise 14.6(b).

14.8 (a) $1° = \frac{2\pi}{360}$ is not constructible by Exercise 14.6, since $360 = 2^3 \cdot 3^2 \cdot 5$.
(b) $3° = \frac{2\pi}{120}$ is constructible.
(c) $5° = \frac{2\pi}{72}$ is not constructible.

14.9 For example, $\alpha = \frac{2\pi}{7}$. Use Exercise 14.6 and the equality $\frac{\alpha}{3} = 2(\frac{\pi}{3} - \alpha)$.

Problems of Chap. 15

15.1 (a) Consider two cases: $f(X)$ reducible or irreducible over K. Notice that $\Delta(f) = g(\alpha_1)^2(\alpha_2 - \alpha_3)^2$, when $\alpha_1, \alpha_2, \alpha_3$ are zeros of $f(X)$. Show that $g(X) = f(X)/(X - \alpha_1) = (X - \alpha_2)(X - \alpha_3)$ has its coefficients in $K(\alpha_1)$. Notice also that the Galois group of an irreducible cubic equation has at least three and at most six elements (as a subgroup of S_3—see Exercise 6.1).
(b) Show that $F(X_1, X_2, X_3)$ has two different images $\pm F(X_1, X_2, X_3)$ under the permutations in $G = S_3$, notice that $G_F = A_3$ and use the definition of $r_{G,F}(f)$. Apply **T. 15.2**.

15.2 Show that $F(X_1, \ldots, X_n)$ has two different images $\pm F(X_1, \ldots, X_n)$ under the permutations in $G = S_n$, notice that $G_F = A_n$ and use the definition of $r_{G,F}(f)$. Apply **T. 15.2**.

15.3 (a) What are the possibilities for a in the factorization $X^4 + pX^2 + qX + r = (X^2 + aX + b)(X^2 + a'X + b')$?

15.4 (a) Check directly that the permutations in G_F do not change $F(X_1, X_2, X_3, X_4)$ and that this polynomial has exactly three different images under the action of the permutations in A_4 (and S_4).
(b) Use the definition $r_{G,F}(f)$ and the three different images of $F(X_1, X_2, X_3, X_4)$ in order to compute the resolvent $r_{G,F}(f)$. Use either Vieta's formulae for the zeros $\alpha_1, \alpha_2, \alpha_3, \alpha_4$ of $f(X)$ or Exercise 15.3.

15.5 Use Exercises 15.1 and 15.3 in order to study K_f through $K_{r(f)}$ (but this needs some effort, so if needed check the solution in the next chapter).

15.6 Consider a factorization $f(X) = X^4 + pX^2 + qX + r = (X^2 - aX + b)(X^2 + aX + b')$, where $a, b, b' \in K$, $\delta = a^2 - 4b$ and $\delta' = a^2 - 4b'$ are not squares in K (since otherwise, the polynomial $f(X)$ has zeros in K). Compute $r(f)$ and its zeros. Discuss two possibilities depending on $K(\sqrt{\delta}) = K(\sqrt{\delta'})$ and $K(\sqrt{\delta}) \neq K(\sqrt{\delta'})$.

15.7 (a) This is true for both types of resolvents discussed in Exercise 15.4. If $r(f)(T) = T^3 + 2pT^2 + (p^2 - 4r)T - q^2 = T^3 + 2pT^2 + (p^2 - 4r)T = T(T^2 + 2pT + p^2 - 4r)$, then one zero is $T = 0$ and the remaining two are $-p \pm 2\sqrt{r}$.

If $r(f)(T) = T^3 - pT^2 - 4rT + 4pr - q^2 = T^3 - pT^2 - 4rT + 4pr = (T-p)(T^2 - 4r)$, then the zeros are p and $\pm 2\sqrt{r}$. Use Exercise 15.5(a) in order to express the discriminant in terms of the coefficients.
(b) Use directly Exercise 15.5(b).

15.8 Use Exercise 15.5 in order to construct suitable examples. For example, the Galois group of X^4+X+1 is S_4, X^4+3X+3 is D_4, X^4+5X+5 is C_4, $X^4+8X+12$ is A_4 and X^4+1 is V_4.

15.9 (a) Use **T. 8.2** to represent $k' = k(\theta)$ for some θ and notice that $K' = K(\theta)$. If $[k' : k] = m$, show that $1, \theta, \ldots, \theta^{m-1}$ form a basis of K' over K. Map any automorphism of k' over k to the automorphism of K' over K having the same effect on θ—use Exercise 6.1.
(b) The resolvent is $r_{G,F}(f)(T) = T^2 + p(Y_1+Y_2)T + (p^2-2q)Y_1Y_2 + q(Y_1^2+Y_2^2)$. The resolvent is reducible if and only if its discriminant $\Delta(r_{G,F}(f)) = (p^2-4q)(Y_1-Y_2)^2 = 0$, that is, $p^2 - 4q = 0$ (which, of course, could be expected).

15.10 Consider reductions modulo prime numbers: (a) $2, 3, 5$; (b) $3, 13, 23$ (14 is a zero modulo 23); (c) $7, 13, 31$ (11 is a zero modulo 31), the discriminant is $6^4 \cdot 10^6$; (d) $3, 13, 17$ (or Eisenstein's criterion instead of reduction modulo 3). The discriminant is $2^8 \cdot 5^6 \cdot 11^4$. Using computer may be recommended.

15.11 (a) Take $n > 3$. Choose a prime number $p > n-2$, $p > 3$ and three polynomials of degree n such that the first is irreducible over $\mathbb{F}_2$, the second is a product of an irreducible polynomial of degree $n-1$ with a first degree polynomial over $\mathbb{F}_3$, and the third is a product of an irreducible quadratic polynomial with $n-2$ different first degree polynomials over $\mathbb{F}_p$. Use the Chinese Remainder Theorem A.7.1 in order to show that there is a polynomial $f(X) \in \mathbb{Z}[X]$ which has the chosen polynomials as reductions modulo 2, 3 and p. Use **T. 15.4** and A.8.5 in order to show that the Galois group of $f(X)$ is S_n.
(b) Follow the prescription in (a) for $n = 7, 9$. As in the former exercise, using a computer package may facilitate the computational tasks.

Chapter 19
Examples and Selected Solutions

Problems of Chap. 1

1.2 Notice that a, b in (1.4) satisfy $ab = -\frac{p}{3}$, so when we choose the values of a, b as the third roots of a^3, b^3, respectively, then we have to satisfy this condition. Hence, if we fix one of the three possible values of a (if $a \neq 0$), then the value of b is unique and given by $b = -\frac{p}{3a}$. Let a denote one fixed value of the third root of a^3 and b the corresponding value satisfying $ab = -\frac{p}{3}$. Then the three solutions are $x_1 = a+b, x_2 = \varepsilon a + \varepsilon^2 b, x_3 = \varepsilon^2 a + \varepsilon b$, where $\varepsilon^3 = 1, \varepsilon \neq 1$, that is, $\varepsilon = \frac{-1+i\sqrt{3}}{2}$ (compare Exercise 1.7 for $n = 3$).

1.3 Using Vieta's formulae and $x_i^3 = -px_i - q$ for $i = 1, 2, 3$, we get (assuming $x_i \neq 0$):

$$\Delta(f) = [(x_1 - x_2)(x_2 - x_3)(x_3 - x_1)]^2 = \left(x_3^2 + \frac{4q}{x_3}\right)\left(x_1^2 + \frac{4q}{x_1}\right)\left(x_2^2 + \frac{4q}{x_2}\right)$$

$$= \frac{(x_1x_2x_3)^3 + 16q^2(x_1^3 + x_2^3 + x_3^3) + 4q(x_1^3x_2^3 + x_2^3x_3^3 + x_3^3x_1^3) + 64q^3}{x_1x_2x_3} = -(27q^2 + 4p^3),$$

since $x_1x_2x_3 = -q$, $x_1^3 + x_2^3 + x_3^3 = x_1^3 + x_2^3 - (x_1 + x_2)^3 = 3x_1x_2x_3 = -3q$ and $x_1^3x_2^3 + x_2^3x_3^3 + x_3^3x_1^3 = p^3 + 3q^2$ (use $x_1^3x_2^3 = (px_1 + q)(px_2 + q)$ and similarly for $x_2^3x_3^3$ and $x_3^3x_1^3$, taking into account that $x_2x_2 + x_2x_3 + x_3x_1 = p$). Notice that if some $x_i = 0$, then $q = 0$ and this case becomes trivial.

Problems of Chap. 2

2.4 Let K be a field of prime characteristic p. According to **T. 2.1**, the prime subfield of K is isomorphic to $\mathbb{F}_p = \{0, 1, \ldots, p-1\}$, where to $r \in \mathbb{F}_p$ corresponds $r \cdot 1$ in K. Thus, the least $n > 0$ such that $n \cdot 1 = 0$ is $n = p$. If K is a field of characteristic 0,

J. Brzeziński, *Galois Theory Through Exercises*, Springer Undergraduate Mathematics Series, https://doi.org/10.1007/978-3-319-72326-6_19

then K contains a subfield isomorphic to $\mathbb{Q}$ and, in particular, the field K contains the images $r \cdot 1$ of all positive integers $r \in \mathbb{Z}$. As all these multiples are different, there is no n such that $n \cdot 1 = 0$.

2.7 **Example 1** (to (a) and (b)). We show that $\mathbb{Q}(\sqrt{2}, i\sqrt{2}) = \mathbb{Q}(i, \sqrt{2})$. It is clear that $\sqrt{2}, i\sqrt{2} \in \mathbb{Q}(i, \sqrt{2})$, so the smallest field $\mathbb{Q}(\sqrt{2}, i\sqrt{2})$ which contains $\mathbb{Q}$ and these two numbers must be contained in the field on the right-hand side. On the other hand, we have $i, \sqrt{2} \in \mathbb{Q}(\sqrt{2}, i\sqrt{2})$ (i is the quotient of $i\sqrt{2}$ by $\sqrt{2}$), so a similar argument shows that the field on the right-hand side is contained in the field on the left-hand side. Thus the two fields are equal.
Example 2 (to (c) and (d)). We show that $\mathbb{Q}(\sqrt{2}, \sqrt{3}) = \mathbb{Q}(\sqrt{2} + \sqrt{3})$. It is clear that $\sqrt{2} + \sqrt{3} \in \mathbb{Q}(\sqrt{2}, \sqrt{3})$, so $\mathbb{Q}(\sqrt{2} + \sqrt{3}) \subseteq \mathbb{Q}(\sqrt{2}, \sqrt{3})$. Let $\alpha = \sqrt{2} + \sqrt{3}$. Hence $\sqrt{3} = \alpha - \sqrt{2}$ and squaring both sides, we get $\sqrt{2} = \frac{\alpha^2 - 1}{2\alpha}$. Hence $\sqrt{2} \in \mathbb{Q}(\alpha)$. But $\sqrt{3} = \alpha - \sqrt{2}$, so $\sqrt{3} \in \mathbb{Q}(\alpha)$ as well. Thus both $\sqrt{2}, \sqrt{3} \in \mathbb{Q}(\alpha)$, so $\mathbb{Q}(\sqrt{2}, \sqrt{3}) \subseteq \mathbb{Q}(\sqrt{2} + \sqrt{3})$. Both inclusions prove that these fields are equal.

2.13 (a) It is clear that M_1M_2, as a field containing M_1 and M_2, must contain every quotient of finite sums $\sum \alpha_i\beta_i$, where $\alpha_i \in M_1$, $\beta_i \in M_2$. On the other hand, all such quotients form a field, since a sum, difference, product and quotient (by a nonzero sum) of two such quotients is a quotient of the same type. Thus, M_1M_2 as the smallest field containing M_1 and M_2, is equal to the field of all such quotients.
(b) If $x \in M_1M_2$ is expressed as a quotient of finite sums of some elements from M_1 and M_2 as in (a), then taking all $\alpha_i \in M_1$ ($i = 1, \ldots, m$ for some m) and all β_j ($j = 1, \ldots, n$ for some n) which appear in such a representation of x, we get that $x \in K(\alpha_1, \ldots, \alpha_m, \beta_1, \ldots, \beta_n)$.

Problems of Chap. 3

3.1 (a) If $f(X)$ is an arbitrary polynomial of degree at least 2 such that $f(x_0) = 0$ for $x_0 \in K$, then by Theorem **T. 3.1**, we have $f(X) = (X - x_0)g(X)$, where $g(X) \in K[X]$ is a polynomial of degree at least 1. Thus $f(X)$ is reducible in $K[X]$. If a polynomial of degree 2 or 3 is reducible in $K[X]$, then $f(X) = g(X)h(X)$, where $g(X), h(X) \in K[X]$ and the degree of at least one of the factors $g(X), h(X)$, say $g(X)$, equals 1. Then $g(X) = aX + b$, $a \neq 0$, and $x_0 = -\frac{b}{a} \in K$ is a zero of $f(X)$. Notice that the same argument doesn't work when $f(X)$ has degree 4.
(b) **Example**. We show that the polynomial $X^4 + 4$ has no rational zeros, but is reducible in $\mathbb{Q}[X]$. For any real x, we have $x^4 + 4 > 0$, so the polynomial has no real zeros. Hence the polynomial has no factors of degree 1 in $\mathbb{Q}[X]$ (or $\mathbb{R}[X]$). We have $X^4 + 4 = X^4 + 4X^2 + 4 - 4X^2 = (X^2 + 2)^2 - (2X)^2 = (X^2 - 2X + 2)(X^2 + 2X + 2)$, so the polynomial is a product of two polynomials of degree 2 with integer coefficients. Notice that in general, we have $X^4 + a^2 = X^4 + 2aX^2 + a^2 - 2aX^2 = (X^2 + a)^2 - (\sqrt{2a}X)^2 = (X^2 - \sqrt{2a}X + a)(X^2 + \sqrt{2a}X + a)$ for any complex number a.

3.3 If

$$f\left(\frac{p}{q}\right) = a_n\left(\frac{p}{q}\right)^n + a_{n-1}\left(\frac{p}{q}\right)^{n-1} + \cdots + a_1\left(\frac{p}{q}\right) + a_0 = 0,$$

then $a_np^n + a_{n-1}p^{n-1}q + \cdots + a_1pq^{n-1} + a_0q^n = 0$, which shows that $p|a_0q^n$ and $q|a_np^n$. Since p, q are relatively prime, we get $p|a_0$ and $q|a_n$.

3.4 Example to (d). We show that $X^4 + 2$ is irreducible in $\mathbb{F}_5[X]$. It has no factor of degree 1, since it has no zeros in $\mathbb{F}_5$ (check it!). If $X^4 + 2 = (X^2 + aX + b)(X^2 + cX + d)$, then $a + c = 0$, $ac + b + d = 0$, $ad + bc = 0$, $bd = 2$. The last equation gives the following possibilities: $(b, d) = (1, 2), (2, 1), (3, 4), (4, 3)$. Since $c = -a$, the second equation gives $a^2 = 2$ or 3. This is impossible, since $a^2 = 0, 1, 4$ in $\mathbb{F}_5$.

3.5 (a) Let $f(X) = a_nX^n + \cdots + a_1X + a_0$, where $a_i \in \mathbb{R}$, and assume that $f(z) = 0$ for a complex number z. Then $f(\bar{z}) = a_n\bar{z}^n + \cdots + a_1\bar{z} + a_0 = \overline{a_nz^n + \cdots + a_1z + a_0} = \overline{f(z)} = 0$, which shows that $\bar{z}$ is a zero of $f(X)$. Observe that we use two properties of complex conjugation: $\overline{z_1 + z_2} = \bar{z}_1 + \bar{z}_2$ and $\overline{z_1z_2} = \bar{z}_1\bar{z}_2$, as well as $\bar{a} = a$ when a is a real number.
(b) The polynomial $f(X)$ of degree n has n zeros. Each time we have a nonreal zero z, we also have a different nonreal zero $\bar{z}$, so we have a pair of nonreal zeros of $f(X)$: $z, \bar{z}$. Denote by $x_1, \ldots, x_r$ all real zeros of $f(X)$. Let $z_1, \bar{z}_1, \ldots, z_s, \bar{z}_s$ be all pairs of nonreal zeros of $f(X)$. Thus, we have

$$f(X) = a_n(X - x_1)\cdots(X - x_r)(X - z_1)(X - \bar{z}_1)\cdots(X - z_s)(X - \bar{z}_s).$$

Now $(X - z_i)(X - \bar{z}_i) = X^2 - p_iX + q_i$, where $p_i = z_i + \bar{z}_i$ and $q_i = z_i\bar{z}_i$ are real numbers. Hence $f(X)$ is a product of r real polynomials $X - x_i$ of first degree and s real polynomials $X^2 - p_iX + q_i$ of second degree. These are irreducible in $\mathbb{R}[X]$ as they do not have real zeros (see Exercise 3.1).

3.6 Example. Using Gauss's Lemma, we show that the polynomial $f(X) = X^4 - 10X^2 + 1$ is irreducible over $\mathbb{Q}$. It has degree 4, so its possible factorization is either a product of a polynomial of degree 1 by a polynomial of degree 3 or a product of two second degree polynomials. But $f(X)$ has no linear factors. In fact, by Exercise 3.3 and **T. 3.1**, we need only check as possible rational (in fact, integer) zeros the numbers ± 1. But $f(\pm 1) \neq 0$. Assume that we have the second possibility, that is, $X^4 - 10X^2 + 1 = (X^2 + aX + b)(X^2 + cX + d)$ with integer coefficients a, b, c, d. Then $a + c = 0$, $ac + b + d = -10$, $ad + bc = 0$, $bd = 1$. Hence the last equation gives $b = d = 1$ or $b = d = -1$. Since $c = -a$, the second equation gives $a^2 = \pm 2$, which cannot be satisfied by an integer. Hence $f(X)$ cannot be factorized in the ring $\mathbb{Z}[X]$. Using Gauss's Lemma, we get that it cannot be factorized in $\mathbb{Q}[X]$.

3.7 (a) According to the assumption concerning the divisibility of the coefficients of $f(X)$, we get $\bar{f}(X) = \bar{a}_nX^n = \bar{g}(X)\bar{h}(X)$ in the polynomial ring $\mathbb{F}_p[X]$. Since we have unique factorization in this polynomial ring, we get $\bar{g}(X) = aX^k$ and $\bar{h}(X) = bX^l$, where $a, b \in \mathbb{F}_p^*$ $(k, l \geq 1)$. Hence $p|g(0)$ and $p|h(0)$, which gives that $p^2|a_0 = f(0) = g(0)h(0)$. This is a contradiction. Hence $f(X)$ is irreducible in $\mathbb{Z}[X]$. Using Gauss's Lemma (see Exercise 3.6), we get irreducibility of $f(X)$ in $\mathbb{Q}[X]$.

3.8 (a) If $f(X)$ is reducible in $\mathbb{Q}[X]$ then, by Gauss's Lemma, it is reducible in $\mathbb{Z}[X]$, that is, $f(X) = g(X)h(X)$, where the degrees of g, h are less than the degree of f. Thus we also have $\bar{f}(X) = \bar{g}(X)\bar{h}(X)$ modulo p so the polynomial $\bar{f}$ in $\mathbb{F}_p[X]$ is reducible, contrary to the assumption.
(b) Suppose $f(X)$ is reducible in $\mathbb{Q}[X]$, that is, $f(X) = g(X)h(X)$, where $g(X), h(X) \in \mathbb{Z}[X]$ (we use Gauss's Lemma). Let d_g and d_h be the degrees of g and h. Then we also have $\bar{f}(X) = \bar{g}(X)\bar{h}(X)$ modulo p and modulo q, with the same degrees $d_{\bar{g}} = d_g$ and $d_{\bar{h}} = d_h$, since $f(X)$ is monic. According to **T. 3.2**, we must have $\{d_{1p}, d_{2p}\} = \{d_g, d_h\} = \{d_{1q}, d_{2q}\}$ contrary to the assumption. Hence $f(X)$ must be irreducible in $\mathbb{Q}[X]$.

3.9 (a) Let $X^4 + pX^2 + qX + r = (X^2 + aX + b)(X^2 + a'X + b')$, where a, b, a', b' are in some field containing K. Comparing the coefficients to the right and left, we get $a' = -a$ and the system:

$$\begin{aligned} b + b' &= a^2 + p, \\ a(b' - b) &= q, \\ bb' &= r. \end{aligned}$$

Multiplying the first equation by a, then squaring the first two equations and finally subtracting the second from the first, we get

$$4a^2bb' = a^2(a^2 + p)^2 - q^2.$$

Replacing bb' by r, we obtain

$$a^6 + 2pa^4 + (p^2 - 4r)a^2 - q^2 = 0,$$

so a^2 is a zero of $r(f)(T) = T^3 + 2pT^2 + (p^2 - 4r)T - q^2$.
Conversely, let a^2 be a zero of $r(f)$, for some value of a. The case $a = 0$ can only happen when $q = 0$ and then we find b, b' from the first and third equation in the system. If $a \neq 0$, then we find b, b' from the first two equations in the system and the third equation is evidently satisfied, since $r(f)(a^2) = 0$.
(b) If $q \neq 0$ and the resolvent has a zero a^2 such that $a \in K$, then we already know from (a) that $a \neq 0$ and using the first two equations of the system in (a), we find $b, b' \in K$, which give a factorization of $f(X)$. Conversely, if we have a factorization of $f(X)$ in which $a \in K$, then the zero of the resolvent a^2 is a square in K.
If $q = 0$, there are two cases. If there is a factorization over K with $a = 0$, then $\delta = p^2 - 4r = (b + b')^2 - 4bb' = (b - b')^2$ is a square in K. Conversely, if δ is a square in K, then the quadratic polynomial $T^2 - pT + r$ has two zeros b, b' in K and we get a factorization $f(X) = X^4 + pX^2 + r = (X^2 + b)(X^2 + b')$ with $a = 0$.

If $q = 0$ and we have a factorization with $a \neq 0$, then the resolvent has two zeros which are squares in K: 0 and a^2. Conversely, assume that the resolvent $r(f)$ has two zeros which are squares in K (both may be equal to 0). We easily find that the zeros of $r(f)(T) = T^3+2pT^2+(p^2-4r)T$ are: $\beta_1 = 0, \beta_2 = -p+2\sqrt{r}, \beta_3 = -p-2\sqrt{r}$, so $\sqrt{r} \in K$. Of course, 0 is one of the squares. If the second is β_2, we check that $f(X)$ has the following factorization in $K[X]$:

$$f(X) = X^4 + pX^2 + r = (X^2 + \sqrt{r})^2 - \beta_2 X^2 = (X^2 - \sqrt{\beta_2}X + \sqrt{r})(X^2 + \sqrt{\beta_2}X + \sqrt{r})$$

and similarly for β_3, when β_3 is a square in K.

Problems of Chap. 4

4.1 Example 1. (to (a), (b), (c)). We prove that the number $\sqrt{2} + \sqrt[5]{2}$ is algebraic. The number $\sqrt{2}$ is algebraic as a zero of the polynomial $X^2 - 2$. The number $\sqrt[5]{2}$ is algebraic as a zero of the polynomial $X^5 - 2$. Thus the number $\sqrt{2} + \sqrt[5]{2}$ is algebraic as a sum of two algebraic numbers according to Theorem **T. 4.6**.
Example 2. (to (d), (e), (f)). We prove that the number $\alpha = \sqrt[3]{3} + \sqrt[5]{\pi}$ is transcendental. If it is algebraic, then according to Theorem **T. 4.6**, the number $\alpha - \sqrt[3]{3} = \sqrt[5]{\pi}$ is also algebraic, since the number $\sqrt[3]{3}$ is algebraic as a solution of the equation $X^3 - 3 = 0$. Hence, by the same theorem, the number $\pi = \left(\sqrt[5]{\pi}\right)^5$ is algebraic, which gives a contradiction (see Exercise 4.17). Thus the number α is transcendental.

4.2 Let $n = \deg(f)$ and $r = [L : K]$. Let α be a zero of $f(X)$ in a field containing L (see **T. 5.1**). Then $[K(\alpha) : K] = n$ (since $f(X)$ is irreducible over K—see **T. 4.2**) and let $[L(\alpha) : L] = m$ for an integer m. Of course, we have $m \leq n$, since $f(X)$ has degree n and may be reducible over L. The field $L(\alpha)$ contains the field $K(\alpha)$ as well as the field L.

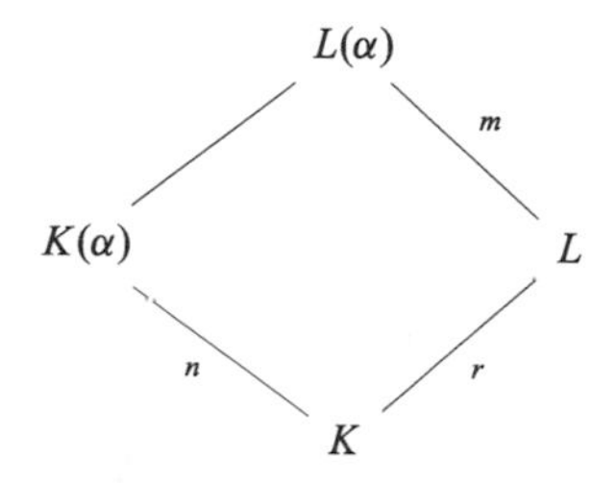

Since n and r are relatively prime and both divide $[L(\alpha) : K]$, the product nr also divides $[L(\alpha) : K] = [L(\alpha) : L][L : K] = mr$ (see **T. 4.3**). Thus n divides m, and since $m \leq n$, we have $m = n$. This tells us that $f(X)$ is also irreducible over L, since $[L(\alpha) : L] = n$ (see **T. 4.2**).

4.3 Example. We find the minimal polynomial (and the degree) of the number $\alpha = \sqrt[5]{3}$ over the field $K = \mathbb{Q}(i)$. We easily see that α is a zero of the polynomial X^5-3 over K. Now this polynomial is certainly irreducible over the rational numbers $\mathbb{Q}$, for example, according to Eisenstein's criterion (see Exercise 3.7) for $p = 3$. But this polynomial remains irreducible over the field $K = \mathbb{Q}(i)$, since its degree 5 is relatively prime to the degree 2 of the extension $K \supset \mathbb{Q}$ (see Exercise 4.2). Thus, the minimal polynomial of α over K is $X^5 - 3$ and its degree over K equals 5.

4.4 Example. We find the degree and a basis for the extension $K \subset L$, where $K = \mathbb{Q}$ and $L = \mathbb{Q}(\sqrt{2}, \sqrt[3]{2})$. Consider the tower: $\mathbb{Q} \subset \mathbb{Q}(\sqrt{2}) \subset \mathbb{Q}(\sqrt{2}, \sqrt[3]{2})$. In the first extension, we adjoin $\sqrt{2}$, which is a zero of the polynomial $X^2 - 2$. It is the minimal polynomial of $\sqrt{2}$ over $\mathbb{Q}$ (it is irreducible over $\mathbb{Q}$). A basis is $1, \sqrt{2}$ by **T. 4.2**. In the second extension, we adjoin $\sqrt[3]{2}$, which is a zero of the polynomial $X^3 - 2$. This polynomial is certainly irreducible over $\mathbb{Q}$, but it is also irreducible over $\mathbb{Q}(\sqrt{2})$ as its degree 3 and the degree of the extension $\mathbb{Q} \subset \mathbb{Q}(\sqrt{2})$ are relatively prime (see Exercise 4.2). Thus $X^3 - 2$ is also the minimal polynomial of $\sqrt[3]{2}$ over the field $\mathbb{Q}(\sqrt{2})$. By **T. 4.2**, a basis is given by $1, \sqrt[3]{2}, \sqrt[3]{4}$. According to the Tower Law (**T. 4.3**), the whole extension has degree 6 and the proof of this theorem shows that we get a basis of L by multiplying the basis elements of the first extension by the basis elements of the second. Thus, we get all the numbers $1, \sqrt{2}, \sqrt[3]{2}, \sqrt[3]{4}, \sqrt[6]{2}, \sqrt[6]{32}$.

4.7 Let $r = \frac{p}{q}$, where p, q are integers and $q > 0$. Consider $z = \cos r\pi + i \sin r\pi = e^{\frac{pi\pi}{q}}$. Hence, $z^{2q} = 1$, that is, z is a solution of the equation $X^{2q} - 1 = 0$. Thus z is an algebraic number. Now use Exercise 4.6.

4.8 (b) Write

$$\frac{1}{1 + \sqrt[3]{2}} = a + b\sqrt[3]{2} + c\sqrt[3]{4},$$

multiply both sides by the denominator to the left and equate the corresponding coefficients of $1, \sqrt[3]{2}, \sqrt[3]{4}$ as a basis of $\mathbb{Q}(\sqrt[3]{2})$ over $\mathbb{Q}$ (use **T. 4.2**). Compute a, b, c by solving a system of linear equations.

4.9 Show that $1, \sqrt{2}, \sqrt{3}, \sqrt{6}$ is a basis of $\mathbb{Q}(\sqrt{2}, \sqrt{3})$ over $\mathbb{Q}$ and use the same method as in the solution of Exercise 4.8(b).

4.10 Show that $1, \alpha, \alpha^2, \alpha^3$ is a basis of $L = \mathbb{F}_2(\alpha)$ over $\mathbb{F}_2$ and use the same method as in the solution of Exercise 4.8(b).

4.11 (b) Assume that α is algebraic over K. Then, it follows from **T. 4.2** that $[K(\alpha) : K]$ is finite. Since $[K(X) : K(\alpha)]$ is also finite by (a), we get that the extension $K(X)$ of K is finite by the Tower Law (**T. 4.3**). However, this is a contradiction, since X is transcendental over K. Thus, α must be transcendental over K as well.
(c) We have to show that the polynomial $f(\alpha, Y) = p(Y) - \alpha q(Y)$ is irreducible over the field $K(\alpha)$, that is, as a polynomial with respect to Y in the polynomial

ring $K(\alpha)[Y]$. But as we know from **T. 4.2**(b), the ring $K[\alpha]$ is isomorphic to the polynomial ring in one variable (α is transcendental over K by (b)). We now consider $f(\alpha, Y)$ as a polynomial in the polynomial ring $K[\alpha, Y]$ in two variables α, Y. If $f(\alpha, Y) = p(Y) - \alpha q(Y)$ is reducible over $K(\alpha)$, then it must already be reducible over $K[\alpha]$ by Gauss's Lemma, which is also true over the polynomial rings in one variable over fields (see the comment after Gauss's Lemma **T. 3.3**). So assume that $f(\alpha, Y) = p(Y) - \alpha q(Y) = g(\alpha, Y)h(\alpha, Y)$ in the polynomial ring $K[\alpha, Y]$, where the degrees of g, h with respect to Y are at least 1. Since $f(\alpha, Y)$ has degree 1 with respect to α, one of the polynomials $g(\alpha, Y), h(\alpha, Y)$ does not contain α. Assume the polynomial g only has terms depending on Y. Then the equality $p(Y) - \alpha q(Y) = g(\alpha, Y)h(\alpha, Y)$ says that both $p(Y)$ and $q(Y)$ have a polynomial factor $g(\alpha, Y)$, which belongs to $K[Y]$ (we consider both sides of the last equality as polynomials of α with coefficients in $K[Y]$). This is a contradiction, since $p(Y)$ and $q(Y)$ are relatively prime (see (a)). Hence the polynomial $f(\alpha, Y)$ is irreducible as a polynomial in Y over $K(\alpha)$ and has degree n with respect to Y (see (a)). Since $Y = X$ is a zero of this polynomial, we have $[K(X) : K(\alpha)] = n$ by **T. 4.2**(a).

4.13 If $q \neq p$, then according to Exercise 4.2, the polynomial f remains irreducible over L.

4.14 As a counterexample take $\alpha = 1 + i$ over $K = \mathbb{Q}$.

4.15 (b) If $L \neq \mathbb{R}$, choose $\alpha \in L \setminus \mathbb{R}$. According to Exercise 3.5(b), the minimal polynomial of α over $\mathbb{R}$ has degree 2, so α is a zero of a polynomial $X^2 + pX + q$, where $p, q \in \mathbb{R}$ and $p^2 - 4q < 0$. Hence, we have $\alpha = a + bi$, where a, b are real and $b > 0$. Thus, we have $\mathbb{R}(\alpha) = \mathbb{R}(i) = \mathbb{C} \subseteq L$. Since the degree of L over $\mathbb{R}$ is finite, it is also finite over $\mathbb{C}$ and by (a), we get $L = \mathbb{C}$.

4.16 Consider $\mathbb{Q}(\alpha) \supset \mathbb{Q}$, where $\alpha^4 + \alpha + 1 = 0$. The polynomial $X^4 + X + 1$ is irreducible over $\mathbb{Q}$ (for example, a reduction modulo 2 gives a proof). Thus the degree of $\mathbb{Q}(\alpha)$ over $\mathbb{Q}$ is 4. Assume that there is a quadratic subfield K in this field. Then α has degree 2 over it, so the polynomial $X^4 + X + 1$ must be divisible by a second degree polynomial over K (the minimal polynomial of α over K). Study a factorization $X^4 + X + 1 = (X^2 + aX + b)(X^2 + cX + d)$, where $a, b, c, d \in K$. Find an irreducible polynomial with rational coefficients having a^2 as its zero using the method applied in Exercise 3.9. Prove that such a polynomial cannot be minimal for an element of K. Try to give your own example!

4.17 Assume that both $e + \pi$ and $e\pi$ are algebraic. Prove that in the tower of fields $\mathbb{Q}(e, \pi) \supset \mathbb{Q}(e + \pi, e\pi) \supset \mathbb{Q}$ both subextensions are algebraic, so both e and π must be algebraic according to Theorem **T. 4.4**. This is a contradiction.

4.18 (b) **Example**. We prove that if α is algebraic over a field K, then $\beta = \sqrt{1/\alpha + \sqrt[3]{\alpha}}$ is also algebraic over K. Consider the field $L = K(\alpha)$. The number $1/\alpha \in L$ is algebraic over K, since L is an algebraic extension of K by **T. 4.4**. The number $\sqrt[3]{\alpha}$ is algebraic over K, since it is a zero of the polynomial $X^3 - \alpha$ (here we use (a)). Thus the sum $1/\alpha + \sqrt[3]{\alpha} = \beta^2$ is algebraic over K by **T. 4.6**. Now β is a zero $X^2 - \beta^2$, so it is algebraic over K by (a).

Problems of Chap. 5

5.1 Example. We find the degree and a basis for the splitting field of $f(X) = X^3 - 2$ over $K = \mathbb{Q}(\sqrt{2})$. The zeros of $f(X)$ are $\sqrt[3]{2}, \varepsilon\sqrt[3]{2}, \varepsilon^2\sqrt[3]{2}$, where $\varepsilon^3 = 1$ and $\varepsilon \neq 1$, so the splitting field of $f(X)$ over K is $L = K(\sqrt[3]{2}, \varepsilon\sqrt[3]{2}, \varepsilon^2\sqrt[3]{2}) = K(\sqrt[3]{2}, \varepsilon)$. Thus, we have a chain of fields:

$$K \subset K(\sqrt[3]{2}) \subset K(\sqrt[3]{2}, \varepsilon).$$

The first extension $K \subset K(\sqrt[3]{2})$ has degree 3, since $f(X) = X^3 - 2$ is the minimal polynomial of $\sqrt[3]{2}$ over K ($f(X)$ is irreducible over $\mathbb{Q}$ as a polynomial of degree 3 without rational zeros and it is still irreducible over K, since its degree 3 and the degree 2 of the extension K over $\mathbb{Q}$ are relatively prime—see Exercise 4.2). According to **T. 4.2**, a basis of $K(\sqrt[3]{2})$ over K is $1, \sqrt[3]{2}, \sqrt[3]{4}$. The second extension $K(\sqrt[3]{2}) \subset K(\sqrt[3]{2}, \varepsilon)$ is generated by ε, whose minimal polynomial is $X^2 + X + 1$. In fact, this quadratic polynomial has no zeros in the (real) field $K(\sqrt[3]{2})$. Hence a basis of L over $K(\sqrt[3]{2})$ is $1, \varepsilon$. By the Tower Law (see **T. 4.3**), the degree of L over K is 6 and a basis is $1, \sqrt[3]{2}, \sqrt[3]{4}, \varepsilon, \varepsilon\sqrt[3]{2}, \varepsilon^2\sqrt[3]{4}$.

5.2 For (a) and (b) use **T. 5.1**. **Example** (to (c)). We show that the fields $\mathbb{Q}(\sqrt[3]{2})$ and $\mathbb{Q}(\sqrt[3]{5})$ are not isomorphic. Assume that there is an isomorphism $\varphi : \mathbb{Q}(\sqrt[3]{2}) \to \mathbb{Q}(\sqrt[3]{5})$. As $(\sqrt[3]{2})^3 = 2$, we have $\varphi((\sqrt[3]{2})^3) = \varphi(2) = 2$, so $\varphi(\sqrt[3]{2})^3 = 2$. Thus $x = \varphi(\sqrt[3]{2})$ is a real number, which solves the equation $x^3 = 2$ in $\mathbb{Q}(\sqrt[3]{5})$. Hence, $x = \sqrt[3]{2} \in \mathbb{Q}(\sqrt[3]{5})$ (which implies that $\mathbb{Q}(\sqrt[3]{2}) = \mathbb{Q}(\sqrt[3]{5})$). Now one can choose different arguments in order to show that $\sqrt[3]{2}$ is not in the field $\mathbb{Q}(\sqrt[3]{5})$. Probably the most direct way is to check that the equality $\sqrt[3]{2} = a + b\sqrt[3]{5} + c\sqrt[5]{25}$ for some rational numbers a, b, c is impossible. Define $r = \sqrt[3]{2} - a = b\sqrt[3]{5} + c\sqrt[3]{25}$ and take the third powers of both sides. We get $2 - a^3 - 3a\sqrt[3]{2}r = 5b^3 + 25c^3 + 15bcr$. Thus, we have $r(15bc + 3a\sqrt[3]{2}) = 2 - 5b^3 - 25c^3$, so $(\sqrt[3]{2} - a)(15bc + 3a\sqrt[3]{2})$ is a rational number. Since $1, \sqrt[3]{2}, \sqrt[3]{4}$ form a basis of $\mathbb{Q}(\sqrt[3]{2})$ over $\mathbb{Q}$, the product $(\sqrt[3]{2} - a)(15bc + 3a\sqrt[3]{2})$ can be a rational number if and only if $a = 0$ and $bc = 0$. Thus, we have $\sqrt[3]{2} = b\sqrt[3]{5}$ or $\sqrt[3]{2} = c\sqrt[3]{25}$. Both equalities give a contradiction (e.g. multiplying both sides of the first one by $\sqrt[3]{25}$ gives that $\sqrt[3]{50}$ is a rational number, which is not the case).

5.5 If a field with p^m elements is contained in a field with p^n elements, then according to **T. 5.4**, the first one, which is a splitting field of $X^{p^m} - X$ over $\mathbb{F}_p$, is contained in a splitting field of $X^{p^n} - X$ over the same field. The degrees of these fields over $\mathbb{F}_p$ are m and n, also by **T. 5.4**. According to **T. 4.3**, the degree m of the first field divides the degree n of the second one.

Conversely, if $m \mid n$, then it is easy to check that any splitting field of $X^{p^n} - X$ contains the splitting field of $X^{p^m} - X$, since every zero of the last polynomial is a zero of the former.

5.6 (a) The quotient ring $K = \mathbb{F}_p[X]/(f(X))$ is a field with p^n elements in which $f(X)$ has a zero (see A.5.7 and A.5.8). According to **T. 5.4**, this field must be

isomorphic to a splitting field of $X^{p^n} - X$. Since every element of K is a zero of $X^{p^n} - X$, in particular, the zeros of $f(X)$ in K are zeros of this polynomial. Since $f(X)$ is an irreducible polynomial and its zero is a zero of $X^{p^n} - X$, it must divide this polynomial (see **T. 4.1**).
(b) Denote by m the degree of the irreducible polynomial $f(X)$. According to (a), we have $f(X) \mid X^{p^m} - X$. If $m \mid n$, then $X^{p^m} - X \mid X^{p^n} - X$, so $f(X) \mid X^{p^n} - X$. This follows easily if we note that every zero of $X^{p^m} = X$ is a zero of $X^{p^n} = X$ (say, in a splitting field of $(X^{p^m} - X)(X^{p^n} - X)$). In fact, if $n = dm$ and α is a zero of $X^{p^m} = X$, then $\alpha^{p^m} = \alpha$ gives $(\alpha^{p^m})^{p^m} = \alpha^{p^m} = \alpha$, so $\alpha^{p^{2m}} = \alpha$ and continuing in this way $d-1$ times, we get $\alpha^{p^{dm}} = \alpha^{p^n} = \alpha$, that is, α is a zero of $X^{p^n} = X$.
If $f(X) \mid X^{p^n} - X$, then a splitting field of $X^{p^n} - X$, which is a field of degree n over $\mathbb{F}_p$, contains a splitting field of $f(X)$. As we know from (a), a splitting field of $f(X)$ is a field with p^m elements whose degree over $\mathbb{F}_p$ is m. As it is a subfield of a field of degree n over $\mathbb{F}_p$, we have $m \mid n$.

5.9 (b) If $\mathbb{F}$ is a field with q elements, where $q = p^m$ for a prime number p and some exponent m, then according to **T. 5.4**, $\mathbb{F}$ is a splitting field over $\mathbb{F}_p$ of the polynomial $X^q - X$. Let K be a splitting field of the polynomial $X^{q^n} - X = X^{p^{mn}} - X$ over $\mathbb{F}_p$. Then K contains $\mathbb{F}$ and has degree mn over $\mathbb{F}_p$. Hence, the field K has degree n over $\mathbb{F}$. By Exercise 5.8(b), we have $K = \mathbb{F}(\gamma)$ for some $\gamma \in K$. The minimal polynomial of γ over $\mathbb{F}$ has degree n and is irreducible.

5.11 (b) If $n = pk$, then $X^n - 1 = (X^k - 1)^p$ over any field of characteristic p, so the number of different zeros of $X^n - 1$ is at most $k < n$. If $p \nmid n$, then it follows from **T. 5.3** that all zeros of $X^n - 1$ are different since this polynomial and its derivative nX^{n-1} do not have common zeros (are relatively prime).
(c) If α and β are solutions of $X^n - a = 0$ in L, then $(\frac{\alpha}{\beta})^n = 1$, so $\beta = \eta\alpha$, where $\eta^n = 1$ and η is a solution of $X^n - 1 = 0$ belonging to L. All such solutions form a cyclic group by Exercise 5.8 and if $\varepsilon \in L$ generates it, then $\eta = \varepsilon^i$ for some i. Thus every zero of $X^n - a$ belongs to $K(\varepsilon, \alpha) \subseteq L$, which means that the splitting field L of $X^n - a$ is contained in $K(\varepsilon, \alpha)$. Thus, we get the equality $L = K(\varepsilon, \alpha)$.

5.12 (a) Let $X^p - a = \prod_{i=0}^{p}(X - \varepsilon^i\alpha)$ in a splitting field $L = K(\alpha, \varepsilon)$ of $X^p - a$ over K, where $\alpha^p = a$ and ε is a generator of the group of all solutions of $X^p = 1$ in L. If $X^p - a = (X^k + \cdots + b)(X^l + \cdots + c)$ is reducible over K, then $b = \varepsilon^{k'}\alpha^k$ and $c = \varepsilon^{l'}\alpha^l$, where $0 \leq k', l' \leq p-1$ and $k + l = p$. Hence k, l are relatively prime, since p is a prime number. Let r, s be integers such that $kr + ls = 1$ (see A.3.4(b)). Then

$$b^r c^s = (\varepsilon^{k'}\alpha^k)^r(\varepsilon^{l'}\alpha^l)^s = \eta\alpha,$$

where $\eta^p = 1$ ($\eta = \varepsilon^{k'r+l's}$). But $b^r c^s \in K$ and $\eta\alpha$ is a zero of $X^p - a$, so this polynomial, if reducible over K, has a zero in K.

5.13 For $p = 2$, we have the factorization $X^4 + 1 = (X + 1)^4$. If $p > 2$, then p is odd, so the number $p^2 - 1$ is divisible by 8 ($p = 2k + 1$ for an integer k, so

$p^2 - 1 = 4k(k+1)$, where k or $k+1$ is even). Consider the field $\mathbb{F}_{p^2}$ (see **T. 5.4**). The group $\mathbb{F}^*_{p^2}$ is cyclic of order $p^2 - 1$, so it contains an element α of order 8. We have $\alpha^4 = -1$, since $\alpha^8 = 1$ and the order of α in the group $\mathbb{F}^*_{p^2}$ is 8. Hence α is a zero of the polynomial $X^4 + 1$. The minimal polynomial of α over $\mathbb{F}_p$ has degree 1 or 2 since the extension $\mathbb{F}_p \subset \mathbb{F}_{p^2}$ has degree 2. Hence $X^4 + 1$ must be divisible modulo p by a polynomial of degree 1 or 2 with coefficients in $\mathbb{F}_p$ (in fact both cases can happen, that is, the polynomial X^4+1 is either a product of two irreducible quadratic polynomials or four linear polynomials modulo p (why not any other possibility?).

Problems of Chap. 6

6.1 (b) Every element $x \in K(\alpha_1, \dots, \alpha_r)$ can be written as a fraction $x = \frac{p(\alpha_1,\dots,\alpha_r)}{q(\alpha_1,\dots,\alpha_r)}$, where $p(X_1, \dots, X_r), q(X_1, \dots, X_r) \in K[X_1, \dots, X_r]$. Since σ and τ are automorphisms, we have

$$\sigma(p(\alpha_1, \dots, \alpha_r)) = p(\sigma(\alpha_1), \dots, \sigma(\alpha_r)) = p(\tau(\alpha_1), \dots, \tau(\alpha_r)) = \tau(p(\alpha_1, \dots, \alpha_r))$$

and similarly for q. Hence, we have

$$\sigma(x) = \frac{\sigma(p(\alpha_1, \dots, \alpha_r))}{\sigma(q(\alpha_1, \dots, \alpha_r))} = \frac{\tau(p(\alpha_1, \dots, \alpha_r))}{\tau(q(\alpha_1, \dots, \alpha_r))} = \tau(x)$$

for every $x \in K(\alpha_1, \dots, \alpha_r)$, so σ and τ are equal as automorphisms of L over K.

6.2 Example. We find the Galois group $G(L/K)$, when $K = \mathbb{Q}$ and $L = \mathbb{Q}(\sqrt{2+\sqrt{3}})$. Denote $\sqrt{2+\sqrt{3}}$ by α. Squaring two times, we get the polynomial $f(X) = X^4 - 4X^2 + 1$ such that $f(\alpha) = 0$. The polynomial $f(X)$ is irreducible over $\mathbb{Q}$, which can be easily checked by one of the methods described in Chap. 3 (for example using Gauss's Lemma). Hence $[L : \mathbb{Q}] = 4$. Let σ be an automorphism of L over $\mathbb{Q}$. Then according to Exercise 6.1(a), the image $\sigma(\alpha)$ must also be a zero of $f(X)$. Thus, we have at most four automorphisms according to Exercise 6.1(b), since the degree of $f(X)$ is 4. How many zeros of $f(X)$ are contained in L? The zeros are of course $\pm\sqrt{2 \pm \sqrt{3}}$. Defining $\beta = \sqrt{2-\sqrt{3}}$, we observe that $\alpha\beta = 1$, so all four zeros of $f(X)$ are in L. According to **T. 5.1**(b), there is an automorphism of L mapping α onto any of these four zeros. Thus, $G(L/\mathbb{Q}) = \{\sigma_0, \sigma_1, \sigma_2, \sigma_3\}$, where $\sigma_0(\alpha) = \alpha$, $\sigma_1(\alpha) = -\alpha$, $\sigma_2(\alpha) = \beta$, $\sigma_3(\alpha) = -\beta$.

6.3 (a) Let σ be an automorphism of a field. Since $\sigma(1) = 1$ for every automorphism of any field, we get $\sigma(2) = \sigma(1+1) = 2$, $\sigma(3) = \sigma(2+1) = 3$ and so on, so $\sigma(n) = n$ for any multiple n of unity 1 (very formally, it is an easy induction). We also have $\sigma(-1) = -1$ for every automorphism of any field (since $(-1) + 1 = 0$ and $\sigma(0) = 0$), so we get that $\sigma(n) = n$ holds for all integer multiples of 1. The

prime field consists of only such multiples when its characteristic is nonzero (the fields $\mathbb{F}_p$ for all prime numbers p). If the prime field is the field of rational numbers $\mathbb{Q}$, then each element is a quotient of two integers, so the automorphism σ, being constant on $\mathbb{Z}$, is also constant on $\mathbb{Q}$.
(b) and (c) Use Exercise 2.12, from which it follows that $\sigma(x) = x^p$ is an automorphism of $\mathbb{F}_p$. In fact, we have $\sigma(x+y) = (x+y)^p = x^p + y^p = \sigma(x) + \sigma(y)$ by this exercise and, trivially, $\sigma(xy) = \sigma(x)\sigma(y)$. Moreover, the kernel of σ is trivial (consists of only 0), since $\sigma(x) = x^p = 0$ gives $x = 0$. Hence, the mapping σ is one-to-one, that is, it maps $\mathbb{F}_p$ onto the whole field $\mathbb{F}_p$ (p different elements map to p different elements). Thus σ really is an automorphism of the field $\mathbb{F}_p$. As we know from (a), it must be the identity, that is, we have $\sigma(x) = x^p = x$ for every $x \in \mathbb{F}_p$. If a is an integer, then for its residue x modulo p, we have $x^p = x$, which means that p divides $a^p - a$ (both a^p and a have the same residue modulo p). One can also formulate this more formally in terms of the natural homomorphism $\varphi : \mathbb{Z} \to \mathbb{F}_p$ of reduction modulo p ($\varphi(a)$ = residue of a modulo p). For the image of $a^p - a$, we have $\varphi(a^p - a) = x^p - x = 0$, so p divides $a^p - a$.

6.5 The field $M = L^{G(L/K)}$ is a subfield of L containing K. By **T. 6.2**, we have $[L : M] = [L : L^{G(L/K)}] = |G(L/K)|$. Thus, by the Tower Law (**T. 4.3**), we have $|G(L/K)| = [L : M] \mid [L : K]$.

6.6 (a) If $\sigma : K(X) \to K(X)$ is an automorphism of $K(X)$, then $\sigma(X) = \frac{p(X)}{q(X)}$, where $p(X), q(X)$ are relatively prime polynomials in X and both cannot be constants. By Exercise 4.11(c), we know that the degrees of these polynomials cannot be bigger than 1 when σ, as an automorphism of the field $K(X)$, maps this field onto itself. Thus $\sigma(X)$ has the required form. The condition $ad - bc \neq 0$ is equivalent to the requirement that $\sigma(X)$ is not a constant.

6.7 By Exercise 6.6, the group G has six elements: $\sigma_1(X) = X, \sigma_2(X) = X + 1, \sigma_3(X) = \frac{1}{X}, \sigma_4(X) = \frac{1}{X+1}, \sigma_5(X) = \frac{X}{X+1}, \sigma_6(X) = \frac{X+1}{X}$. This group is nonabelian and as such isomorphic to the group S_3 of permutations on the set $\{1, 2, 3\}$.

Hence, by **T. 6.2**, we get $[L : L^G] = |G| = 6$, where $L = \mathbb{F}_2(X)$. The function $\alpha = \frac{(X^3+X+1)(X^3+X^2+1)}{(X^2+X)^2}$ has degree 6 and it is not too difficult to check that every automorphism σ_i for $i = 1, \ldots, 6$ maps it onto itself. Thus $\alpha \in L^G$ and we have a chain of fields: $\mathbb{F}_2(\alpha) \subseteq L^G \subset \mathbb{F}_2(X) = L$. Since $[L : L^G] = 6$ and $[L : \mathbb{F}_2(\alpha)] = 6$ by Exercise 4.11(c), the Tower Law (**T. 4.3**) implies $L^G = \mathbb{F}_2(\alpha)$.

How is it possible to find the element α? The automorphism group G of order 6 has a (normal) subgroup of order 3 generated by an element of this order, for example, σ_4 (in fact, we have $\sigma_4^2(X) = \sigma_4(\frac{1}{X+1}) = \frac{X+1}{X} = \sigma_6(X)$ and $\sigma_4^3(X) = \sigma_4(\sigma_6(X)) = \sigma_4(\frac{X+1}{X}) = X$, that is, $\sigma_4^3(X) = X$). For the cyclic group $\langle\sigma_4\rangle$ of order 3 generated by σ_4, we get $L^{\langle\sigma_4\rangle} = \mathbb{F}_2(\beta)$, where

$$\beta = \mathrm{Tr}_{\langle\sigma_4\rangle}(X) = X + \sigma_4(X) + \sigma_4^2(X) = X + \frac{1}{X+1} + \frac{X+1}{X} = \frac{X^3+X+1}{X^2+X},$$

since $\mathbb{F}_2(\beta) \subseteq L^{\langle\sigma_4\rangle} \subset L$ and the degree of L over both smaller subfields in the chain is 3 by **T. 6.2** and Exercise 4.11 correspondingly. Now we note that the automorphism σ_2 is an automorphism of $L^{\langle\sigma_4\rangle}$ when restricted to this field. First, we observe that $\sigma_2\sigma_4 = \sigma_4^2\sigma_2$. Then we check that if $\varphi \in L^{\langle\sigma_4\rangle}$, that is, $\sigma_4(\varphi) = \varphi$, then $\sigma_4(\sigma_2(\varphi)) = \sigma_2(\sigma_4^2(\varphi)) = \sigma_2(\varphi)$, that is, $\sigma_2(\varphi) \in L^{\langle\sigma_4\rangle}$.

Since we know that $\beta \in L^{\langle\sigma_4\rangle}$, we now take the group $\langle\sigma_2\rangle$ of order 2 acting on $L^{\langle\sigma_4\rangle}$ and choose

$$\alpha = \mathrm{Nr}_{\langle\sigma_2\rangle}(\beta) = \beta\sigma_2(\beta) = \frac{(X^3 + X^2 + 1)(X^2 + X + 1)}{(X^2 + X)^2}$$

for which not only $\sigma_4(\alpha) = \alpha$, but also $\sigma_2(\alpha) = \alpha$. Thus α is fixed by all σ_i in G, since σ_2, σ_4 generate this group (in terms of S_3, or the group of symmetries of an equilateral triangle, σ_4 is a rotation and σ_2 a symmetry). Thus (as noted earlier), we have a chain of fields $\mathbb{F}_2(\alpha) \subseteq L^G \subset L$ in which the degree of L over the two smaller fields is 6 by **T. 6.2** and Exercise 4.11 correspondingly. We get $L^G = \mathbb{F}_2(\alpha)$ by the Tower Law.

6.8 Example. Consider $L = \mathbb{F}_3(X)$ and let G be the group of all automorphisms of L such that $\sigma(X) = aX + b$, where $a, b \in \mathbb{F}_3$ and $a \neq 0$. We want to find L^G.

It follows from Exercise 6.6 that we have the following automorphisms in G: $\sigma_1(X) = X, \sigma_2(X) = X+1, \sigma_3(X) = X+2, \sigma_4(X) = 2X, \sigma_5(X) = 2X+1, \sigma_6(X) = 2X + 2$. Thus $|G| = 6$, so according to **T. 6.2**, we have $[L : L^G] = |G| = 6$, where $L = \mathbb{F}_3(X)$. We choose $\alpha = \mathrm{Nr}_G(X) = X(X + 1)(X + 2)2X(2X + 1)(2X + 2) = X^6 + X^4 + X^2 \in L^G$, which gives the chain of fields $\mathbb{F}_3(\alpha) \subseteq L^G \subset L$. It follows from Exercise 4.11(c) that $[L : \mathbb{F}_3(\alpha)] = 6$, so the Tower Law (**T. 4.3**) implies $L^G = \mathbb{F}_3(\alpha) = \mathbb{F}_3(X^6 + X^4 + X^2)$.

6.9 (a) We have $\mathbb{R}(X^2, Y) \subseteq L^G \subset L$, where $L = \mathbb{R}(X, Y)$, since both X^2 and Y are fixed by the automorphisms σ_1, σ_2. But $[\mathbb{R}(X, Y) : \mathbb{R}(X^2, Y)] \leq 2$, since $X \in \mathbb{R}(X, Y)$ satisfies the equation $T^2 - X^2 = 0$ over the field $\mathbb{R}(X^2, Y)$. Now $[L : L^G] = |G| = 2$ (by **T. 6.2**) and the Tower Law imply that $L^G = \mathbb{R}(X^2, Y)$.
(b) Notice that the condition $f(-X, Y) = -f(X, Y)$ implies that

$$\sigma_2\left(\frac{f(X, Y)}{X}\right) = \frac{f(-X, Y)}{-X} = \frac{f(X, Y)}{X}.$$

This says that $\frac{f(X,Y)}{X} \in \mathbb{R}(X^2, Y)$, that is, there is a rational function $h \in \mathbb{R}(X, Y)$ such that $f(X, Y)/X = h(X^2, Y)$. Thus $f(\sin x, \cos x)/\sin x = h(1 - \cos^2 x, \cos x)$ and using standard substitutions, we get:

$$\int f(\sin x, \cos x)dx = \left\{\begin{array}{l} t = \cos x, \\ dt = -(\sin x)dx \end{array}\right\} = \int -h(1 - t^2, t)dt = \int f(t)dt,$$

where $f(t) = -h(1 - t^2, t)$.

6.10 (a) Notice that $X^2, X/Y \in L^G$ and consider the chain $\mathbb{R}(X^2, X/Y) \subseteq L^G \subset L$, where $L = \mathbb{R}(X, Y)$. The degree of L over both subfields in the chain is two (similar arguments to those in Exercise 6.9(a)), so $L^G = \mathbb{R}(X^2, X/Y)$ (we could take XY instead of X/Y but our choice depends on (b)).
(b) Similarly to Exercise 6.9(b), we note that the condition $f(-X, -Y) = f(X, Y)$ implies that $f(X, Y) \in L^G = \mathbb{R}(X^2, X/Y)$, that is, there is a rational function $h \in \mathbb{R}(X, Y)$ such that $f(X, Y) = h(X^2, X/Y)$. Hence:

$$\int f(\sin x, \cos x)dx = \int h(\sin^2 x, \tan x)dx = \left\{ \begin{array}{l} t = \tan x, \\ dt = (1+t^2)dx \end{array} \right\}$$

$$= \int h\left(\frac{t^2}{1+t^2}, t\right) \frac{1}{1+t^2} dt = \int f(t)dt,$$

where $f(t) = h\left(\frac{t^2}{1+t^2}, t\right) \frac{1}{1+t^2}$.

6.11 It is clear that $X^2, Y^2 \in L^G$. Thus we have a chain of fields $\mathbb{Q}(X^2, Y^2) \subseteq L^G \subset L = \mathbb{Q}(X, Y)$. Show that $[L : \mathbb{Q}(X^2, Y^2)] \leq 4$ and prove that $L^G = \mathbb{Q}(X^2, Y^2)$.

6.12 Notice that $\sigma(X) = X + 1$ generates an infinite group (what are $\sigma^2(X), \sigma^3(X), \ldots$?). Use **T. 6.2** and Exercise 4.11(a) in order to conclude that $\mathbb{Q}(X)^G$ cannot contain any nonconstant functions.

6.13 (b) It is clear that all the elementary symmetric polynomials $s_1, s_2, \ldots, s_n$ belong to L^G, since they do not change under any permutation of $X_1, X_2, \ldots, X_n$. Thus, we have a chain of fields $K(s_1, s_2, \ldots, s_n) \subseteq L^G \subset K(X_1, X_2, \ldots, X_n)$ and we have to show that $L^G = K(s_1, s_2, \ldots, s_n)$. Since $[L : L^G] = n!$ by **T. 6.2** and the fact that $|S_n| = n!$, we only need to prove that $[L : K(s_1, s_2, \ldots, s_n)] \leq n!$ and use the Tower Law (**T. 4.3**).
It is clear that $L = K(X_1, X_2, \ldots, X_n)$ is a splitting field of the polynomial $(X - X_1)(X - X_2) \cdots (X - X_n) = X^n - s_1 X^{n-1} + s_2 X^{n-2} - \cdots + (-1)^n s_n$ of degree n over the field $K(s_1, s_2, \ldots, s_n)$ as the zeros of this polynomial are $X_1, X_2, \ldots, X_n$. Thus, it is clear that the degree of this splitting field over the field containing its coefficients cannot be bigger than $n!$ (see Exercise 5.3).

6.15 (a) Since ε is a primitive m-th root of unity, every other root of unity is its power. If $\varepsilon_i = \varepsilon^{k_i}$ for an integer $0 \leq k_i < m$, we take $m_i = m - k_i$ for $i = 2, \ldots, n$. Then, we have

$$\sigma(X_1^{m_i} X_i) = \varepsilon^{m_i} \varepsilon_i X_1^{m_i} X_i = \varepsilon^{m_i + k_i} X_1^{m_i} X_i = X_1^{m_i} X_i,$$

which shows that $X_1^{m_i} X_i \in L^G$. Hence, we have a chain of fields

$$\mathbb{C} \subset \mathbb{C}(X_1^m, X_1^{m_2} X_2, \ldots, X_1^{m_n} X_n) \subseteq L^G \subset L = \mathbb{C}(X_1, \ldots, X_n).$$

We have $[L : L^G] = m$ by **T. 6.2**, since $|G| = m$. Next X_1 is an algebraic element of degree at most m over the field $\mathbb{C}(X_1^m, X_1^{m_2}X_2, \ldots, X_1^{m_n}X_n)$ and adjoining it to this field gives L. Thus, we have $[L : \mathbb{C}(X_1^m, X_1^{m_2}X_2, \ldots, X_1^{m_n}X_n)] \leq m$. The conclusion is that $L^G = \mathbb{C}(X_1^m, X_1^{m_2}X_2, \ldots, X_1^{m_n}X_n)$ by the Tower Law (**T. 4.3**).
(b) We use an inductive argument. The claim is of course true for a group with one element. Assume that it is true for finite abelian groups of order less than the order of a finite abelian group G. If G is cyclic, then we have the case considered in (a). If G is not cyclic, then we take $\sigma \in G, \sigma \neq 1$, and consider the subgroup $N = \langle \sigma \rangle$ of some order $m > 1$. The claim is true for N by (a), so we have $L^N = \mathbb{C}(Z_1, \ldots, Z_n)$, where $Z_1 = X_1^m, Z_2 = X_1^{m_2}X_2, \ldots, Z_n = X_1^{m_n}X_n$. Now we take the quotient group G/N. The finite abelian group G/N has order less than the order of G, and acts as an automorphism group of the field L^N (see Exercise 6.14). This action on the variables $Z_1, \ldots, Z_n$ is of the same type as the action of G on L, that is, $\sigma(Z_i) = \sigma(X_1^{m_i}X_i) = \varepsilon_{\sigma,1}^{m_i}\varepsilon_{\sigma,i}X_1^{m_i}X_i = \varepsilon_{\sigma,1}^{m_i}\varepsilon_{\sigma,i}Z_i$. Thus, applying induction, we get (see Exercise 6.14) that $\left(L^N\right)^{G/N} = \mathbb{C}(Y_1, \ldots, Y_n)$, where Y_i are products of suitable powers of X_i.

6.16 Let $H_1 = \langle \sigma_1 \rangle$, where $\sigma_1(X) = -\frac{1}{X}$ and $H_2 = \langle \sigma_2 \rangle$, where $\sigma_2(X) = -\frac{1}{X+1}$. Notice that $\sigma_1^2 = 1$ and $\sigma_2^3 = 1$, so $|\langle \sigma_1 \rangle| = 2$ and $|\langle \sigma_2 \rangle| = 3$. Thus $[L : L^{\langle \sigma_1 \rangle}] = 2$ and $[L : L^{\langle \sigma_2 \rangle}] = 3$ by **T. 6.2**. But $\sigma_3(X) = \sigma_1(\sigma_2(X)) = X + 1$, so σ_3 has infinite order and $L^{\langle \sigma_3 \rangle} = K$ (see Exercise 6.12). Taking $K_1 = L^{\langle \sigma_1 \rangle}$ and $K_2 = L^{\langle \sigma_2 \rangle}$, we get a required example, since $K_1 \cap K_2 = L^{\langle \sigma_1 \rangle} \cap L^{\langle \sigma_2 \rangle} \subseteq L^{\langle \sigma_3 \rangle} = K$, so $[L : K_1 \cap K_2] = \infty$ but $[L : K_1] \neq \infty$ and $[L : K_2] \neq \infty$.

Problems of Chap. 7

7.1 Example 1. Let $L = \mathbb{Q}(\sqrt{2 + \sqrt{6}})$. Is L normal over $\mathbb{Q}$? Squaring two times, we find that $\alpha = \sqrt{2 + \sqrt{6}}$ is a zero of the polynomial $f(X) = X^4 - 4X^2 - 2$, which is irreducible by Eisenstein's criterion for $p = 2$ (see Exercise 3.7). All zeros of $f(X)$ are $\pm\sqrt{2 \pm \sqrt{6}}$. Do they belong to L? Define $\beta = \sqrt{2 - \sqrt{6}}$. We have $\alpha\beta = \sqrt{-2}$, so β cannot belong to L, which is a real field. Thus L is not normal over $\mathbb{Q}$.
Example 2. Consider $L = \mathbb{Q}(\sqrt{5 + \sqrt{3}}, \sqrt{2})$. Defining $\alpha = \sqrt{5 + \sqrt{3}}$ and squaring two times, we find that α is a zero of the polynomial $f(X) = X^4 - 10X^2 + 22$, which is irreducible by Eisenstein's criterion for $p = 2$. All zeros of $f(X)$ are $\pm\sqrt{5 \pm \sqrt{3}}$. Defining $\beta = \sqrt{5 - \sqrt{3}}$, we have $\alpha\beta = \sqrt{2}$, so all four zeros of $f(X)$ are in L. Of course, the field L contains the splitting field of $g(X) = X^2 - 2$, so L is a splitting field of $f(X)g(X)$. As such, it is a normal extension of $\mathbb{Q}$.

7.2 Example. We find a normal closure of $L = \mathbb{Q}(\sqrt[3]{5}, i)$ over $\mathbb{Q}$. According to **T. 7.2**, we can do it, finding the minimal polynomials of the generators and taking a splitting field of their product. The minimal polynomial of $\sqrt[3]{5}$ over $\mathbb{Q}$ is $X^3 - 5$ and the minimal polynomial of i over $\mathbb{Q}$ is $X^2 + 1$. Thus the normal closure is the splitting field of $f(X) = (X^3 - 5)(X^2 + 1)$. Since $X^3 - 5$ has zeros $\varepsilon^i\sqrt[3]{5}$, where

$i = 1, 2, 3$ and $\varepsilon^3 = 1, \varepsilon \neq 1$, we only need to extend L by ε in order to get a splitting field of $f(X)$. Of course, the splitting field of $X^3 - 5$ must contain ε as a quotient of two of its zeros. Thus a normal closure of $L \supset \mathbb{Q}$ is $N = \mathbb{Q}(\sqrt[3]{5}, i, \varepsilon)$.

7.3 Take $K = \mathbb{Q}, M = \mathbb{Q}(\sqrt{2})$ and $L = \mathbb{Q}(\sqrt[4]{2})$. Then $K \subset M$ is normal (M is the splitting field of $X^2 - 2$ over K), $M \subset L$ is normal (L is the splitting field of $X^2 - \sqrt{2}$ over M), but $K \subset L$ is not normal, since the polynomial $X^4 - 2$, which is irreducible over K, has one zero (even two) in L but not all its zeros (L is real and the two remaining zeros $\pm i\sqrt[4]{2}$ are not real).
(b) The answer is yes, since the same polynomial which gives L as a splitting field over K gives L as a splitting field over M.
(c) For example, take Exercise 7.1(g) ($M = \mathbb{Q}(\sqrt[4]{2})$).

7.4 (a) According to **T. 5.1**(b), there is an isomorphism of the fields $K(\alpha)$ and $K(\beta)$ over K. By **T. 5.2** it is possible to find an automorphism of L over K which extends this isomorphism.
(b) Since L is a finite and normal extension of K, it is a splitting field of a polynomial $f(X) \in K[X]$ and also a splitting field of the same polynomial over both M and M' (see **T. 7.1**). Thus for $\tau : M \to M'$, there exists a $\sigma : L \to L$ which restricted to M is equal to τ by **T. 5.2**.

7.5 (a) Let $f(X)$ be a minimal polynomial over K of an element $\alpha \in M$. If $\sigma \in G(L/K)$, then by Exercise 6.1, we know that $\sigma(\alpha)$ is a zero of $f(X)$ and by the definition of a normal extension $\sigma(\alpha) \in M$. Hence $\sigma(M) \subseteq M$, but taking into account that M and $\sigma(M)$ have the same dimension over K, we get $\sigma(M) = M$.
(b) Given (a), it remains to prove that if $\sigma(M) = M$ for each $\sigma \in G(L/K)$, then $M \supseteq K$ is normal. Let $f(X)$ be an irreducible polynomial over K having a zero α in M. Thus all zeros of $f(X)$ are in L. Let β be any of them. By Exercise 7.4, there exists an automorphism $\sigma \in G(L/K)$ such that $\sigma(\alpha) = \beta$. But $\sigma(M) = M$, so $\beta \in M$, that is, all zeros of $f(X)$ are in M. This proves that M is normal over K.

7.6 (a) Let M_1 be a splitting field of $f_1(X)$ over K and M_2 a splitting field of $f_2(X)$ over K. Then M_1M_2 is a splitting field of $f_1(X)f_2(X)$ over K. For $M_1 \cap M_2$ use the definition of the normal extensions. In (b), we get M_1M_2 as a splitting field of $f_1(X)$ over M_2.

7.7 If $\beta \in K$ then the result is trivial. Assume that $\beta \notin K$. Let $f(X)$ be the minimal polynomial of β over K. Then all zeros of $f(X)$ are in $K(\beta)$. If this polynomial is reducible over $K(\alpha)$, then the coefficients of its irreducible factors belong to $K(\alpha)$ and to $K(\beta)$. Thus these coefficients must belong to $K(\alpha) \cap K(\beta) = K$, which leads to a contradiction, since $f(X)$ is irreducible over K. Hence $f(X)$ must be irreducible over $K(\alpha)$ and $[K(\alpha, \beta) : K(\alpha)] = \deg f(X) = [K(\beta) : K]$ by **T. 4.2**. By the Tower Law (**T. 4.3**), we get $[K(\alpha, \beta) : K] = [K(\alpha, \beta) : K(\alpha)][K(\alpha) : K] = [K(\beta) : K][K(\alpha) : K]$.

7.8 Let α_i and α_j be zeros of two irreducible factors $f_i(X)$ and $f_j(X)$ of $f(X)$ over L. Since α_i and α_j are zeros of the same irreducible polynomial $f(X)$ over K, we know by **T. 5.1** that there is an isomorphism of the fields $K(\alpha_i)$ and $K(\alpha_j)$ mapping α_i onto α_j. Let L be a splitting field of a polynomial $g(X)$ over K. Then $L(\alpha_i)$ is a splitting field of $g(X)$ over $K(\alpha_i)$, and $L(\alpha_j)$ is a splitting field of $g(X)$ over $K(\alpha_j)$. According to **T. 5.2**, there is an isomorphism σ_{ij} of the fields $L(\alpha_i)$ and $L(\alpha_j)$, which extends the isomorphism of $K(\alpha_i)$ and $K(\alpha_j)$ mapping α_i onto α_j.

This isomorphism maps each element $\beta \in L$ to an element of the same field, since β is a zero of an irreducible polynomial with coefficients in K and $\sigma_{ij}(\beta)$ is a zero of the same polynomial (see Exercise 6.1(a)). But all the zeros of the minimal polynomial of β over K are in L, since L is normal over K. Hence σ_{ij} is an automorphism of L, which maps α_i onto α_j. Hence the degrees of $L(\alpha_i)$ and $L(\alpha_j)$ over K are equal. Thus, the degrees of $L(\alpha_i)$ and $L(\alpha_j)$ over L are also equal, which shows that the irreducible factors of $f(X)$ corresponding to α_i and α_j over L have the same degree by **T. 4.2**. (Notice that $[L(\alpha) : L] = [L(\alpha) : K]/[L : K]$.)

Now $\sigma_{ij}(f_i(X))$ is a polynomial in $L[X]$ having α_j as its zero. Hence it is divisible by the minimal polynomial of α_j over L. But both have the same degree and are monic, so they are equal, that is, $\sigma_{ij}(f_i(X)) = f_j(X)$.

Problems of Chap. 8

8.1 (a) The minimal polynomial of X over the field $\mathbb{F}_2(X^2)$ is $T^2 - X^2$. This polynomial is irreducible, since $X \notin \mathbb{F}_2(X^2)$ and is not separable, since its zero X has multiplicity 2. In fact, if $X \in \mathbb{F}_2(X^2)$, then $X = \frac{p(X^2)}{q(X^2)}$, where $p(T), q(T)$ are polynomials over $\mathbb{F}_2$. Then $Xq(X^2) = p(X^2)$ and we get a contradiction, since the polynomial to the left has odd degree, and to the right, even degree.
(b) We want to prove that $L = \mathbb{F}_p(X)$, $K = \mathbb{F}_p(X^p)$ is a non-separable field extension. The minimal polynomial of X over K is $T^p - X^p$, which shows that X is not separable over K using the definition of separability (the polynomial has a multiple zero X) or **T. 8.1**(b). In order to show that $T^p - X^p$ is irreducible over $\mathbb{F}_p(X^p)$, we can use Exercise 4.11(c), which gives $[\mathbb{F}_p(X) : \mathbb{F}_p(X^p)] = p$. Thus $T^p - X^p$ is irreducible over $\mathbb{F}_p(X^p)$ by **T. 4.2**.

8.2 Assume first that the characteristic of K is different from 2. If $\gamma \in L \setminus K$, then $L = K(\gamma)$ and $\gamma^2 + p\gamma + q = 0$ for some $p, q \in K$. Hence, $\alpha = \gamma + p/2$ satisfies $\alpha^2 = a \in K$, where $a = (p/2)^2 - q$ and $L = K(\gamma) = K(\alpha)$. According to **T. 8.1**(b), the extension $K \subset L$ is separable.

If the characteristic of K is 2 and $\gamma \in L \setminus K$, then we have two possibilities: Either γ is not separable and then according to **T. 8.1**(b) its minimal polynomial is $X^2 + a$ for some $a \in K$ or γ is separable and has a minimal polynomial $X^2 + pX + q$ where $p, q \in K$ and $p \neq 0$ by the same theorem. In the first case, we have $L = K(\alpha)$, where $\alpha^2 = a \in K$ if we take $\alpha = \gamma$. In the second case, we can choose $\alpha = (1/p)\gamma$ and then $\alpha^2 = \alpha + a$, where $a = q/p^2$. Of course, we have $L = K(\alpha)$, since $\alpha \in L \setminus K$.

8.3 (a) If α is not separable over K and has minimal polynomial $f(X)$, then by **T. 8.1** (b), we have $f(X) = g(X^p)$ for a polynomial g over K. If $\deg f = n$, then $\deg g = n/p$ and $g(\alpha^p) = 0$, so $[K(\alpha^p) : K] \leq n/p$ and $[K(\alpha) : K] = n$, thus $K(\alpha^p) \subsetneq K(\alpha)$.
Assume now that α is separable and suppose that $K(\alpha^p) \subsetneq K(\alpha)$. Let $f(X)$ and $g(X)$ be the minimal polynomials of α, respectively, over K and $K(\alpha^p)$. Of course, we have $\deg g(X) \geq 2$. Since α is a zero of $X^p - \alpha^p$ over $K(\alpha^p)$, we get that $g(X)$ divides both $f(X)$ and $X^p - \alpha^p$. But the former polynomial has only the multiple zero α (with multiplicity p). Thus $g(X)$ has a multiple zero α, which implies that $f(X)$ has a multiple zero α, which contradicts the assumption that α is separable over K. Thus the separability of α implies $K(\alpha^p) = K(\alpha)$.
(b) If $K \subseteq L$ is separable and $\alpha \in L$, then $\alpha \in K(\alpha^p) \subseteq KL^p$ by (a), so $L \subseteq KL^p \subseteq L$. Hence, we have $KL^p = L$.
Conversely, assume that $KL^p = L$. This equality implies that if $\alpha_1 = \alpha, \ldots, \alpha_r \in L$ are linearly independent over K, then $\alpha_1^p, \ldots, \alpha_r^p$ are also linearly independent over K. In fact, we can assume that $r = [L : K]$ (if not, we can extend these elements to a basis of L over K). Then L is the set of all linear combinations of $\alpha_1, \ldots, \alpha_r$ with coefficients in K. Hence, L^p consists of all linear combinations of the elements $\alpha_1^p, \ldots, \alpha_r^p$ with coefficients in K^p, and $L = KL^p$ is the set of all linear combinations of $\alpha_1^p, \ldots, \alpha_r^p$ with coefficients in K. Thus $\alpha_1^p, \ldots, \alpha_r^p$ generate L over K and since the number of them is equal to the dimension of L over K, they form a basis of this space. Hence they are linearly independent.
Now we get that any $\alpha \in L$ must be separable over K. If not, then the minimal polynomial $f(X)$ of α over K has the form $f(X) = g(X^p)$ for a polynomial $g(X) \in K[X]$ by **T. 8.1**(b). Let $f(X) = X^{pm} + a_{m-1}X^{p(m-1)} + \cdots + a_1X^p + a_0$, $a_i \in K$. Since $g(\alpha^p) = 0$, we get that $(\alpha^m)^p, (\alpha^{m-1})^p \ldots, \alpha^p, 1$ are linearly dependent but $\alpha^m, \alpha^{m-1}, \ldots, \alpha, 1$ are linearly independent, since by **T. 4.2** all powers $\alpha^{pm-1}, \ldots, \alpha, 1$ are linearly independent.
(c) Assume that K is perfect. If $a \in K$, take any irreducible factor $f(X)$ of $X^p - a$. If $\deg f(X) > 1$, then a splitting field of $f(X)$ over K cannot be separable, since $f(X)$ has only one zero with multiplicity equal to its degree (the same is true of the polynomial $X^p - a$). Thus all irreducible factors of $X^p - a$ have degree 1, which means that $X^p - a$ has a zero $b \in K$, that is, we have $K \subseteq K^p$. Since always, $K^p \subseteq K$, we get $K = K^p$.
Conversely, if $K = K^p$, then use the same argument as in the proof of **T. 8.1**(a) on p. 121 in order to show that K is perfect.
(d) By (b), we know that L is separable over K if and only if $KL^p = L$. But of course, we have $L^p = L$, since L is algebraically closed (for each $a \in L$, the equation $X^p - a = 0$ has a solution in L). Now if $f(X) \in K[X]$ is an irreducible polynomial, then it splits in $L[X]$ as L is algebraically closed. Thus, the extension $K \subseteq L$ is normal.

8.4 (a) By **T. 8.1**(a) every field extension of a field K of characteristic 0 is separable, so we can assume that K is a field of finite characteristic p.

We have to show that if the elements $\alpha_1, \ldots, \alpha_n$ are separable over K, then every $\beta \in K(\alpha_1, \ldots, \alpha_n)$ is separable over K. Notice that by Exercise 8.3(a), we have $K(\alpha_1^p, \ldots, \alpha_n^p) = K(\alpha_1, \ldots, \alpha_n)$ as $K(\alpha_i^p) = K(\alpha_i)$ for every $i = 1, \ldots, n$.
According to **T. 8.2**, there exists an $\alpha \in K(\alpha_1, \ldots, \alpha_n)$ such that $K(\alpha_1, \ldots, \alpha_n) = K(\alpha)$. Notice that α is separable over K, since otherwise, we have $K(\alpha^p) \subsetneq K(\alpha)$ and noting that $\alpha_i^p \in K(\alpha^p)$ for every $i = 1, \ldots, n$, we get a contradictory inclusion: $K(\alpha_1^p, \ldots, \alpha_n^p) \subseteq K(\alpha^p) \subsetneq K(\alpha) = K(\alpha_1^p, \ldots, \alpha_n^p)$.
Let $\beta \in K(\alpha)$ and let $f(X)$ be the minimal polynomial of α over $K(\beta)$. Then $f(X)^p = f_1(X^p)$, where $f_1(X)$ is a polynomial whose coefficients are p-powers of the coefficients of $f(X)$. Thus α^p is a zero of the polynomial f_1 with coefficients in $K(\beta^p)$ whose degree, say m, is equal to the degree of $f(X)$. Hence $[K(\alpha) : K(\beta^p)] \leq m$ and $[K(\alpha) : K(\beta)] = m$, so $K(\beta^p) = K(\beta)$, by the Tower Law, since $K(\beta^p) \subseteq K(\beta)$. According to Exercise 8.3(a), the element β is separable over K.

8.5 Let $\alpha, \beta \in L$ be separable over K. By Exercise 8.4, the field $K(\alpha, \beta)$ is a separable extension of K. Thus $\alpha \pm \beta, \alpha\beta$ and α/β, when $\beta \neq 0$, are separable over K as elements belonging to $K(\alpha, \beta)$.

8.6 Let $f(X)$ be the minimal polynomial of $\alpha \in L$ over K. If α is separable over K, then $r = 0$, since $\alpha \in L_s$. Assume that α is not separable over K. Then by **T. 8.1**(b), there is a polynomial $f_1(X) \in K[X]$ such that $f(X) = f_1(X^p)$. Of course, the polynomial $f_1(X)$ is irreducible over K, since if it is reducible, then $f(X)$ is also reducible, which is not the case. Let r be the biggest exponent such that $f(X) = f_r(X^{p^r})$ for some polynomial $f_r(X) \in K[X]$ (we have $r \geq 1$). We claim that α^{p^r} is separable over K, that is, $\alpha^{p^r} \in L_s$. If not, then for the polynomial $f_r(X)$, which is irreducible over K and minimal for α^{p^r}, we have $f_r(X) = g(X^p)$ by **T. 8.1**(b). This gives $f(X) = f_r(X^p) = g(X^{p^{r+1}})$, which contradicts the definition of r. Hence, we have $\alpha^{p^r} \in L_s$.

8.7 (a) Since L_s is a finite extension of K, we have $L_s = K(\alpha)$ for a separable element $\alpha \in L$ by **T. 8.2**. Let $f(X)$ be the minimal polynomial of α over K and let $\alpha_1 = \alpha, \ldots, \alpha_m$ be all zeros of $f(X)$ in N. Notice that all zeros of $f(X)$ are in N, since N is normal over K. Moreover, all these zeros are different since α is separable over K. Of course, we have $[K(\alpha) : K] = m$ by **T. 4.2**. It follows from Theorem **T. 5.1** that for every α_i $(i = 1, \ldots, m)$, there exists an isomorphism $\tau_i : K(\alpha) \to K(\alpha_i)$ such that $\tau(\alpha) = \alpha_i$. Using Exercise 7.4(b), we get an automorphism $\sigma_i : N \to N$ such that σ_i restricted to $K(\alpha)$ is equal to τ_i. Since every isomorphism of $K(\alpha)$ over K maps α onto a zero of $f(X)$, we see that there are exactly m images of $K(\alpha)$ under the automorphisms of N over K, that is, the elements of $G(N/K)$. What remains to show is that m is exactly the number of restrictions of these automorphisms to L. If the characteristic of K is zero, then L is separable over K, so $L = L_s$ and there is nothing to prove. So assume that K has a finite characteristic p. By Exercise 8.6, for every $\alpha \in L$, there is an exponent r such that $\alpha^{p^r} \in L_s$. If now $\sigma|_{L_s} = \sigma'|_{L_s}$ for $\sigma, \sigma' \in G(N/K)$, so $\sigma(\alpha^{p^r}) = \sigma'(\alpha^{p^r})$, which gives $(\sigma(\alpha) - \sigma'(\alpha))^{p^r} = 0$ (see Exercise 2.12), so $\sigma(\alpha) = \sigma'(\alpha)$). Thus $\sigma|_L = \sigma'|_L$, which means that there are exactly m different restrictions of automorphisms of N over K to L.

(b) Let $K' = \tau(K) \subseteq N$. The proof is almost the same as in (a)—we only have to take the zeros of the polynomial $\tau(f(X))$ in the field N, which contains all the zeros of this polynomial, since it is algebraically closed (in fact, it is normal over K'). These zeros are all different, since $\tau(f(X))$, like $f(X)$, is separable (see **T. 8.1**). This time, we only count the number of embeddings of L into N, which are defined by m different images of α.

8.8 If $L \supseteq M \supseteq K$ and L is separable over K, then clearly M is separable over K. Every element $\alpha \in L$ is separable over K, so its minimal polynomial $f(X)$ over K has different zeros. The minimal polynomial $g(X)$ of α over M divides $f(X)$ (see **T. 4.1**) so it also has different zeros, that is, the element α is also separable over M.

If both $L \supseteq M$ and $M \supseteq K$ are separable, then by Exercise 8.3(b), we have $KM^p = M$ and $ML^p = L$. Hence $L = ML^p = KM^pL^p \subseteq KL^p \subseteq L$, which implies $KL^p = L$, that is, the extension $K \subseteq L$ is separable by Exercise 8.3(b).

8.9 We want to use Exercise 8.7(b). Let N be any algebraically closed field in which we can embed the field K. For example, we take $N = \overline{K}$, the algebraic closure of K with the identity as an embedding of K into $\overline{K}$. Let $[L : K]_s = n$, $[M : K]_s = m$ and $[L : M]_s = q$. By Exercise 8.7(b), there are s embeddings $\tau_1, \dots, \tau_s$ of M into $\overline{K}$ which restrict to the identity on K. For every τ_i ($i = 1, \dots, s$), there exists exactly q embeddings of L into $\overline{K}$, which restricted to M are equal to τ_i. Thus, we have qm different embeddings of L into $\overline{K}$, which restricted to K give the identity on K. In this way, we have at least qm different embeddings of L into $\overline{K}$, which restricted to K give the identity. But each embedding of L into $\overline{K}$, which is the identity on K, when restricted to M must be equal to one of the τ_i. Hence, it must be one of the qm embeddings, so $qm = n$.

8.10 Example. We find a primitive element of $L = \mathbb{Q}(i, \sqrt[3]{2})$ over $K = \mathbb{Q}$. We have $[L : K] = 6$ (see Exercise 4.4(c)). Hence, according to the proof of **T. 8.2**, we have to find $c \in K$ such that $\alpha = i + c\sqrt[3]{2}$ has six different images under the isomorphisms of L into the complex numbers. These isomorphisms map i onto $\pm i$ and $\sqrt[3]{2}$ onto $\varepsilon\sqrt[3]{2}$, where ε is one of the three third roots of unity ($\varepsilon = 1, \frac{-1\pm\sqrt{3}}{2}$). It is easy to check that $c = 1$ is a good choice. Hence, we have $L = \mathbb{Q}(i + \sqrt[3]{2})$. Notice that also $L = \mathbb{Q}(i\sqrt[3]{2})$, since $\beta = i\sqrt[3]{2}$ also has six different images under the isomorphisms of L into $\mathbb{C}$. The minimal polynomial of β is $X^6 + 4$ (just take the smallest power of β which gives an integer). Notice also that a normal closure of L is $L(\varepsilon) = \mathbb{Q}(i\sqrt[3]{2}, \varepsilon)$, where $\varepsilon \neq 1$ is a third root of unity (compare Exercise 7.2 on p. 190). The field $\mathbb{Q}(i\sqrt[3]{2}, \varepsilon)$, is the splitting field of $X^6 + 4$ over $\mathbb{Q}$, has degree 12 and Galois group D_6 (see Exercise 10.16(e)).

8.11 Let $K = \mathbb{F}_2(X^2, Y^2)$ and $L = \mathbb{F}_2(X, Y)$. First show that $[L : K] = 4$ considering the chain $L \supset \mathbb{F}_2(X^2, Y) \supset K$. Then show that for each $\gamma \in L$, we have $\gamma^2 \in K$, so the degree $[K(\gamma) : K]$ is at most 2. Thus $L \neq K(\gamma)$.

8.12 (b) We prove that if the number of fields M between K and L is finite, then L is a simple algebraic extension of K. First of all, the field L is algebraic over K, since a transcendental element $X \in L$ generates a subfield $M = K(X)$ which has infinitely many subfields by Exercise 4.11. Next note that L is finitely generated over K, since otherwise, we could construct an infinite chain

$$K \subset K(\alpha_1) \subset K(\alpha_1, \alpha_2) \subset \cdots \subset K(\alpha_1, \alpha_2, \ldots, \alpha_n) \subset \cdots \subset L$$

of subfields of L choosing each time an element in L which does not belong to a finitely generated subfield $K(\alpha_1, \alpha_2, \ldots, \alpha_n)$ for consecutive n. Thus L is a finitely generated algebraic extension of K and as such, it is a finite extension (see **T. 4.4**). Hence L is a vector space of finite dimension over K.
Here, we split our proof into two cases (even if there exist proofs where this is not necessary). If K is finite, then also L is finite and the group L^* is cyclic by Exercise 5.8. Any generator γ of this group generates the field L over K, that is, we have $L = K(\gamma)$.
If K is infinite, we use the fact that a linear space over K cannot be a union of a finite number of its subspaces. Knowing this, we can take the union of all proper subfields of L containing K (a subfield of L containing K is a linear space over K). Since such a union is not equal to L, we choose any $\gamma \in L$ outside this union. Then $L = K(\gamma)$, since $K(\gamma)$ is a subfield of L and it is not proper depending on the choice of γ.
Thus what remains is a proof of the following auxiliary result:

Lemma. *A linear space over an infinite field K is not a union of a finite number of its subspaces.*

Proof. Let a linear space V over an infinite field K be a union of its proper subspaces V_i for $i = 1, \ldots, n$ and assume that n has the minimal value in such presentations of V (that is, no V_i can be removed from the union when it is equal to V). Take $x \in V_1$ and choose $y \in V \setminus V_1$. Now form infinitely many vectors $x + ay$ where $a \in K$ and $a \neq 0$. Notice that all these vectors are not in V_1, since $y \notin V_1$. Hence, there exists an index $i > 1$ such that infinitely many vectors $x + ay \in V_i$. If $x + ay, x + a'y \in V_i$ for $a \neq a'$, then the difference of these vectors is in V_i, which gives that $y \in V_i$. Hence, we have $x \in V_i$. Since x was an arbitrary vector in V_1, this shows that V_1 is contained in the union of V_i for $i > 1$, which contradicts our choice of the presentation of V as a minimal union of its proper subspaces. Hence such a presentation of V is impossible. □

8.13 Take $K = \mathbb{F}_p(X^2, Y^2)$, $L = \mathbb{F}_p(X, Y)$ and consider the fields $M_c = \mathbb{F}_p(X^2, Y^2, X + cY)$, where $c \in K$. Notice that $X + cY \notin K$ and $(X + cY)^2 \in K$, so $M_c \supset K$ is a quadratic field extension. In order to show that the fields M_c are different for different c, use Exercise 2.5(b).

8.14 According to **T. 8.2**, there is an $\alpha \in L$ such that $L = \mathbb{Q}(\alpha)$. Let $f(X)$ be the minimal polynomial of α over $\mathbb{Q}$. Since the degree of L over $\mathbb{Q}$ is odd, the polynomial $f(X)$ has odd degree. Hence, it has a real zero and we can choose it in order to generate L over $\mathbb{Q}$, since L is normal. Thus $L \subset \mathbb{R}$ (and all the zeros of $f(X)$ are real).

8.15 (c) Take $n = 2$. The cyclic group $\mathbb{F}_9^*$ has eight elements and the number of its generators is $\varphi(8) = 4$ (see A.2.2). Since the field $\mathbb{F}_9$ has nine elements and the field $\mathbb{F}_3$ has three elements, the number of elements in the first and not in the second is equal to 6. Each of these six elements generates the field $\mathbb{F}_9$ over $\mathbb{F}_3$.

Problems of Chap. 9

9.1 We look at two examples—a Galois field extension and an extension which is not Galois.
Example 1. Let $K = \mathbb{Q}(\sqrt{2}, \sqrt{3})$. Is $\mathbb{Q} \subset K$ a Galois extension? We notice that $\sqrt{2}$ and $\sqrt{3}$ have minimal polynomials X^2-2 and X^2-3. The zeros of these polynomials are $\pm\sqrt{2}$ and $\pm\sqrt{3}$, so all these numbers are in the field K. Hence, the field K is a splitting field of the polynomial $(X^2-2)(X^2-3)$, so it is normal according to **T. 7.1** and separable (as a field of characteristic zero—see **T. 8.1**). By **T. 9.1**(d), the field K is Galois over $\mathbb{Q}$.
Example 2. Consider now the field $K = \mathbb{Q}(\sqrt{2}, \sqrt[4]{2})$. In this case, the minimal polynomial of $\sqrt[4]{2}$ is $X^4 - 2$ and its zeros are $\pm\sqrt[4]{2}, \pm i\sqrt[4]{2}$. Not all of them are in K, since $i\sqrt[4]{2}$ is not real, but the field K consists of only real numbers. Hence K is not normal over $\mathbb{Q}$ by **T. 7.1**, so it is not Galois by **T. 9.1**(d).

9.2 We consider two examples similar to those in the exercise.
Example 1. Let $f(X) = X^4 - 4X^2 + 1$. We want to find the splitting field of the polynomial, find its Galois group, all its subgroups and the corresponding subfields. We find the zeros of the polynomial: $\pm\sqrt{2 \pm \sqrt{3}}$. Let $\alpha = \sqrt{2+\sqrt{3}}$ and $\beta = \sqrt{2-\sqrt{3}}$. The splitting field is $L = \mathbb{Q}(\alpha, \beta)$, since the zeros are $\pm\alpha, \pm\beta$. Now notice that $\alpha\beta = 1$, so $L = \mathbb{Q}(\alpha)$. The polynomial $f(X)$ is irreducible (check this by one of the methods from Chap. 3). Thus $[L : \mathbb{Q}] = 4$ and the Galois group has order 4. The automorphisms σ_i ($i = 0, 1, 2, 3$) are given by the four possible images of α according to the table below, which defines each automorphism by its effect on α. In the last column, we write the order of the automorphism as an element in the group $G(L/\mathbb{Q})$ (this information helps to list all the subgroups and also to find the corresponding subfields):

G	α	o
σ_0	α	1
σ_1	$-\alpha$	2
σ_2	β	2
σ_3	$-\beta$	2

The subgroups of G: $G, I = \{\sigma_0\}$,
$H_1 = \{\sigma_0, \sigma_1\}, H_2 = \{\sigma_0, \sigma_2\}, H_3 = \{\sigma_0, \sigma_3\}$.
The fields between $\mathbb{Q}$ **and** $L = \mathbb{Q}(\sqrt{2+\sqrt{3}}) = \mathbb{Q}(\sqrt{2}, \sqrt{3})$:
$L^G = \mathbb{Q}, L^I = L$,
$L^{H_1} = \mathbb{Q}(\sqrt{3}), L^{H_2} = \mathbb{Q}(\sqrt{6}), L^{H_3} = \mathbb{Q}(\sqrt{2})$.

It is easy to find the order of σ_1. For σ_2, we have $\alpha \xrightarrow{\sigma_2} \beta = \frac{1}{\alpha} \xrightarrow{\sigma_2} \frac{1}{\beta} = \alpha$, that is, we get the identity after two compositions. Similarly for σ_3. Having the orders of the elements, it is easy to find all subgroups (their orders are $1, 2, 4$ by

Lagrange's theorem—see A.2.6) choosing the identity and one of the elements of order 2 in subgroups of order 2.

In order to find the fixed fields corresponding to nontrivial subgroups H_i for $i = 1, 2, 3$, we take the norms $\mathrm{Nr}_{H_i}(\alpha) = \alpha \cdot \sigma_i(\alpha)$ and the traces $\mathrm{Tr}_{H_i}(\alpha) = \alpha + \sigma_i(\alpha)$ for $i = 1, 2, 3$ (see p. 36), which are fixed by σ_i. We have $\mathrm{Nr}_{H_1}(\alpha) = \alpha \cdot \sigma_i(\alpha) = -\alpha^2$, so α^2 is fixed by σ_1. But $\alpha^2 = 2 + \sqrt{3}$, so $\sqrt{3} = \alpha^2 - 2$ is fixed by σ_1. Moreover, we have $\mathrm{Tr}_{H_2}(\alpha) = \alpha + \sigma_2(\alpha) = \alpha + \beta$. Notice that $(\alpha + \beta)^2 = 6$, so $\alpha + \beta = \pm\sqrt{6}$ (in fact, the sum is equal to $\sqrt{6}$, since it is positive). Thus $\sqrt{6}$ is fixed by σ_2. Finally, we have $\mathrm{Tr}_{H_3}(\alpha) = \alpha + \sigma_3(\alpha) = \alpha - \beta$. But $(\alpha - \beta)^2 = 2$, so $\alpha - \beta = \pm\sqrt{2}$ (in fact, $\sqrt{2}$), which shows that $\sqrt{2}$ is fixed by σ_3. Now $[L : L^{H_i}] = 2$ by **T. 9.1**(a), so $[L^{H_i} : \mathbb{Q}] = 2$ by **T. 4.3** for $i = 1, 2, 3$. Hence $L^{H_1} = \mathbb{Q}(\sqrt{3})$, $L^{H_2} = \mathbb{Q}(\sqrt{6})$, $L^{H_3} = \mathbb{Q}(\sqrt{2})$.

Example 2. Consider the polynomial $f(X) = X^4 - 4X^2 + 2$. The zeros of the polynomial are $\pm\sqrt{2 \pm \sqrt{2}}$. Let $\alpha = \sqrt{2 + \sqrt{2}}$ and $\beta = \sqrt{2 - \sqrt{2}}$. The splitting field is $L = \mathbb{Q}(\alpha, \beta)$, since the zeros are $\pm\alpha, \pm\beta$. Now notice that $\alpha\beta = \sqrt{2}$, so $L = \mathbb{Q}(\alpha)$ as $\beta = \alpha/\sqrt{2}$ and $\sqrt{2} \in \mathbb{Q}(\alpha)$. The polynomial $f(X)$ is irreducible by Eisenstein's criterion (for $p = 2$). Thus $[L : \mathbb{Q}] = 4$ and the Galois group has order 4. As in Example 1 above, the automorphisms σ_i $(i = 0, 1, 2, 3)$ are given by the four possible images of α according to the table below, which defines each automorphism by its effect on α. As before, we write the order of each automorphism as an element in the group $G(L/\mathbb{Q})$. Notice the essential difference between this case and the previous one.

G	α	o
σ_0	α	1
σ_1	$-\alpha$	2
σ_2	β	4
σ_3	$-\beta$	4

The subgroups of G: $G, I = \{\sigma_0\}$,
$H_1 = \{\sigma_0, \sigma_1\}, H_2 = \{\sigma_0, \sigma_2\}, H_3 = \{\sigma_0, \sigma_3\}$.
The fields between $\mathbb{Q}$ **and** $L = \mathbb{Q}(\sqrt{2 + \sqrt{3}}) = \mathbb{Q}(\sqrt{2}, \sqrt{3})$**:**
$L^G = \mathbb{Q}, L^I = L$,
$L^{H_1} = \mathbb{Q}(\sqrt{3}), L^{H_2} = \mathbb{Q}(\sqrt{6}), L^{H_3} = \mathbb{Q}(\sqrt{2})$.

The order of σ_1 is of course 2. To find the orders of σ_3 and σ_4, we need their effect on $\sqrt{2}$, since $\beta = \sqrt{2}/\alpha$. But $\sqrt{2} = \alpha^2 - 2$, so we have $\sigma_2(\sqrt{2}) = \sigma_2(\alpha^2) - 2 = \beta^2 - 2 = -\sqrt{2}$ and similarly, $\sigma_4(\sqrt{2}) = -\sqrt{2}$. Hence

$$\alpha \xrightarrow{\sigma_2} \beta = \frac{\sqrt{2}}{\alpha} \xrightarrow{\sigma_2} \frac{-\sqrt{2}}{\beta} = -\alpha \ .$$

Thus, we have $\sigma_2^2(\alpha) = -\alpha$ (that is, $\sigma_2^2 = \sigma_1$), which means that $\sigma_2^4 = I$, that is, the order of σ_2 is 4. Similarly, we get that σ_3 has order 4. Therefore, the Galois group $G(L/\mathbb{Q})$ is this time cyclic of order 4 (and generated by σ_2 or σ_3). We have three subgroups: $I = \{\sigma_0\}$, $G = G(L/\mathbb{Q})$ and the only nontrivial subgroup $H = \{\sigma_0, \sigma_1\}$. The field L^H corresponding to H is rather evident, since it must be the only quadratic extension contained in L, and such a field is $\mathbb{Q}(\sqrt{2})$, since we already know that $\sqrt{2} \in L$. But we could find L^H using the same method as in Example 1 above by, for example, taking $\mathrm{Nr}_H(\alpha) = \alpha\sigma_1(\alpha) = -(2 + \sqrt{2})$, which is fixed by σ_1 (so $\sqrt{2}$ is also fixed).

9.3 (a) We denote by $X_f = \{\alpha_1, \ldots, \alpha_n\}$ the set of all zeros of the polynomial $f(X)$ $(L = K(\alpha_1, \ldots, \alpha_n))$ and note that according to Exercise 6.1, for every automorphism $\sigma \in G(L/K)$, we get a mapping $\sigma : X_f \to X_f$, where $\sigma(\alpha_i) = \alpha_j$ for some $j \in \{1, \ldots, n\}$. As in the text, it is convenient to denote the permutation of the indices i of the α_i by the same symbol σ even if formally it is a mapping on the set $\{1, \ldots, n\}$. Thus $\sigma(\alpha_i) = \alpha_j$ means that $\sigma(i) = j$. In this way, we get a mapping $\Phi : G(L/K) \to S_n$ defined by $\Phi(\sigma) = \sigma$, where σ on the right-hand side denotes a permutation of the indices $\{1, \ldots, n\}$ of the zeros α_i, that is, $\sigma(i)$ is defined by $\sigma(\alpha_i) = \alpha_{\sigma(i)}$. It is clear that Φ is injective (if $\Phi(\sigma)$ is the identity permutation, then σ is the identity in the Galois group $G(L/K)$). But we also have $\Phi(\sigma\tau) = \Phi(\sigma)\Phi(\tau)$, since $\Phi(\sigma\tau)(i) = \sigma\tau(i) = \Phi(\sigma)\Phi(\tau)(i) = \sigma(\tau(i))$ as $\sigma\tau(\alpha_i) = \alpha_{\sigma\tau(i)} = \sigma(\tau(\alpha_i)) = \sigma(\alpha_{\tau(i)}) = \alpha_{\sigma(\tau(i))}$.

9.4 (a) If $\sigma \in G(L/K)$ and $\alpha_i \in X_f = \{\alpha_1, \ldots, \alpha_n\}$ (X_f is the set of zeros of $f(X)$) is a zero of an irreducible factor $f_i(X)$ of $f(X)$, then the image $\sigma(\alpha_i)$ is also a zero of the same irreducible factor $f_i(X)$ by Exercise 6.1. Hence the permutation of the indices of the zeros, which corresponds to σ according to Exercise 9.3, splits into k orbits, since $f(X)$ has k irreducible factors (see p. 267). This means that the set $X = \{1, \ldots, n\}$ of the indices of the zeros is partitioned into k subsets so that every σ permutes the elements of every such subset.
(b) If $f(X)$ is irreducible and α_i, α_j are two zeros of it, then by **T. 5.1**(b) there exists an isomorphism $\tau : K(\alpha_i) \to K(\alpha_j)$ over K. By **T. 5.2**(b), this isomorphism τ is a restriction of an automorphism $\sigma \in G(K_f/K)$, since K_f is a splitting field of $f(X)$ over both $K(\alpha_i)$ and $K(\alpha_j)$. Thus, $G(K_f/K)$ acts transitively on the zeros of $f(X)$ (see Exercise 9.4).
Conversely, if $G(K_f/K)$ acts transitively on the zeros of $f(X)$, then $f(X)$ cannot be reducible over K, since its reducibility implies, according to (a), that there are at least two orbits ($k > 1$).
(c) Let the set $X_f = \{\alpha_1, \ldots, \alpha_n\}$ of the zeros of $f(X)$ be partitioned into k subsets so that each $\sigma \in G(K_f/K)$ permutes only the elements of each of these sets. Let $X = \{\beta_1, \ldots, \beta_r\}$ be one of these k subsets of X_f. Consider the polynomial $g(X) = (X - \beta_1)\cdots(X - \beta_r)$. The coefficients of this polynomial are the elementary symmetric functions of $\beta_1, \ldots, \beta_r$ (see section "Symmetric Polynomials" in Appendix), so every $\sigma \in G(K_f/K)$ maps every coefficient of $g(X)$ to itself. Thus, these coefficients belong to $K_f^{G(K_f/K)} = K$ (the equality by **T. 9.1**(b)). This shows that $g(X) \in K[X]$ and $g(X)$ is irreducible by (b). Thus $f(X)$ has k irreducible factors corresponding to the given k subsets of X_f.

9.6 The group $V = \mathbb{Z}_2 \times \mathbb{Z}_2$ has three different subgroups of order 2 (every nonzero $(a, b) \in \mathbb{Z}_2 \times \mathbb{Z}_2$ together with $(0, 0)$ form a subgroup). Take two of these subgroups H_1, H_2. Then according to **T. 9.2**, the fields L^{H_1} and L^{H_2} are different and are quadratic extensions of $\mathbb{Q}$ by **T. 9.1**(a) and **T. 4.3**. Now using Exercise 4.5, we find d_1, d_2 such that $L^{H_1} = \mathbb{Q}(\sqrt{d_1})$ and $L^{H_2} = \mathbb{Q}(\sqrt{d_2})$. The product d_1d_2 is not a square, since otherwise $\mathbb{Q}(\sqrt{d_1}) = \mathbb{Q}(\sqrt{d_2})$, which contradicts the choice of different subgroups of V. The field $\mathbb{Q}(\sqrt{d_1d_2})$ corresponds to the third subgroup of order 2 in V, since it is quadratic and different from the fields $\mathbb{Q}(\sqrt{d_1}), \mathbb{Q}(\sqrt{d_2})$ (for example, by Exercise 2.5(b)).

9.7 Take $K = \mathbb{Q}$, $M = \mathbb{Q}(\sqrt{2})$ and $L = \mathbb{Q}(\sqrt[4]{2})$. L is not Galois over $\mathbb{Q}$ as the irreducible polynomial $X^4 - 2$ has one of its zeros $\sqrt[4]{2}$ in L, but not all of them. Hence, the extension $\mathbb{Q} \subset L$ is not Galois (see **T. 9.1**(a)). The extensions $\mathbb{Q} \subset M$ and $M \subset L$ are Galois.

9.10 (b) Let $\tau : M \to M'$ be a K-isomorphism of M with a subfield M' of L containing K. By Exercise 7.4(b), there is a $\sigma \in G(L/K)$ such that τ is a restriction of σ to M. According to (a), those $\sigma \in G(L/K)$ for which the restriction maps M onto M form a subgroup $H(L/M)$. If $M_1 = M, M_2, \ldots, M_r$ are all different subfields of L, which are K-isomorphic to M, then $M_i = \sigma_i(M)$ for some $\sigma_i \in G(L/K)$ and $\sigma_i(M) = \sigma_j(M)$ if and only if $\sigma_i^{-1}\sigma_j(M) = M$, that is, $\sigma_i^{-1}\sigma_j \in H(L/M)$, which is equivalent to $\sigma_i(H(L/M)) = \sigma_j(H(L/M))$. Thus, we get that the number of different subfields M_i K-isomorphic to M is equal to the index of $H(L/M)$ in $G(L/K)$, that is, $r = |G(L/K)|/|H(L/M)|$. Consider a group homomorphism $\varphi : H(L/M) \to G(M/K)$, where $\varphi(\sigma) = \sigma|_M$. As we already know, this is a surjective homomorphism whose kernel is equal to $G(L/M)$. Thus, the group $G(L/M)$ is normal in $H(L/M)$, the group $G(M/K)$ is isomorphic to the quotient $H(L/M)/G(L/M)$ and we have:

$$|G(M/K)| = \frac{|H(L/M)|}{|G(L/M)|} = \frac{|G(L/K)|}{r|G(L/M)|} = \frac{[L:K]}{r[L:M]} = \frac{[M:K]}{r},$$

so $r = [M : K]/|G(M/K)|$. In the computations above, we use **T. 9.1**(a) for Galois extensions $L \supseteq K$ and $L \supseteq M$.
(c) It follows immediately from (b) that $G(M/K) = [M : K]$ if and only if $r = 1$, which is equivalent to $H(L/M) = G(L/K)$. This equality says that each automorphism of L over K restricted to M gives an automorphism of M.
(d) Since $G(L/M)$ is normal in $H(L/M)$ by (b), we get that $H(L/M) \subseteq \mathcal{N}(G(L/M))$ by the definition of the normaliser (see p. 241). Conversely, if $\sigma \in \mathcal{N}(G(L/M))$, then $\sigma^{-1}\tau\sigma \in G(L/M)$ for any $\tau \in G(L/M)$. Thus, for $m \in M$, we have $\sigma^{-1}\tau\sigma(m) = m$, that is, $\tau(\sigma(m)) = \sigma(m)$ for any $\tau \in G(L/M)$. Hence, we have $\tau(m) \in M$. Thus $\tau(M) \subseteq M$, that is, $\tau \in H(L/M)$. This proves that $\mathcal{N}(G(L/M)) \subseteq H(L/M)$ and together with the converse inclusion noted earlier, we get the equality. The isomorphism of $G(M/K)$ to the quotient $H(L/M)/G(L/M)$ was proved in (*b*) and now we can replace $H(L/M)$ by $\mathcal{N}(G(L/M))$.

9.11 (a) We have $\sigma_i(\alpha) = \sigma_j(\alpha)$ if and only if $\sigma_i^{-1}\sigma_j(\alpha) = \alpha$, that is, $\sigma_i^{-1}\sigma_j \in G_\alpha$, which is equivalent to $\sigma_i G_\alpha = \sigma_j G_\alpha$. Thus the number of different images of α by elements of $G(L/K)$ is equal to the index of G_α in $G(L/K)$. This gives the equality $|G(L/K)\alpha| = |G(L/K)|/|G_\alpha|$. Since $G_\alpha = G(L/K(\alpha))$, we get the equality $|G(L/K)\alpha| = [K(\alpha) : K]$ by **T. 9.1**(a) and **T. 4.3**.
(b) Let $X = \{\beta_1, \ldots, \beta_r\}$ be the set of all different elements in the orbit $G(L/K)\alpha$. The coefficients of the polynomial $f_\alpha(X) = (X - \beta_1)\cdots(X - \beta_r)$

are the elementary symmetric functions of $\beta_1, \ldots, \beta_r$ (see section "Symmetric Polynomials" in Appendix), so every $\sigma \in G(L/K)$ maps every coefficient of $f_\alpha(X)$ to itself. Thus, these coefficients belong to $L^{G(L/K)} = K$ (see **T. 9.1**(b)). This shows that $f_\alpha(X) \in K[X]$ and $f_\alpha(X)$ is irreducible by Exercise 9.4(b).
(c) The number of different images $\beta = \sigma(\alpha)$ for $\sigma \in G(L/K)$ is equal to $[K(\alpha) : K]$, so every such β appears $|G(L/K)|/[K(\alpha) : K] = |G_\alpha|$ times. This says that $f_\alpha(X)$ appears as a factor of $F_\alpha(X)$ exactly $|G_\alpha|$ times.

9.12 (a) We have $\mathrm{Tr}_H(\alpha) = \sum_{\sigma \in H} \sigma(\alpha)$. By **T. 6.3**, there exists an $\alpha \in L$ such that $\mathrm{Tr}_H(\alpha) \neq 0$, since otherwise, $a_1 = \cdots = a_n = 1$ in **T. 6.3** contradicts the claim. Consider now L and L^H as vector spaces over L^H. Then Tr_H is a linear function from L to L^H (of vector spaces over L^H). We see that the image of L, which is a linear subspace of L^H, contains nonzero elements, so it must be equal to the whole one-dimensional space L^H.
(b) Since $\alpha_1, \ldots, \alpha_n$ generate L over L^H, their images $\mathrm{Tr}_H(\alpha_1), \ldots, \mathrm{Tr}_H(\alpha_n)$ generate L^H, since there is at least one among these images which is nonzero, by (a).

9.13 (b) According to Exercise 2.12, the function $\sigma(x) = x^q$ is an automorphism of L. Let $[L : K] = n$. Then L has q^n elements. We have $\sigma^i(x) = x^{q^i}$, so σ^i, for $i = 1, \ldots, n$, are n automorphisms and $\sigma^n(x) = x^{q^n} = x$ is the identity, since the field L has q^n elements (see **T. 5.4**). Moreover, these automorphism are all different, since $\sigma^i = \sigma^j$ for $1 \leq i < j \leq n$ should mean $x^{q^i} = x^{q^j}$ for every $x \in L$, that is, $x^{q^j - q^i} = 1$ for all $x \neq 0$ in L, which contradicts the fact that L has $q^n - 1$ different nonzero elements (of course, $q^j - q^i < q^n - 1$). Thus L has at least n different automorphisms over K and $[L : K] = n$. Hence the powers of σ really are all automorphisms and the extension $K \subseteq L$ is Galois with the cyclic Galois group $G(L/K) = \{\sigma, \ldots, \sigma^{n-1}, \sigma^n = 1\} = \langle\sigma\rangle$, since an extension of degree n has at most n automorphisms (see Exercise 6.5).
(c) Since the group L^* is cyclic of order $q^n - 1$ and the kernel of $\mathrm{Nr}_{L/K}$ has $\frac{q^n-1}{q-1}$ elements, the image of L^* in K^* has $q-1$ elements (see A.2.11). Hence each element of K^* is in the image of the norm map.

9.14 Example. We solve a similar problem for the polynomial $X^4 + X + 1$. It is irreducible over $\mathbb{F}_2$, so $[L : \mathbb{F}_2] = 4$. According to Exercise 9.13(b), the Galois group $G(L/\mathbb{F}_2)$ is cyclic of order 4 and it is generated by the automorphism $\sigma(\alpha) = \alpha^2$. The Galois group contains exactly one subgroup H of order 2 and $H = \langle\sigma^2\rangle$. Thus, by **T. 9.2**, there is exactly one quadratic field extension and it equals L^H. In order to find β such that $L^H = \mathbb{F}_2(\beta)$, we take $\beta = \mathrm{Tr}_H(\alpha) = \alpha + \alpha^2$. Since $\alpha^4 + \alpha + 1 = 0$, we easily check that $\beta^2 + \beta + 1 = 0$ (the polynomial $X^2 + X + 1$ is the only irreducible quadratic polynomial over $\mathbb{F}_2$, so we know which polynomial should be checked). Since $X^2 + X + 1$ is irreducible over $\mathbb{F}_2$, we have $[\mathbb{F}_2(\beta) : \mathbb{F}_2] = 2$ and $\mathbb{F}_2(\beta) \subseteq L^H$, so $L^H = \mathbb{F}_2(\beta)$.

9.15 (a) By Exercise 7.6, the field M_1M_2 is normal over K. It is separable over K by Exercise 8.16. Thus M_1M_2 is a Galois extension of K by **T. 9.1**(d).

(b) For $\sigma \in G(M_1M_2/M_2)$ define $\varphi(\sigma) = \sigma|_{M_1}$. Since $\sigma(M_1) = M_1$ by Exercise 7.5(a), we have a group homomorphism $\varphi : G(M_1M_2/M_2) \to G(M_1/K)$. By Exercise 7.4(b), φ is surjective and if σ is in its kernel, then it is the identity on both M_2 and M_1, so it is simply the identity on M_1M_2. Thus φ is an isomorphism of $G(M_1M_2/M_2)$ with $G(M_1/K)$.

9.17 Choose a prime p such that $G \subseteq S_p$ (G can be embedded in such a group S_p by A.8.3). Let $\mathbb{Q} \subset L (\subset \mathbb{C})$ be a Galois extension with Galois group S_p (such an extension exists by Exercise 13.4(c)). Take $K = L^G$ and apply **T. 9.1**(c).

9.18 (b) Choose a prime number $p = nk + 1$ for some integer k, which is possible by Dirichlet's Theorem. Let $L = \mathbb{Q}(\varepsilon)$, where $\varepsilon = e^{\frac{2\pi i}{p}}$. Then L is a splitting field of $X^p - 1$, $[L : \mathbb{Q}] = p - 1$ (see Exercise 3.7(d)) and $G(L/\mathbb{Q})$ is a cyclic group of order $p - 1$ (see Exercise 5.9). This group contains a cyclic subgroup H of order k since k divides $p-1$ (see A.2.2). Let $K = L^H$. Then K is a Galois field extension of $\mathbb{Q}$ (since L is a Galois field extension of $\mathbb{Q}$ with an abelian Galois group, all its subfields are Galois extensions of $\mathbb{Q}$). Moreover, we have $[L^H : \mathbb{Q}] = (p-1)/k = n$, so the Galois group $G(L^H/\mathbb{Q}) = G(L/\mathbb{Q})/G(L/L^H)$ is cyclic as a quotient of a cyclic group and has order n.

9.19 We show that $L = \mathbb{Q}(\sqrt[3]{3+\sqrt{3}})$ is such a field with a subfield $M = \mathbb{Q}(\sqrt{3})$, that is, there are no nontrivial automorphisms of L over $\mathbb{Q}$ but, of course, the field $\mathbb{Q}(\sqrt{3})$ has a nontrivial automorphism mapping $\sqrt{3}$ onto $-\sqrt{3}$. In order to show that there are no nontrivial automorphisms of the field L, we have to show that no other zeros of the minimal polynomial of $\alpha = \sqrt[3]{3+\sqrt{3}}$ are in L. We have $\alpha^3 = 3 + \sqrt{3}$ and squaring $\sqrt{3} = \alpha^3 - 3$, we get that $X^6 - 6X^3 + 6$ is the minimal polynomial of α (for example by Eisenstein's criterion—see Exercise 3.6). All its zeros are $\varepsilon^i\alpha$ and $\varepsilon^i\beta$, where $\beta = \sqrt[3]{3-\sqrt{3}}$, $i = 0, 1, 2$ and $\varepsilon^3 = 1, \varepsilon \neq 1$ is a primitive third root of unity. We have $[L : \mathbb{Q}] = 6$.

The field $L' = \mathbb{Q}(\varepsilon\alpha)$ is isomorphic to L and we prefer to show that L' contains only one zero $\varepsilon\alpha$ of the polynomial $f(X)$ (which shows that L also contains only one zero α of this polynomial).

Of course, the real numbers α, β are not in L' since they generate real fields of degree 6 over $\mathbb{Q}$ and L' is not real.

If $\varepsilon^2\alpha \in L'$, then $\varepsilon = \varepsilon^2\alpha/\varepsilon\alpha \in L'$, so L' contains a subfield $\mathbb{Q}(\sqrt{3}, \varepsilon)$ whose degree over $\mathbb{Q}$ is 4. This is impossible, since L' has degree 6 over $\mathbb{Q}$.

If $\varepsilon\beta \in L'$, then $\gamma = \varepsilon\alpha/\varepsilon\beta = \sqrt[3]{2+\sqrt{3}} \in L'$. We have $\gamma^3 = 2 + \sqrt{3}$, so $X^6 - 4X^3 + 1$ is the minimal polynomial of the real number γ—it is not difficult to check that this polynomial is irreducible using, for example, Gauss's Lemma (see Chap. 3) or by replacing X by $X+2$ and using Eisenstein's criterion for $p = 3$. Now we have a contradiction, since $\mathbb{Q}(\gamma)$ is a real field of degree 6 over $\mathbb{Q}$ but L' is not real.

If $\varepsilon^2\beta \in L'$, then $\varepsilon\alpha\varepsilon^2\beta = \sqrt[3]{6} \in L'$, so L' contains the real field $\mathbb{Q}(\sqrt{3}, \sqrt[3]{6})$ whose degree over $\mathbb{Q}$ is 6, which is impossible, as before. Thus, the field L' has no nontrivial automorphisms.

9.21 Since L is an abelian Galois extension of K, each subfield M of L containing K is a Galois extension of K. In fact, the subgroup $G(L/M)$ of $G(L/K)$ is normal, so M is Galois over K by **T. 9.3**. Since $\alpha \in M = K(\alpha)$ is one of the zeros of $f(X)$ and $K(\alpha)$ is Galois over K, all other zeros of this polynomial must belong to $K(\alpha)$ as well.

9.22 (a) Let L be a splitting field of a polynomial $f(X) \in K[X]$. It is clear that LM is a splitting field of the same polynomial considered as a polynomial with coefficients in M (in fact, in K). If σ is an automorphism of LM over M, then its restriction to L maps this field into itself, since L over K is normal. Thus, the image of any element $x \in L$ must be a zero of the minimal polynomial of x over K (and all zeros of this polynomial are in L by the normality of L over K). This shows that the restriction defines a homomorphism of the Galois group $G(LM/M)$ into $G(L/K)$. If $\sigma \in G(LM/M)$ belongs to the kernel of this restriction, that is, σ gives the identity on L, then σ is the identity on both L and M, so it is the identity on LM. Hence the restriction of automorphisms from LM to L is an injection of $G(LM/M)$ into $G(L/K)$.
(b) Of course, we have $L \cap M \subseteq L^{G(LM/M)}$, since any element in $L \cap M$ is fixed by $\sigma \in G(LM/M)$. On the other hand, if $x \in L^{G(LM/M)}$, then x is in LM and is fixed by the group $G(LM/M)$, so it belongs to M and L, that is, to the intersection $L \cap M$. Thus $L^{G(LM/M)} \subseteq L \cap M$. Thus, we have $L^{G(LM/M)} = L \cap M$.

9.23 If L is not real, then the complex conjugation is a nontrivial automorphism of L, since for any $\alpha \in L$, we have $\overline{\alpha} \in L$. In fact, both α and its complex conjugate $\overline{\alpha}$ have the same minimal polynomial over K (see Exercise 3.5(a)). The automorphism $\sigma(\alpha) = \overline{\alpha}$ of L is over K and has order 2, so $M = L^H$, where $H = \langle\sigma\rangle$, is a real field and by **T. 6.2**, we have $[L : M] = |H| = 2$.

9.25 (a) Every transposition of two zeros α_i, α_j of $f(X)$ changes the sign of $\alpha_i - \alpha_j$, so the square of such a factor in $\Delta(f)$ is fixed by a transposition. Since every element of the Galois group $\mathrm{Gal}(L/K)$ is a composition of such transpositions of two fixed elements (in S_n which contains $\mathrm{Gal}(L/K)$), the discriminant $\Delta(f)$ is fixed by all automorphisms in $\mathrm{Gal}(L/K)$. Hence by **T. 9.1**(b), we get $\Delta(f) \in L^{\mathrm{Gal}(L/K)} = K$.
(b) First observe that $\sqrt{\Delta(f)} \in L$ is fixed by each even permutation, since such a permutation is a composition of even number of transpositions, that is, an even permutation gives a sign change an even number of times. Conversely, if an automorphism in $\mathrm{Gal}(L/K)$ does not change the sign of $\sqrt{\Delta(f)}$ then it must be an even permutation of zeros, since odd permutations are compositions of an odd number of transpositions, so they change the sign of $\sqrt{\Delta(f)}$. Thus $\sqrt{\Delta(f)} \in K$ if and only if $\mathrm{Gal}(L/K)$ has only even permutations, that is, $\mathrm{Gal}_0(L/K) = \mathrm{Gal}(L/K)$. If there are odd permutations in $\mathrm{Gal}(L/K)$, then the index of $\mathrm{Gal}_0(L/K)$ is 2, since the composition of two odd permutations is an even permutation (see A.8.2). Thus [Gal(L/K): $\mathrm{Gal}_0(L/K)$] $= 1$ or 2 and the second possibility occurs if and only if $\sqrt{\Delta} \notin K$.

Of course, $K(\sqrt{\Delta(f)}) \subseteq L^{\mathrm{Gal}_0(L/K)}$, since every even permutation does not change $\sqrt{\Delta(f)}$.
If all permutations in Gal(L/K) are even, then Gal(L/K)=Gal$_0(L/K)$ and $\sqrt{\Delta} \in K$, so $L^{\mathrm{Gal}_0(L/K)} = L^{\mathrm{Gal}(L/K)} = K = K(\sqrt{\Delta})$.
If there are odd permutations in Gal(L/K), then the index of Gal$_0(L/K)$ in Gal(L/K) is 2 and $\sqrt{\Delta} \notin K$. Since $L^{\mathrm{Gal}_0(L/K)} \supseteq K(\sqrt{\Delta})$ and both these fields have degree 2 over K, we have $L^{\mathrm{Gal}_0(L/K)} = K(\sqrt{\Delta})$ in this case.

9.26 (b) The minimal polynomials of $\alpha_i = \alpha + i$ over K all have the same degree, since this degree equals $[K(\alpha_i) : K]$ by **T. 4.2**. Since all of them divide $f(X) = X^p - X + 1$, their common degree must divide p, which shows that there is only one such divisor and it must be $f(X)$ itself. Thus $f(X)$ is irreducible. The field L is its splitting field and as such is normal over K. It is also separable, since the zeros of $f(X)$ are different. Hence, the field L is Galois over K and $G(L/K)$ is a group of prime order p, so it is cyclic. In fact, we can easily describe all its elements. These map α onto any other zero of $f(X)$, so we have p automorphisms $\sigma_i(\alpha) = \alpha + i$ for $i = 0, 1, \ldots, p-1$.

9.30 Example. We construct an irreducible polynomial of degree 15 over $\mathbb{F}_2$ by using irreducible polynomials of degrees 3 and 5 whose splitting fields are subfields of $\mathbb{F}_{2^{15}}$.

As we know, the degree of $L = \mathbb{F}_{2^{15}}$ over $\mathbb{F}_2$ is 15 and the Galois group $G(L/\mathbb{F}_2)$ is cyclic. Thus, the field L contains subfields of degrees 3 and 5 over $\mathbb{F}_2$. This follows from the fundamental theorem of Galois theory **T. 9.2**, since the Galois group, as a cyclic group, contains subgroups of orders 5 and 3 and, in fact, these subgroups are unique (see A.2.2(b)). Let $K = \mathbb{F}_2(a)$ be the unique subfield of L of degree 3 over $\mathbb{F}_2$, and let $K' = \mathbb{F}_2(b)$ be the unique subfield of L of degree 5 over $\mathbb{F}_2$. We can assume that the minimal polynomial of a is any irreducible polynomial of degree 3, and the minimal polynomial of b is any irreducible polynomial of degree 5 both in $\mathbb{F}_2[X]$ (all finite fields with the same number of elements are isomorphic). We define

```
>alias(a=RooOf(X^3+X+1) mod 2)
```

$$a$$

```
>alias(b=RooOf(X^5+X^2+1) mod 2)
```

$$a, b$$

We claim that the element ab is a generator of L over $\mathbb{F}_2$. In fact, this product is neither in K nor K' (notice that $K \cap K' = \mathbb{F}_2$ depending on the relatively prime degrees of K and K' over $\mathbb{F}_2$). Thus, we have $\mathbb{F}_2(ab) = L$, since the subfield generated by ab is one of four possible subfields of L and L is the only possibility. We could find the minimal polynomial for ab over $\mathbb{F}_2$ by computing the product $X - (ab)^{2^i}$ for $i = 0, \ldots, 14$ as all automorphisms in the Galois group $G(L/\mathbb{F}_2)$ are $\sigma(x) = x^{2^i}$ (see Exercise 9.13). Unfortunately, Maple may be unable to find the

product as the task can be too demanding for your computer. But one can split the computation in a simple way. First, we find the minimal polynomial of ab over K':
>simplify($(X-ab)(X-a^{32}b^{32})(X-a^{1024}b^{1024})$) **mod** 2

$$X^3 + b^2X + b^3$$

as the Galois group $G(L/K')$ consists of automorphisms $x \mapsto x, x^{2^5}, x^{2^{10}}$ (this is a subgroup of order 3 in the cyclic group $G(L/\mathbb{F}_2)$ of order 15 generated by the automorphism $x \mapsto x^2$). The polynomial above is one of the five factors of the minimal polynomial of ab (over $\mathbb{F}_2$), when this polynomial (of degree 15) is factorized over the field K (see Exercise 7.8). The automorphisms of L over K are $x \mapsto x, x^{2^3}, x^{2^6}, x^{2^9}, x^{2^{12}}$, so the minimal polynomial of ab over $\mathbb{F}_2$ is
>simplify($(X^3+b^2X+b^3)(X^3+b^{16}X+b^{24})(X^3+b^{128}X+b^{192})(X^3+b^{1024}X+b^{1536})(X^3+b^{8192}X+b^{12288})$) **mod** 2

$$1 + X^2 + X^6 + X^7 + X^{10} + X^{12} + X^{15}$$

Notice that the field L over $\mathbb{F}_2$ can described by a somewhat simpler irreducible polynomial of degree 15, namely, the polynomial $X^{15} + X + 1$, which can be easily found using Maple. But our construction of an irreducible polynomial using subfields of a finite field as intermediatory steps is instructive and has some advantages, providing an alternative to random search. Irreducible polynomials are useful in different mathematical applications (e.g. coding theory or cryptography), but, unfortunately, it is not easy to find such polynomials of high degrees.

Problems of Chap. 10

10.2 Example. We compute the cyclotomic polynomial $\Phi_{300}(X)$. Since $n = 300 = 2^2 \cdot 3 \cdot 5^2$, we have $r(n) = 2 \cdot 3 \cdot 5 = 30$. It follows from (b), that $\Phi_{300}(X) = \Phi_{30}(X^{10})$. Now we could use (c) twice but we prefer to start with $p = 5$ and use the already computed polynomial $\Phi_6(X) = X^2 - X + 1$: $\Phi_{30}(Y) = \Phi_6(Y^5)/\Phi_6(Y)$, so $\Phi_{300}(X) = \Phi_{30}(X^{10}) = \Phi_6(X^{50})/\Phi_6(X^{10}) = (X^{100} - X^{50} + 1)/(X^{20} - X^{10} + 1) = X^{80} + X^{70} - X^{50} - X^{40} - X^{30} + X^{10} + 1$.

10.3 (a) Let $\varepsilon_j = e^{\frac{2\pi ij}{k}}$, for $0 < j < k, (j,k) = 1$, be all $\varphi(k)$ primitive k-th roots of unity. Let $d = n/k$ and consider all d-th roots of all ε_j. These numbers are $e^{\frac{\frac{2\pi ij}{k}+2\pi il}{d}} = e^{\frac{2\pi i(j+kl)}{n}}$ for j as before and $l = 0, 1, \ldots, d-1$. Now we note that all these numbers are n-th roots of unity and all $j + kl$ are relatively prime with n, since any prime dividing n and $j + kl$ must divide k (by the assumption $r(k) = r(n)$), so it must divide j, which is impossible as $(j,k) = 1$. The number of sums $j + kl$ is equal to $\varphi(k)d = \varphi(n)$ (see A.11.4), so the numbers $e^{\frac{2\pi i(j+kl)}{n}}$ are all the zeros of $\Phi_n(X)$ and at the same time all zeros of $\Phi_k(X^{\frac{n}{k}})$, which proves the equality.

(b) We use **T. 10.2**(a) several times. We start with $\Phi_s(X) = \prod_{a|s}(X^a - 1)^{\mu(\frac{s}{a})}$, so $\Phi_s(X^d) = \prod_{a|s}(X^{ad} - 1)^{\mu(\frac{s}{a})}$. Therefore,

$$\prod_{d|r} \Phi_s(X^d)^{\mu(\frac{r}{d})} = \prod_{d|r}\prod_{a|s}(X^{ad} - 1)^{\mu(\frac{s}{a})\mu(\frac{r}{d})} = \prod_{d|r}\prod_{a|s}(X^{ad} - 1)^{\mu(\frac{rs}{ad})}.$$

But the products $b = ad$ are exactly all divisors of rs, so the products above are equal to $\prod_{b|rs}(X^b - 1)^{\mu(\frac{rs}{b})} = \Phi_n(X)$, which shows the required equality. Notice that we use the equality $\mu(x)\mu(y) = \mu(xy)$, when x, y are relatively prime (see A.11.1).

10.4 The polynomial $\Phi_{2m}(X)$ is characterized as a monic irreducible polynomial having a primitive $2m$-th root of unity as its zero. We want to show that $\Phi_m(-X)$ is such a polynomial. Of course, this is a monic irreducible polynomial (if $m > 1$), since $\Phi_m(X)$ is irreducible by **T. 10.2**(d) and its degree is even (if $m > 1$).

Let ε be a primitive m-th root of unity. We want to show that $-\varepsilon$ is a primitive $2m$-th root of unity. Since $-\varepsilon$ is a zero of $\Phi_m(-X)$, we get the required equality $\Phi_{2m}(X) = \Phi_m(-X)$.

Assume that the order of $-\varepsilon$ is n. Since we have $(-\varepsilon)^{2m} = 1$, it follows that $n \mid 2m$ (see p. 237).

Since $\varepsilon^{2n} = (-\varepsilon)^{2n} = 1$ we have $m \mid 2n$, and as m is odd, we get $m \mid n$. Let $n = qm$ for an integer q. Thus $qm \mid 2m$ shows that q is 1 or 2. But $m = n$ is impossible (as $(-\varepsilon)^m = -1$), so $n = 2m$.

10.5 (a) If $\varepsilon_{mn} = e^{\frac{2\pi i}{mn}}$, then $\varepsilon_m = \varepsilon_{mn}^n$ and $\varepsilon_n = \varepsilon_{mn}^m$, which shows that $\mathbb{Q}(\varepsilon_m)\mathbb{Q}(\varepsilon_n) \subseteq \mathbb{Q}(\varepsilon_{mn})$. On the other hand, since m, n are relatively prime, we have $1 = km + ln$ for suitable integers k, l, so $\varepsilon_{mn} = \varepsilon_{mn}^{km+ln} = \varepsilon_n^k\varepsilon_m^l$. Thus, we have $\mathbb{Q}(\varepsilon_{mn}) \subseteq \mathbb{Q}(\varepsilon_m)\mathbb{Q}(\varepsilon_n)$, so $\mathbb{Q}(\varepsilon_m)\mathbb{Q}(\varepsilon_n) = \mathbb{Q}(\varepsilon_{mn})$.
By **T. 10.1**, we have $[\mathbb{Q}(\varepsilon_{mn}) : \mathbb{Q}] = \varphi(mn)$, $[\mathbb{Q}(\varepsilon_m) : \mathbb{Q}] = \varphi(m)$, $[\mathbb{Q}(\varepsilon_n) : \mathbb{Q}] = \varphi(n)$. Thus $[\mathbb{Q}(\varepsilon_{mn}) : \mathbb{Q}(\varepsilon_m)] = \varphi(n)$ and $[\mathbb{Q}(\varepsilon_{mn}) : \mathbb{Q}(\varepsilon_n)] = \varphi(m)$.
Let $M = \mathbb{Q}(\varepsilon_m) \cap \mathbb{Q}(\varepsilon_n)$. Using Exercise 9.15(b) with $M_1 = \mathbb{Q}(\varepsilon_m)$ and $M_2 = \mathbb{Q}(\varepsilon_n)$, we get $[\mathbb{Q}(\varepsilon_{mn}) : \mathbb{Q}(\varepsilon_n)] = [\mathbb{Q}(\varepsilon_m) : M]$. Thus $[\mathbb{Q}(\varepsilon_m) : M] = \varphi(m)$ as well as $[\mathbb{Q}(\varepsilon_m) : \mathbb{Q}] = \varphi(m)$ imply $[M : \mathbb{Q}] = 1$, that is, $M = \mathbb{Q}$.
(b) The polynomial $\Phi_n(X)$, of degree $\varphi(n)$, is irreducible over $\mathbb{Q}(\varepsilon_m)$, since $\mathbb{Q}(\varepsilon_{mn}) = \mathbb{Q}(\varepsilon_m)(\varepsilon_n)$ has degree $\varphi(n)$ over the field $\mathbb{Q}(\varepsilon_m)$ (see **T. 4.2**). Similarly, the polynomial $\Phi_m(X)$, of degree $\varphi(m)$ is irreducible over $\mathbb{Q}(\varepsilon_n)$.
(c) This is a generalization of (a). The proof below follows almost exactly the argument given in (a), but it it necessary to make use of a relation between the least common multiple $[m, n]$ and the greatest common divisor (m, n) of two integers m, n, which introduces a (small) complication. So we prefer to present a solution again.
The relation is $[m, n](m, n) = mn$ (see A.3.4 for $R = \mathbb{Z}$). We have $\varepsilon_{[m,n]} = e^{\frac{2\pi i}{[m,n]}}$, $\varepsilon_m = \varepsilon_{[m,n]}^{\frac{n}{(m,n)}}$ and $\varepsilon_n = \varepsilon_{[m,n]}^{\frac{m}{(m,n)}}$, which shows that $\mathbb{Q}(\varepsilon_m)\mathbb{Q}(\varepsilon_n) \subseteq \mathbb{Q}(\varepsilon_{[m,n]})$. On the other hand, since $(m, n) = km + ln$ for suitable integers k, l, we get $\varepsilon_n^k\varepsilon_m^l = \varepsilon_{[m,n]}$. Thus, we have $\mathbb{Q}(\varepsilon_{[m,n]}) \subseteq \mathbb{Q}(\varepsilon_m)\mathbb{Q}(\varepsilon_n)$, which, together with the converse inclusion, gives $\mathbb{Q}(\varepsilon_m)\mathbb{Q}(\varepsilon_n) = \mathbb{Q}(\varepsilon_{[m,n]})$.

We also have $\varphi([m,n])\varphi((m,n)) = \varphi(m)\varphi(n)$ (see A.11.4(a)). By **T. 10.1**, we have $[\mathbb{Q}(\varepsilon_{[m,n]}) : \mathbb{Q}] = \varphi([m,n])$, $[\mathbb{Q}(\varepsilon_m) : \mathbb{Q}] = \varphi(m)$, $[\mathbb{Q}(\varepsilon_n) : \mathbb{Q}] = \varphi(n)$. Thus $[\mathbb{Q}(\varepsilon_{[m,n]}) : \mathbb{Q}(\varepsilon_n)] = \varphi([m,n])/\varphi(n) = \varphi(m)/\varphi((m,n))$.
Let $M = \mathbb{Q}(\varepsilon_m) \cap \mathbb{Q}(\varepsilon_n)$. Of course, we have $\mathbb{Q}(\varepsilon_{(m,n)}) \subseteq M$. Using Exercise 9.15(b) with $M_1 = \mathbb{Q}(\varepsilon_m)$ and $M_2 = \mathbb{Q}(\varepsilon_n)$, we get $[\mathbb{Q}(\varepsilon_{[m,n]}) : \mathbb{Q}(\varepsilon_n)] = [\mathbb{Q}(\varepsilon_m) : M]$. Thus $[\mathbb{Q}(\varepsilon_m) : M] = \varphi(m)/\varphi((m,n))$ as well as $[\mathbb{Q}(\varepsilon_m) : \mathbb{Q}(\varepsilon_{(m,n)})] = \varphi(m)/\varphi((m,n))$ imply $[M : \mathbb{Q}(\varepsilon_{(m,n)})] = 1$, that is, $M = \mathbb{Q}(\varepsilon_{(m,n)})$.

10.6 Let $\sigma(x) = \bar{x}$ be the complex conjugation on the field $K = \mathbb{Q}(\varepsilon_n)$. The subgroup $H = \{1, \sigma\}$ of the Galois group of K has order 2. The maximal real subfield of K is the field K^H (see Exercise 9.23). Of course, we have $\sigma(\eta) = \sigma(\varepsilon_n + \overline{\varepsilon_n}) = \eta$, so $\mathbb{Q}(\eta) \subseteq K^H \subset K$. We have $[K : K^H] = 2$ by **T. 6.2**. We also have $[K : \mathbb{Q}(\eta)] \leq 2$, since ε_n is a zero of the polynomial $X^2 - \eta X + 1$. Thus, we have $K^H = \mathbb{Q}(\eta)$ by **T. 4.3**.

10.7 (b) Notice that $[K^{G_d} : \mathbb{Q}] = m$ by **T. 6.2** and $\mathbb{Q}(\theta_d) \subseteq K^{G_d}$, since $\sigma^m(\theta_d) = \theta_d$, where σ^m generates the group G_d. According to **T. 4.1**, the elements $1, \varepsilon_p, \ldots, \varepsilon_p^{p-2}$ are linearly independent over $\mathbb{Q}$. Consider

$$\sigma^r(\theta_d) = \sigma^r(\varepsilon_p) + \sigma^{m+r}(\varepsilon_p) + \cdots + \sigma^{(d-1)m+r}(\varepsilon_p) = \varepsilon_p^{g^r} + \varepsilon_p^{g^{m+r}} + \cdots + \varepsilon_p^{g^{(d-1)m+r}}$$

for $r = 0, 1, \ldots, m-1$. All these m elements are different, since any equality $\sigma^r(\theta_d) = \sigma^{r'}(\theta_d)$ for $r \neq r'$ gives a relation of linear dependence between $2d$ of the elements $1, \varepsilon_p, \ldots, \varepsilon_p^{p-2}$ (notice that the elements $\varepsilon_p^{g^k}$, where $k = qm + r$ for $q = 0, 1, \ldots, d-1$ and $r = 0, 1, \ldots, m-1$ are all $dm = p-1$ elements $1, \varepsilon_p, \ldots, \varepsilon_p^{p-2}$). By Exercise 9.24, we have the equality $K^{G_d} = \mathbb{Q}(\theta_d)$, since the number of images of θ_d is equal to the degree of K^{G_d} over $\mathbb{Q}$.

10.8 Example. We find all the quadratic subfields in the cyclotomic extension $K = \mathbb{Q}(\varepsilon_{12})$ of $\mathbb{Q}$. We know that the Galois group $G(K/\mathbb{Q}) \cong (\mathbb{Z}/12\mathbb{Z})^* = \{1, 5, 7, 11\}$. If ε denotes a primitive 12-th root of unity (say, $\varepsilon = e^{\frac{2\pi i}{12}}$), then the automorphisms in the Galois group are $\sigma_0(\varepsilon) = \varepsilon$, $\sigma_1(\varepsilon) = \varepsilon^5$, $\sigma_2(\varepsilon) = \varepsilon^7$, $\sigma_3(\varepsilon) = \varepsilon^{11}$. The order of each automorphism different from the identity σ_0 is equal to 2, which is a consequence of the equalities $5^2 \equiv 7^2 \equiv 11^2 \equiv 1 \pmod{12}$. By **T. 9.2**, the quadratic subfields of K correspond to subgroups of order 2 in the Galois group $G(K/\mathbb{Q})$. Thus there are three such subgroups: $H_i = \{\sigma_0, \sigma_i\}$ for $i = 1, 2, 3$. In order to find all these subfields, we note that all 12-th roots of unity are zeros of the polynomial $X^{12} - 1 = (X^6 - 1)(X^6 + 1) = (X-1)(X+1)(X^2+X+1)(X^2-X+1)(X^2+1)(X^4-X^2+1)$. Of course, the primitive roots $\varepsilon, \varepsilon^5, \varepsilon^7, \varepsilon^{11}$ are zeros of the last factor. The quadratic factor X^2+X+1 with zeros $\varepsilon^4, \varepsilon^8$ (since $X^3 - 1 = (X-1)(X^2+X+1)$ and these numbers are 3-rd roots of unity) has the splitting field $\mathbb{Q}(\sqrt{-3})$. The other quadratic factor $X^2 + 1$ has zeros $\varepsilon^3, \varepsilon^9$ (since $\varepsilon^6 = -1$) and the splitting field $\mathbb{Q}(i)$. The third quadratic subfield is, of course, $\mathbb{Q}(\sqrt{3})$ (since $i\sqrt{-3} = \sqrt{3}$). Finally, the cyclotomic field $\mathbb{Q}(\varepsilon_{12})$ has three quadratic subfields: $\mathbb{Q}(\sqrt{i})$, $\mathbb{Q}(\sqrt{3})$ and $\mathbb{Q}(\sqrt{-3})$. Which of the automorphisms is the complex conjugation? How to express $\sqrt{3}$ in terms of powers of ε? An alternative possibility to find the quadratic subfields could be to study suitable traces or norms of ε with respect to the subgroups H_i. Notice that $K^{H_1} = \mathbb{Q}(i)$, $K^{H_2} = \mathbb{Q}(\sqrt{-3})$ and $K^{H_3} = \mathbb{Q}(\sqrt{3})$.

10.9 Example. We find all subfields of the cyclotomic field $K = \mathbb{Q}(\varepsilon_{11})$ using the Gaussian periods defined in Exercise 10.7. The Galois group $G(K/\mathbb{Q}) \cong (\mathbb{Z}/11\mathbb{Z})^*$ is cyclic of order 10. It has four subgroups corresponding to the divisors 1, 2, 5, 10 of its order. The trivial subgroups of orders 1 and 10 define, correspondingly, the subfields K and $\mathbb{Q}$. Let $\sigma(\varepsilon_{11}) = \varepsilon_{11}^2$. It is easy to check that σ generates the Galois group $G(K/\mathbb{Q})$. In fact, we have $2^2 = 4$ and $2^5 \equiv -1 \pmod{11}$, so the order of 2 in $(\mathbb{Z}/11\mathbb{Z})^*$ is 10 (it could be 2, 5 or 10 and it is neither 2 or 5), which shows that σ has order 10 in $G(K/\mathbb{Q})$. Thus, we have $G_2 = \langle\sigma^5\rangle$ and $G_5 = \langle\sigma^2\rangle$, the subgroups of $G(K/\mathbb{Q})$ of orders 2 and 5, respectively.

Now we want to find the subfields of K: $K^{G_2} = \mathbb{Q}(\theta_2)$ and $K^{G_5} = \mathbb{Q}(\theta_5)$, which correspond to the subgroups G_2 and G_5.

By Exercise 10.7(c), we have $\theta_2 = \varepsilon_{11} + \sigma^5(\varepsilon_{11}) = \varepsilon_{11} + \varepsilon_{11}^{10}$ and $[K^{G_2} : \mathbb{Q}] = 5$, since $[K : K^{G_2}] = 2$ by **T. 6.2**. We have $K^{G_2} = \mathbb{Q}(\theta_2)$ and the minimal polynomial of θ_2 is

$$f_{\theta_2}(X) = \prod_{i=0}^{4}(X - \sigma^i(\theta_2)) = X^5 + X^4 - 4X^3 - 3X^2 + 3X + 1.$$

In order to describe K^{G_5}, we consider $\theta_5 = \varepsilon + \sigma^2(\varepsilon) + \sigma^4(\varepsilon) + \sigma^6(\varepsilon) + \sigma^8(\varepsilon) = \varepsilon + \varepsilon^4 + \varepsilon^5 + \varepsilon^9 + \varepsilon^3$. We have $[K^{G_5} : \mathbb{Q}] = 2$, since $[K : K^{G_5}] = 5$ by **T. 6.2**. Thus θ_5 is a zero of a quadratic polynomial and we can find it using its second zero $\sigma(\theta_5) = \varepsilon^2 + \varepsilon^8 + \varepsilon^{10} + \varepsilon^7 + \varepsilon^6$. Thus $K^{G_5} = \mathbb{Q}(\theta_5)$, where

$$f_{\theta_5}(X) = (X - \theta_5)(X - \sigma(\theta_5)) = X^2 + X + 3.$$

Notice that $K^{G_5} = \mathbb{Q}(\sqrt{-11})$. Compare Exercise 10.10 and its solution.

Remark. In order to carry out the computation, we can use the following commands in Maple:
`>alias(`ε `=RootOf(`$x^{11} - 1)/(x - 1)$`)`
which defines a primitive 11-th root of unity and
$>\theta := \varepsilon + \varepsilon^{10}$
$>\sigma := x \to x^2$
`>simplify(`$\prod_{i=0}^{4}(X - (\sigma^i(\varepsilon) + \sigma^i(\varepsilon^{10}))$`))`
which is the sought minimal polynomial of θ_2. Similarly for θ_5 even if it is easy to compute the coefficients of f_{θ_5} without a computer.

10.10 According to Exercise 10.7(c), the quadratic subfield of $\mathbb{Q}(\varepsilon)$, where ε is a p-th root of unity ($\varepsilon \neq 1$) is defined as an extension of $\mathbb{Q}$ by θ_2 corresponding to $m = 2$ and $d = (p-1)/2$. Thus $\theta_2 = \sum_{i=0}^{\frac{p-3}{2}} \sigma^{2i}(\varepsilon) = \sum_{i=0}^{\frac{p-3}{2}} \varepsilon^{g^{2i}}$, where σ is a generator of the Galois group $G(\mathbb{Q}(\varepsilon)/\mathbb{Q})$ and $\sigma(\varepsilon) = \varepsilon^g$. Notice that g is a generator of the cyclic group $\mathbb{F}_p^*$, so all nonzero residues modulo p are g^i for $i = 1, \dots, p-1$ ($g^{p-1} \equiv 1 \pmod{p}$).

The automorphism σ restricted to the quadratic subfield gives the nontrivial automorphism of this subfield and maps θ_2 onto $\theta_2' = \sum_{j=0}^{\frac{p-3}{2}} \sigma^{2j+1}(\varepsilon) = \sum_{j=0}^{\frac{p-3}{2}} \varepsilon^{g^{2j+1}}$. We want to find the minimal polynomial of θ_2, which is quadratic, and which must have θ_2' as the second zero. Thus this polynomial is $X^2 - (\theta_2 + \theta_2')X + \theta_2\theta_2'$.

It is easy to see that $\theta_2 + \theta_2' = \sum_{i=0}^{p-2} \sigma^i(\varepsilon) = \sum_{i=0}^{p-2} \varepsilon^{g^i} = -1$, since the sum of all p-th roots of unity is equal to 0, and the last sum has all these roots with the exception of 1. Thus, we have to compute the product $\theta_2\theta_2'$, and this is more difficult.

First, we observe that it is necessary to find all possible products of the terms, which are all of the form ε to a power of $g^{2i} + g^{2j+1} \pmod{p}$. The number of such terms is $(p-1)^2/4$ and they can be considered as values of the quadratic form $X^2 + gY^2$, where X, Y are nonzero elements of the finite field $\mathbb{F}_p$.

In our solution, we need to count the number of solutions $(x, y) \in \mathbb{F}_p^2$ of the equation $X^2 + gY^2 = r$, where $r \in \mathbb{F}_p$. These numbers are the following:

$$\{(x, y) \in \mathbb{F}_p^2 | x^2 + gy^2 = r\} = \begin{cases} 0 & \text{if } p \equiv 1 \pmod{4} \text{ and } r = 0 \\ p+1 & \text{if } p \equiv 1 \pmod{4} \text{ and } r \neq 0 \\ 2p-1 & \text{if } p \equiv 3 \pmod{4} \text{ and } r = 0 \\ p-1 & \text{if } p \equiv 3 \pmod{4} \text{ and } r \neq 0 \end{cases}$$

We prove this auxiliary result at the end of the solution below.

There are two cases depending on the residue of p modulo 4. We start with the case $p \equiv 1 \pmod{4}$. In this case, the quadratic form $X^2 + gY^2$ is non-isotropic, which means that the equation $X^2 + gY^2 \equiv 0 \pmod{p}$ has only the solution $X = Y = 0$. Thus, the sum $g^{2i} + g^{2j+1} \pmod{p}$ is never 0. In how many ways can we obtain a term ε^r with $r \not\equiv 0$, $r \equiv g^{2i} + g^{2j+1} \pmod{p}$, in the product $\theta_2\theta_2'$? The equation $X^2 + gY^2 = r$ has $p+1$ different solutions. We have to exclude those with $X = 0$ or $Y = 0$. It is easy to see that there are two such solutions: either $X^2 = r$ has two solutions if r is a square, or $gY^2 = r$ has two solutions if r is not a square (since g is not a square in $\mathbb{F}_p^*$). Thus, we have $p-1$ solutions such that $XY \neq 0$. These solutions give $(p-1)/4$ possibilities for X^2, since exactly $(\pm X, \pm Y)$ give the same X^2. This means that in the product $\theta_2\theta_2'$, we get every term ε^r exactly $(p-1)/4$ times, so $\theta_2\theta_2' = -\frac{p-1}{4}$ and the minimal polynomial of θ_2 is $X^2 + X - \frac{p-1}{4}$ when $p \equiv 1 \pmod{4}$.

Assume now that $p \equiv 3 \pmod{4}$. In this case, the quadratic form $X^2 + gY^2$ is isotropic, which means that the equation $X^2 + gY^2 \equiv 0 \pmod{p}$ has nontrivial solutions. The number of solutions is $2p-1$ but we have to exclude the trivial solution $(X, Y) = (0, 0)$. In the remaining solutions, we have $XY \neq 0$ and as before, the solutions $(\pm X, \pm Y)$ give the same X^2. Thus, in the product $\theta_2\theta_2'$, we get ε^r with $r = g^{2i} + g^{2j-1} \equiv 0 \pmod{p}$ exactly $(p-1)/2$ times. Every such term contributes $\varepsilon^0 = 1$ to the sum, so we have $(p-1)/2$ terms equal to 1. If $r \neq 0$, then the total number of solutions is equal to $p-1$. As before, we have to exclude the two solutions with $X = 0$ or $Y = 0$ and we get $p-3$ solutions with $XY \neq 0$. This time, we get $(p-3)/4$ different possibilities for X^2, so in the product $\theta_2\theta_2'$, we get every term ε^r with $r \neq 0$ exactly $(p-3)/4$ times. Each sum of all different ε^r (r nonzero) is equal to -1, so we get $(p-3)/4$ times a term -1. Thus, finally, we get $\theta_2\theta_2' = (p-1)/2 - (p-3)/4 = (p+1)/4$ and the minimal polynomial of θ_2 is $X^2 + X + \frac{p+1}{4}$ when $p \equiv 3 \pmod 4$.

Now we show how to compute the number of solutions to $X^2 + gY^2 = r$ over a field $\mathbb{F}_p$. If $p \equiv 1 \pmod 4$, then the equation $X^2 + gY^2 \equiv 0 \pmod p$ has only the solution $X = Y = 0$. In fact, if we had a solution with $Y \neq 0$, then the equality $g = -(X^2/Y^2)$ would imply that $g^{\frac{p-1}{2}} = (X/Y)^{p-1} = 1$, which contradicts the assumption that g is a generator of the cyclic group $\mathbb{F}_p^*$ (and has order $p-1$). Thus the quadratic equation $T^2 + g = 0$ has no solutions in $\mathbb{F}_p$ and when we extend this field by a zero α of this equation, we get a quadratic field $\mathbb{F}_p[\alpha]$ over $\mathbb{F}_p$. Every element of this field has shape $a + b\alpha$, where $a, b \in \mathbb{F}_p$ and the nontrivial automorphism maps this element onto $a - b\alpha$. The norm Nr of such an element is $(a + b\alpha)(a - b\alpha) = a^2 + gb^2$, so it is precisely a value of the quadratic form $X^2 + gY^2$ and the norm mapping $\mathrm{Nr} : \mathbb{F}_p(\alpha) \to \mathbb{F}_p$ is a homomorphism of the group of nonzero elements of $\mathbb{F}_p(\alpha)$ onto the group of nonzero elements of $\mathbb{F}_p$. That we really get a solution of $X^2 + gY^2 = r$ for any $r \neq 0$ follows from the fact that either $X^2 = r$ has two solutions if r is a square in $\mathbb{F}_p$ or $gY^2 = r$ has two solutions if r is not a square in $\mathbb{F}_p$ (since g is not a square and the quotient of two non-squares is a square). The norm maps a group of order $p^2 - 1$ onto a group of order $p-1$, so the kernel (the group of elements of norm 1) has $p+1$ elements. Thus the norm assumes each value $p+1$ times, that is, for every $r \in \mathbb{F}_p^*$ the equation $X^2 + gY^2 = r$ has $p+1$ solutions.

Assume now that $p \equiv 3 \pmod 4$. In this case, the quadratic form $X^2 + gY^2$ is isotropic, which means that the equation $X^2 + gY^2 \equiv 0 \pmod p$ has nonzero solutions in $\mathbb{F}_p$. This means that $-g$ is a square in $\mathbb{F}_p$, that is, there is an $\alpha \in \mathbb{F}_p$ such that $\alpha^2 = -g$. We have $X^2 + gY^2 = (X - \alpha Y)(X + \alpha Y) = 0$ if and only if $X = \alpha Y$ or $X = -\alpha Y$, so we get $2p-1$ solutions when $Y = 0, 1, \ldots, p-1$. Consider the ring $\mathbb{F}_p[\alpha]$ consisting of all $a + b\alpha$, $a, b \in \mathbb{F}_p$. We have a mapping $\mathrm{Nr} : \mathbb{F}_p[\alpha] \to \mathbb{F}_p$ such that $\mathrm{Nr}(a + b\alpha) = (a + b\alpha)(a - b\alpha) = a^2 + gb^2$. The elements $a + b\alpha$ such that $\mathrm{Nr}(a + b\alpha) = a^2 + gb^2 \neq 0$ form a multiplicative group, since a product of two elements in $\mathbb{F}_p[\alpha]$ with norm different from 0 has norm different from 0 (see A.2.1). Since the number of elements in $\mathbb{F}_p[\alpha]$ with norm different from zero is $p^2 - (2p-1) = (p-1)^2$ (p^2 is the number of all elements in $\mathbb{F}_p[\alpha]$), so Nr is a homomorphism of the group of all elements with a nonzero norm in $\mathbb{F}_p[\alpha]$ onto the group $\mathbb{F}_p^*$. That the mapping is surjective follows from the same argument as in the case $p \equiv 1 \pmod 4$. Thus Nr maps a group of order $(p-1)^2$ onto a group of order $p-1$, which means that the kernel has order $p-1$. Thus the norm assumes each value $p-1$ times, that is, for every $r \in \mathbb{F}_p^*$ the equation $X^2 + gY^2 = r$ has $p-1$ solutions.

10.11 (b) Let ε be any primitive n-th root of unity over $\mathbb{F}_p$ and let $f_\varepsilon(X)$ be its minimal polynomial over $\mathbb{F}_p$.

Let k be the degree of $f_\varepsilon(X)$. Since $p \nmid n$, there are $\varphi(n)$ primitive roots and, of course, for each of them the field $\mathbb{F}_p(\varepsilon)$ is a splitting field of $\Phi_{n,\mathbb{F}_p}$. Thus the degrees of all irreducible factors of $\Phi_{n,\mathbb{F}_p}$ are equal to the degree k of this splitting field over $\mathbb{F}_p$ (notice that this is a special case of **T. 10.2**(c)).
The group of nonzero elements in $\mathbb{F}_p(\varepsilon)$ has $p^k - 1$ elements, so $\varepsilon^{p^k-1} = 1$. But the order of ε in a splitting field of $\Phi_{n,\mathbb{F}_p}$ is equal to n, so $n \mid p^k - 1$. Hence, we have $p^k = 1$ in $(\mathbb{Z}/n\mathbb{Z})^*$, which shows that the order of p in this group divides k. But this order must be equal to k. In fact, if the order of p is $l < k$, then $n \mid p^l - 1$, which means that the field $\mathbb{F}_{p^l}$ of degree l over $\mathbb{F}_p$ contains an element of order n, that is, it contains a splitting field of the polynomial $X^n - 1$. Thus, it must also contain a splitting field of the cyclotomic polynomial $\Phi_{n,\mathbb{F}_p}$. This is a contradiction, since the degree of this splitting field is $k > l$. Thus, we have proved that all irreducible factors of $\Phi_{n,\mathbb{F}_p}$ have the same degree k, which is equal to the order of p in $(\mathbb{Z}/n\mathbb{Z})^*$.

10.12 (a) Both polynomials $\Phi_d(X)$ and $\Phi_n(X)$ have a zero x modulo p. The identity $X^n - 1 = \prod_{d|n} \Phi_d(X)$ from **T. 10.2**(a), taken modulo p, shows that $X^n - 1$ has a double zero x modulo p in the field $\mathbb{F}_p$. Hence, it follows from **T. 5.3** (or **T. 8.1**) that the derivative of $X^n - 1$ has a common zero with $X^n - 1$ in $\mathbb{F}_p$, which happens only if $p \mid n$.
(b) Let p be a prime divisor of $\Phi_n(x)$. Notice that $p \nmid x$, since $p \mid \Phi_n(x) \mid x^n - 1$. Denote by k the order of x in the group $\mathbb{Z}_p^*$ of order $p - 1$. Thus $k \mid p - 1$, that is, $p \equiv 1 \pmod{k}$. Since $x^n \equiv 1 \pmod{p}$, we get $k \mid n$.
If $k = n$, then $p \equiv 1 \pmod{n}$. If $k < n$, then the identity $X^k - 1 = \prod_{d|k} \Phi_d(X)$ implies that $p \mid \Phi_d(x)$ for some divisor d of k, since $p \mid x^k - 1$. Thus, we have $p \mid \Phi_d(x)$ and $p \mid \Phi_n(x)$, so (a) implies that $p \mid n$.
(c) Since $\Phi_1(X) = X - 1$, we have $\Phi_1(0) = -1$. By Exercise 10.2(a), we have $\Phi_p(0) = 1$ for any prime p. Now we use induction with respect to n and assume the equality $\Phi_m(0) = 1$ for all m such that $1 < m < n$. If n is not a prime and p is a prime factor of n, then by Exercise 10.3(a), we get $\Phi_n(0) = \Phi_{n/p}(0) = 1$.
(d) Assume that there are only finitely many prime numbers congruent to 1 modulo n. Let N be the product of all such primes and n. Of course, $N > 1$. Consider the cyclotomic polynomial $\Phi_n(X)$. Since its leading coefficient is 1, we can find an exponent k such that $\Phi_n(N^k) > 1$ (of course, $\Phi_n(x) \to \infty$ when $x \to \infty$). Take any prime number p dividing $\Phi_n(N^k)$. Such a prime cannot divide N, since $N \mid \Phi_n(N^k) - 1$ by (c). Thus $p \nmid n$ (n is a factor of N). According to (b), we have $p \equiv 1 \pmod{n}$ and p is different from all the primes dividing N, that is, also all those which are congruent to 1 modulo n. This contradiction shows that the number of primes congruent to 1 modulo n is infinite.

10.13 (a) G is a product of cyclic groups of some orders $n_1, \ldots, n_k$ (see A.6.2). By Dirichlet's Theorem (see Exercise 10.12), for every n_i ($i = 1, \ldots, k$), we can find infinitely many primes congruent to 1 modulo n_i. Let us choose such primes $p_i = n_i k_i + 1$, for some integer k_i, so that all p_i are different. Since the group $\mathbb{Z}_{p_i}^*$ is cyclic of order $p_i - 1 = n_i m_i$, it contains a subgroup of order m_i and the quotient of $\mathbb{Z}_{p_i}^*$ by this subgroup is isomorphic to $\mathbb{Z}_{n_i}$. Hence, we have a surjection

$\varphi_i : \mathbb{Z}_{p_i}^* \to \mathbb{Z}_{n_i}$. Putting these surjections together, we get a surjection $\varphi : \mathbb{Z}_{p_1}^* \times \cdots \times \mathbb{Z}_{p_k}^* \to \mathbb{Z}_{n_1} \times \cdots \times \mathbb{Z}_{n_k}$, where $\varphi((r_1, \ldots, r_k)) = (\varphi_1(r_1), \ldots, \varphi_k(r_k))$. But $\mathbb{Z}_{p_1}^* \times \cdots \times \mathbb{Z}_{p_k}^* \cong \mathbb{Z}_{p_1 \cdots p_k}^*$ (see A.3.11) and $\mathbb{Z}_{n_1} \times \cdots \times \mathbb{Z}_{n_k} = G$, so we have a surjection of $\mathbb{Z}_{p_1 \cdots p_k}^*$ onto G.
(b) Let G be a finite abelian group. By (a), we can find n such that $\mathbb{Z}_n^*$ has a surjective homomorphism onto G. Let H be the kernel of this surjection. By **T. 10.1**, the group $\mathbb{Z}_n^*$ is the Galois group of the cyclotomic field $L = \mathbb{Q}(\varepsilon)$, where ε is a primitive n-th root of unity. By **T. 9.3**, the field $M = L^H$ is Galois over $\mathbb{Q}$ (since L is a Galois abelian extension of $\mathbb{Q}$) and its Galois group is $G(L/\mathbb{Q})/H = \mathbb{Z}_n^*/H \cong G$.

10.14 (a) Consider a tower $K \subseteq K(\varepsilon) \subseteq K(\varepsilon, \alpha)$, where $\varepsilon^n = 1$, $K(\varepsilon)$ is a splitting field of $X^n - 1$ and $\alpha^n = a$. All zeros of $X^n - a$ are $\varepsilon^i \alpha$, where $i = 0, 1, \ldots, n-1$ and ε is any primitive n-th root of unity. Of course, $L = K(\varepsilon, \alpha)$ is a splitting field of $X^n - a$ over K (see Exercise 5.11).
If $\sigma, \tau \in G(L/K(\varepsilon))$, then $\sigma(\alpha) = \varepsilon^i \alpha$ and $\tau(\alpha) = \varepsilon^j \alpha$ for some $i, j \in \mathbb{Z}_n$. We have, $\sigma\tau(\alpha) = \varepsilon^{i+j}\alpha$, so we have an injection of $G(L/K(\varepsilon))$ into $\mathbb{Z}_n$ when we map an automorphism onto the exponent of ε which corresponds to it. Thus $H = G(L/K(\varepsilon))$ is isomorphic to a subgroup of $\mathbb{Z}_n$. It is, of course, a normal subgroup of $G(L/K)$ by **T. 9.3**(a).
If $\sigma, \tau \in G(K(\varepsilon)/K)$, then $\sigma(\varepsilon) = \varepsilon^i$ and $\tau(\varepsilon) = \varepsilon^j$, where i, j are relatively prime to n, since the image of a generator of the group of n-th roots of unity must be a generator of this group. We have $\sigma\tau(\varepsilon) = \varepsilon^{ij}$, so we have an injection of $G(K(\varepsilon)/k)$ into $\mathbb{Z}_n^*$ when we map an automorphism onto the exponent of ε which corresponds to it. Thus $G(K(\varepsilon)/K)$ is isomorphic to a subgroup of $\mathbb{Z}_n^*$. By **T. 9.3**(b), we have $G(K(\varepsilon)/K) = G(L/K)/H$.
(b) Let $\sigma \in G(L/K)$ and let $\sigma(\varepsilon) = \varepsilon^a$, $a \in \mathbb{Z}_n^*$ and $\sigma(\alpha) = \varepsilon^b \alpha$, $b \in \mathbb{Z}_n$. Then $\sigma(\varepsilon^i \alpha) = \varepsilon^{ai+b}\alpha$. We define a mapping $\varphi : G(L/K) \to GL_2(\mathbb{Z}_n)$ such that

$$\varphi(\sigma) = \begin{pmatrix} a & b \\ 0 & 1 \end{pmatrix}.$$

If, in a similar way, we have $\tau(\varepsilon^i \alpha) = \varepsilon^{ci+d}\alpha$, then it is easy to check that $\varphi(\sigma\tau) = \varphi(\sigma)\varphi(\tau)$, so φ is an injection of $G(L/K)$ into the group $GL_2(\mathbb{Z}_n)$.

10.15 (a) Let n be the order of the group of all roots of unity in $K = \mathbb{Q}(\varepsilon_m)$ and let $\varepsilon_n \in K$ be its generator. Of course, we have $m \mid n$, since all m-th roots of unity form a subgroup of order m in the group of order n of all roots of unity. Moreover, the number ε_m is a power of ε_n, so $K = \mathbb{Q}(\varepsilon_m) = \mathbb{Q}(\varepsilon_n)$. Thus $[K : \mathbb{Q}] = \varphi(m) = \varphi(n)$ by **T. 10.1**(a). Using the formula A.11.4(b), the equality $\varphi(m) = \varphi(n)$ together with $m \mid n$ imply $n = m$ or $n = 2m$ and m is odd. If $n = m$, then all roots of unity in K are ε_m^k for $k = 1, \ldots, m$. If $n = 2m$, where m is odd, then there are $2m$ roots of unity in $K = \mathbb{Q}(\varepsilon_n)$: $\varepsilon_n^k =$ for $k = 1, \ldots, n = 2m$. We claim that this set is equal to the set of all powers $\pm\varepsilon_m^k$ for $k = 1, \ldots, m$. We have an inclusion of two cyclic groups: $\langle \varepsilon_m \rangle \subset \langle \varepsilon_n \rangle$. The first group has an odd order m, and the second, the order $n = 2m$. Thus the index of the first group in the second is equal to 2. The number -1 is in

the group of roots of unity in K, but it is not in the first group, since -1 has order 2 while the order of the first group m is odd. Thus the group of roots of unity in K has a splitting in the two cosets $\langle\varepsilon_n\rangle = \langle\varepsilon_m\rangle \cup (-1)\langle\varepsilon_m\rangle$, which proves that it consists of elements $\pm\varepsilon_m^k$ for $k = 1, \ldots, m$. Notice that in the second case ($n = 2m$, m odd), the number $-\varepsilon_m$ is a generator of the group of roots of unity in $\mathbb{Q}(\varepsilon_m)$ (its order is $2m$).
(b) If $n = m$ or $n = 2m$, where m is odd, then $\mathbb{Q}(\varepsilon_m) = \mathbb{Q}(\varepsilon_n)$. This is evident when $n = m$. In the second case, the number $-\varepsilon_m$ generates the group of n-th roots of unity by (a). Hence $\mathbb{Q}(\varepsilon_n) = \mathbb{Q}(-\varepsilon_m) = \mathbb{Q}(\varepsilon_m)$.
Conversely, let $K = \mathbb{Q}(\varepsilon_m) = \mathbb{Q}(\varepsilon_n)$, where $m \leq n$, and denote by N the order of the group of roots of unity in K. Then $m \leq n \leq N$ and $m \mid n$ (say, $n = dm$ for an integer d). As we know from (a), we have $N = m$ or $N = 2m$ and m is odd. Thus either $m = n$ or $m < n$. The second case is only possible when $N = 2m$ and m is odd. Since $m < n = dm \leq N = 2m$, we get $n = 2m$, where m is odd.
(c) We have $\mathbb{Q}(\varepsilon_m) \subseteq \mathbb{Q}(\varepsilon_n)$ if and only if $\mathbb{Q}(\varepsilon_m) \cap \mathbb{Q}(\varepsilon_n) = \mathbb{Q}(\varepsilon_m)$ if and only if $\mathbb{Q}(\varepsilon_{(m,n)}) = \mathbb{Q}(\varepsilon_m)$ (see the solution of Exercise 10.5(c)). By (b), the last equality holds if and only if $(m,n) = m$, that is, $m \mid n$ or $m = 2(m,n)$ and (m,n) is odd (since $m \geq (m,n)$). In the second case, we have that m is even, $\frac{m}{2} \mid n$ and n is odd (since (m,n) is odd and m is even).

10.16 Example 1. Consider the Galois group of $X^{12} + 1$. Let ε be a zero of this polynomial, that is, $\varepsilon^{12} = -1$. Then $\varepsilon^{24} = 1$ so a splitting field of $X^{12} + 1$ is contained in a splitting field of $X^{24} - 1$. Conversely, if ε is a primitive 24-th root of unity, then $\varepsilon^{12} = -1$. Thus $\mathbb{Q}(\varepsilon)$, where ε is a primitive 24-th root of unity is a splitting field of $X^{12} + 1$. By **T. 10.1**, the order of the Galois group is $\varphi(24) = 8$.
Example 2. Consider the Galois group of the binomial $X^{12} - 3$. As in Exercise 10.14, we get the splitting field as $K = \mathbb{Q}(\varepsilon, \sqrt[12]{3})$, where ε is a primitive 12-th root of unity. The field $\mathbb{Q}(\sqrt[12]{3})$ has degree 12 over $\mathbb{Q}$, and $\mathbb{Q}(\varepsilon) = \mathbb{Q}(i, \sqrt{3})$ has degree $\varphi(12) = 4$ over $\mathbb{Q}$ (see Exercise 10.8 in this chapter). Hence $K = \mathbb{Q}(\sqrt[12]{3}, i)$ is the sought splitting field and its degree over $\mathbb{Q}$ is 24, since its degree over $\mathbb{Q}(\sqrt[12]{3})$ is 2.

Problems of Chap. 11

11.1 In order to find a normal basis in a Galois extension $K \subset L$, one can follow the proof of **T. 11.1**, that is, start with any basis $\alpha_1 \ldots, \alpha_n$, write down the images of a "general element" $a_1\alpha_1 + \cdots + a_n\alpha_n$ under all automorphisms σ_i, $i = 1, \ldots, n$, in the Galois group and try to find a_i such that these images are linearly independent, that is, $\alpha = a_1\alpha_1 + \cdots + a_n\alpha_n$ is normal. In the proof of **T. 11.1**, we show that such a choice always exists, but it is not too practical to follow the general argument (in the proof of **T. 11.1**) and usually it is not difficult to find a normal element, "designing" it in a suitable way. We consider one example.
Example. Let L be a splitting field of the polynomial $f(X) = X^3 - 2$ over $\mathbb{Q}$, so $L = \mathbb{Q}(\varepsilon, \sqrt[3]{2})$, where $\varepsilon^2 + \varepsilon + 1 = 0$. The Galois group $G(L/\mathbb{Q})$ consists of six automorphism defined by $\sigma(\varepsilon) = \pm\varepsilon^i$, $i = 1, 2$ and $\sigma(\sqrt[3]{2}) = \varepsilon^j\sqrt[3]{2}$, $j = 0, 1, 2$.

Let us choose $\alpha = (1+\varepsilon)(1+\sqrt[3]{2})$. We want to show that the six images $\sigma(\alpha) = (1+\varepsilon^i)(1+\varepsilon^j\sqrt[3]{2})$ are linearly independent over $\mathbb{Q}$. First note that $1+\varepsilon, 1+\varepsilon^2$ are linearly independent over $\mathbb{Q}$, so they form a basis of $\mathbb{Q}(\varepsilon)$ over $\mathbb{Q}$. Next observe that $1+\varepsilon^j\sqrt[3]{2}$ for $j = 0, 1, 2$ are linearly independent in L over $\mathbb{Q}(\varepsilon)$, since $1, \varepsilon\sqrt[3]{2}, \varepsilon^2\sqrt[3]{2}$ form a basis of L over $\mathbb{Q}(\varepsilon)$, so $a(1+\sqrt[3]{2})+b(1+\varepsilon\sqrt[3]{2})+c(1+\varepsilon^2)\sqrt[3]{2} = 0$ implies $a+b+c = 0, a+b\varepsilon+c(-1-\varepsilon) = 0$, which implies $a = b = c = 0$, since the second equality gives $a-c = b-c = 0$. Now we can use the argument from the proof of **T. 4.3** that a basis of L over $\mathbb{Q}$ can be obtained by multiplying any basis of L over an intermediate field M (here $M = \mathbb{Q}(\varepsilon)$) by a basis of M over $\mathbb{Q}$. A similar argument works when such an intermediate field exists in the extension for which we want to construct a normal basis. See also Exercise 11.2.

11.2 (a) We denote by σ the nontrivial automorphism of L over K. If $\alpha \in K$ or $\mathrm{Tr}(\alpha) = \alpha + \sigma(\alpha) = 0$, then α is not normal, since α and $\beta = \sigma(\alpha) = \pm\alpha$ are linearly dependent.
Conversely, let $\alpha \notin K$ and $\mathrm{Tr}(\alpha) \neq 0$. We want to show that α and $\sigma(\alpha)$ are linearly independent. If not, then $\sigma(\alpha) = a\alpha$ for some $a \in K$. Hence $\sigma^2(\alpha) = a\sigma(\alpha) = a^2\alpha$. Since $\sigma^2 = 1$ and $\alpha \neq 0$, we get $a^2 = 1$. This gives either $\sigma(\alpha) = \alpha$, so $\alpha \in K$, or $\sigma(\alpha) = -\alpha$, so $\mathrm{Tr}(\alpha) = \alpha + \sigma(\alpha) = 0$. Both consequences are impossible, which implies that α and $\sigma(\alpha)$ are linearly independent.
Notice that the solution of (b) below gives another method of treating (a).
(b) Exactly as in (a), if $\alpha \in K$ or $\mathrm{Tr}(\alpha) = \alpha + \sigma(\alpha) + \sigma^2(\alpha) = 0$, then α is not normal, since α, $\beta = \sigma(\alpha)$ and $\gamma = \sigma(\beta)$ are linearly dependant.
Conversely, let $\alpha \notin K$ and $\mathrm{Tr}(\alpha) \neq 0$. Using Lemma 11.2 on p. 129, we get that α, β, γ are linearly dependent if and only if

$$\begin{vmatrix} \alpha & \beta & \gamma \\ \beta & \gamma & \alpha \\ \gamma & \alpha & \beta \end{vmatrix} = (\alpha+\beta+\gamma)[(\alpha-\beta)^2 + (\beta-\gamma)^2 + (\gamma-\alpha)^2] = 0.$$

The last equality implies $\mathrm{Tr}(\alpha) = \alpha+\beta+\gamma = 0$ or $\alpha = \beta = \gamma$, in which case, $\beta = \sigma(\alpha) = \alpha$, so $\alpha \in K$. Thus, we get a contradiction, which shows that α, β, γ are linearly independent.

11.3 (a) Let $G(L/K) = \{\sigma_1, \ldots, \sigma_n\}$ and let $\alpha \in L$ be a normal element. Then $\sigma_1(\alpha), \ldots, \sigma_n(\alpha)$ form a basis of L over K and, in particular, are all different. Hence α is a generator of L over K by Exercise 9.24. The converse is not true, since a generator of L over K need not be normal. For example, the generator i of $\mathbb{Q}(i)$ over $\mathbb{Q}$ is not normal, since $i, -i$ is not a basis of $\mathbb{Q}(i)$ over $\mathbb{Q}$.
(b) Take a finite field generated over $\mathbb{F}_2$ by an element whose trace is 0. For example, let $L = \mathbb{F}_2(\alpha)$, where $\alpha^3+\alpha+1 = 0$. The group L^* has order 7, so 7 is the order of α. The zeros of the minimal polynomial $f(X) = X^3+X+1$ are α, α^2 and $\alpha^4 = \alpha^2+\alpha$, which are the images of α under all automorphisms of $G(L/\mathbb{F}_2)$. They are linearly dependent as $\alpha+\alpha^2+\alpha^4 = 0$. Thus α is a group generator of L^*, but it is not a normal element.

A normal element need not be the group generator for the group of nonzero elements in the field. Consider a quadratic extension $\mathbb{F}_7 \subset \mathbb{F}_7(\alpha) = L$, where $\alpha^2 + 1 = 0$. The nontrivial automorphism in the Galois group is defined by $\sigma(\alpha) = -\alpha$ and an element $x = a + b\alpha$, $a, b \in \mathbb{F}_7$, is normal if and only if $x = a + b\alpha, \sigma(x) = a - b\alpha$ are linearly in dependent over $\mathbb{F}_7$, which happens if and only if $ab \neq 0$. Thus, we have 36 normal elements. Since the field L has 49 elements, the nonzero elements L^* form a group of order 48. The number of its generators is $\varphi(48) = 16$. Thus, certainly there are normal elements in L which do not generate the group L^*. Since there are 12 non-normal elements in L^*, at least 4 elements in L^* must be both normal and group generators. Which ones?

Remark. Summarizing, we see that both group primitive and normal elements are field primitive, but a field primitive element need not be normal or group primitive. On the other hand, the two subsets of field primitive elements—group primitive and normal elements—may intersect, but a normal element need not be group primitive and a group primitive element need not be normal. It is known that in a finite field extension there exist elements which are both normal and group primitive (see [LS]).

11.4 (a) We assume that X, Y, Z are positive integers. Since $\mathrm{Nr}(\alpha) = 1$, where $\alpha = \frac{X+Yi}{Z}$, Hilbert's Theorem 90 says there exists a $\beta = m + ni$ such that $\alpha = \beta/\bar{\beta}$. We can assume that m, n are relatively prime integers (β and $\bar{\beta}$ can be divided by the greatest common divisor of m and n). Thus, we have

$$\alpha = \frac{X}{Z} + \frac{Y}{Z}i = \frac{\beta}{\bar{\beta}} = \frac{m + ni}{m - ni} = \frac{m^2 - n^2}{m^2 + n^2} + \frac{2mn}{m^2 + n^2}i$$

so

$$\frac{X}{Z} = \frac{m^2 - n^2}{m^2 + n^2} \quad \text{and} \quad \frac{Y}{Z} = \frac{2mn}{m^2 + n^2}.$$

Since m, n are chosen to be relatively prime, a common divisor of $m^2 - n^2$ and $m^2 + n^2$ can only be 2 and this happens only when m, n are both odd. The same is true of $2mn$ and $m^2 + n^2$. If it happens, then we can find integers k, l such that $k + l = m$ and $k - l = n$, that is, $k = (m + n)/2$ and $l = (m - n)/2$. Of course k, l are of different parities and

$$\frac{X}{Z} = \frac{2kl}{k^2 + l^2} \quad \text{and} \quad \frac{Y}{Z} = \frac{k^2 - l^2}{k^2 + l^2}.$$

Thus, we can assume that already from the beginning, the numbers m, n are of different parities, so the fractions $\frac{m^2-n^2}{m^2+n^2}$ and $\frac{2mn}{m^2+n^2}$ are in the lowest terms. Since we assume that X, Y, Z are relatively prime positive integers, the fractions $\frac{X}{Z}, \frac{Y}{Z}$ are also in the lowest terms. Thus, we have the equalities $X = m^2 - n^2, Y = 2mn, Z = m^2 + n^2$, m, n are relatively prime positive integers of different parities and $m > n$.

(b) Using similar arguments as in (a), but in the field $\mathbb{Q}(\sqrt{-2})$, the formulae are $X = m^2 - 2n^2$, $Y = 2mn$, $Z = m^2 + 2n^2$, m, n are relatively prime positive integers, m is odd and $m > n\sqrt{2}$.

11.5 (a) A cyclic extension of degree n over K is a Kummer extension of exponent n and by **T. 11.3**, it is of the form $K(\sqrt[n]{a})$ for some $a \in K$, $a \neq 0$. Conversely, every extension of this form is cyclic and has degree n. According to **T. 11.4**(c), the extension $K(\sqrt[n]{a})$ of K corresponds to a subgroup of K^*/K^{*n} generated by aK^{*n}. The extensions $K(\sqrt[n]{a})$ and $K(\sqrt[n]{b})$ of K are isomorphic if and only if aK^{*n} and bK^{*n} generate the same subgroup of K^*/K^{*n}. Since all the generators of the cyclic group generated by aK^{*n} are a^rK^{*n}, where $\gcd(r, n) = 1$ (see A.2.2), we get $bK^{*n} = a^rK^{*n}$ for some r relatively prime to n, that is, $ba^{-r} \in K^{*n}$.
(b) If L is a quadratic field extension of $\mathbb{Q}$, then by Exercise 8.2, we have $L = \mathbb{Q}(\sqrt{a})$, where $a \in \mathbb{Q}$. We can always choose $a \in \mathbb{Z}$, since $a = p/q$, where $p, q \in \mathbb{Z}, q > 0$ and $\sqrt{a} = (1/q)\sqrt{pq}$. Moreover, we can always choose p, q relatively prime and without factors which are squares of integers (we have $\mathbb{Q}(r\sqrt{a}) = \mathbb{Q}(\sqrt{a})$ for any nonzero rational number r). Hence, we can choose pq as a product of different prime numbers or as -1 in the special case when p, q are units (that is, ± 1). If now $\mathbb{Q}(\sqrt{a})$ and $\mathbb{Q}(\sqrt{b})$ are two quadratic extensions such that a, b are both positive or negative products of different prime numbers or -1, then by (a), we get that two such extensions are equal if and only if ba^{-1} is a square of a rational number. This is the same as saying that ab is a square of an integer, since $ba^{-1} = ab \cdot (1/a)^2$. But a, b are square free, so ab is a square if and only if $a = b$. Thus every quadratic extension of $\mathbb{Q}$ can be uniquely written as $\mathbb{Q}(\sqrt{a})$, where a is a product of different factors which are prime numbers or -1. Notice that it is easy to prove (b) directly without the general result in (a).

11.6 (b) If $K \supset \mathbb{Q}$ is a cyclic cubic extension, then by Exercise 9.15, the field EK is a Galois extension of E and Galois extension of $\mathbb{Q}$ with Galois group $\mathbb{Z}_2 \times \mathbb{Z}_3 = \mathbb{Z}_6$. Conversely, if L is a Galois extension of $\mathbb{Q}$ with Galois group $\mathbb{Z}_6$ containing E, then the fixed field of the only subgroup of $\mathbb{Z}_6$ of order 2 (the identity and the complex conjugation) is a cyclic cubic extension K of $\mathbb{Q}$. It is uniquely defined by L.
(c) The minimal polynomial of $\sqrt[3]{\alpha}$ over $\mathbb{Q}$ is $\varphi(X) = (X^3 - \alpha)(X^3 - \bar{\alpha}) = X^6 - \mathrm{Tr}(\alpha)X^3 + \mathrm{Nr}(\alpha)$, since it has rational coefficients and the degree 6 is equal to the degree $[L : \mathbb{Q}] = [L : E][E : \mathbb{Q}] = 6$. Thus, if L over $\mathbb{Q}$ is Galois, we have $\sqrt[3]{\bar{\alpha}} \in L$, since it is a zero of the same irreducible polynomial, which has $\sqrt[3]{\alpha}$ as its zero. Hence, we have $L = E(\sqrt[3]{\alpha}) = E(\sqrt[3]{\bar{\alpha}})$. Conversely, if these equalities hold, then the field L is a splitting field of the polynomial $\varphi(X)$, so it is a Galois extension of $\mathbb{Q}$. The fields $E(\sqrt[3]{\alpha})$ and $E(\sqrt[3]{\bar{\alpha}})$ are Kummer extensions of E of exponent 3 and by **T. 11.4**(c), the equality $E(\sqrt[3]{\alpha}) = E(\sqrt[3]{\bar{\alpha}})$ says that α and $\bar{\alpha}$ generate the same subgroup of E^* modulo E^{*3}. This is equivalent to $\alpha^2E^{*3} = \bar{\alpha}E^{*3}$ or $\alpha E^{*3} = \bar{\alpha}E^{*3}$, that is, $\alpha\bar{\alpha} = \mathrm{Nr}(\alpha) \in E^{*3}$ or $\alpha^2\bar{\alpha} = \alpha\mathrm{Nr}(\alpha) \in E^{*3}$.
Assume that the first case occurs. As a splitting field of $\varphi(X)$, the field L has six automorphisms mapping its zero $\sqrt[3]{\alpha}$ to one of the six others: $\varepsilon^i\sqrt[3]{\alpha}$, $\varepsilon^i\sqrt[3]{\bar{\alpha}}$ for $i = 0, 1, 2$. It is easy to check that the automorphism mapping $\sqrt[3]{\alpha}$ onto $\varepsilon\sqrt[3]{\bar{\alpha}}$

has order 6 (it follows from the assumption that $\sqrt[3]{\alpha}\sqrt[3]{\bar{\alpha}} = \sqrt[3]{\mathrm{Nr}(\alpha)}$ is a rational number). Hence, the Galois group $G(L/\mathbb{Q})$ is cyclic.
Conversely, assume that $G(L/\mathbb{Q})$ is cyclic. We want to prove that $\mathrm{Nr}(\alpha)$ is a rational cube. In fact, we have $\sqrt[3]{\alpha}\sqrt[3]{\bar{\alpha}} = \sqrt[3]{\mathrm{Nr}(\alpha)} \in L$. Hence L contains a splitting field of the polynomial $X^3 - \mathrm{Nr}(\alpha)$ with rational coefficients if this polynomial is irreducible. But the splitting field of such an irreducible polynomial is of degree 6 over $\mathbb{Q}$ and its Galois group is the symmetric group S_3 (see Exercise 10.14(c)). This contradicts our assumption about $G(L/\mathbb{Q})$. Hence the polynomial $X^3 - \mathrm{Nr}(\alpha)$ must be reducible, which shows that $\mathrm{Nr}(\alpha)$ is a rational cube.
Thus, we have proved that the Galois group $G(E(\sqrt[3]{\alpha})/\mathbb{Q})$ is cyclic if and only if $\mathrm{Nr}(\alpha)$ is a rational cube. Consequently, it must be the symmetric group S_3 if and only if $E(\sqrt[3]{\alpha})$ is Galois over $\mathbb{Q}$ and $\alpha\mathrm{Nr}(\alpha)$ is a cube in $\mathbb{Q}$, since there are only two non-isomorphic groups with six elements.
(d) Using the formula $(a+b)^3 = a^3 + b^3 + 3ab(a+b)$ with $a = \sqrt[3]{f\bar{f}^2}$, $b = \sqrt[3]{\bar{f}f^2}$, we get $\gamma^3 = f\bar{f}^2 + \bar{f}f^2 + 3\mathrm{Nr}(f)\gamma = \mathrm{Tr}(f)\mathrm{Nr}(f) + 3\mathrm{Nr}(f)\gamma$. Thus γ has the minimal polynomial:

$$X^3 - 3\mathrm{Nr}(f)X - \mathrm{Tr}(f)\mathrm{Nr}(f), \tag{19.1}$$

over $\mathbb{Q}$. The cubic roots in $\gamma = \sqrt[3]{f\bar{f}^2} + \sqrt[3]{\bar{f}f^2}$ are so chosen that their product is equal to $\mathrm{Nr}(f)$.

11.7 (b) If $K \supset \mathbb{Q}$ is a cyclic quartic extension, then by Exercise 9.15, the field $L = EK$ is a Galois extension of E and a Galois extension of $\mathbb{Q}$ with Galois group $\mathbb{Z}_2 \times \mathbb{Z}_4$. In fact, the quadratic subfield of K is real by Exercise 9.23, so $E \cap K = \mathbb{Q}$. Conversely, let L be a Galois extension of $\mathbb{Q}$ containing E with Galois group $\mathbb{Z}_2 \times \mathbb{Z}_4$. It is not difficult to check that the group $G = \mathbb{Z}_2 \times \mathbb{Z}_4$ contains exactly two elements of order 2, which generate subgroups H such that G/H is cyclic of order 4 (the subgroups H generated by $(1,0)$ or $(1,2)$). The fixed fields K_1 and K_2 of these two subgroups are the only two cyclic quartic extensions of $\mathbb{Q}$ which are contained in L. The third element of order 2 in G generates a subgroup H such that G/H is the group $\mathbb{Z}_2 \times \mathbb{Z}_2$ (the element $(0,2)$). Its fixed field K contains three quadratic extensions of $\mathbb{Q}$ and among them, the field E, since all quadratic extensions of $\mathbb{Q}$ in L are contained in K (the group G contains exactly three subgroups of order 4, whose fixed fields are quadratic over $\mathbb{Q}$, and the field K contains exactly three quadratic subfields). The intersection of K_1 and K_2 must be a real quadratic subfield of both (see Exercises 9.15 and 9.23) and K is not the real subfield of L, since it contains E. Thus one of the fields K_1, K_2 must be real, but not both, since $K_1K_2 = L$ and L is not real (it contains E). Hence, the field L uniquely defines a pair of its subfields K_1 and K_2 (one real and one nonreal), which are cyclic quartic over $\mathbb{Q}$ and such that $EK_1 = EK_2 = L$.
(c) The minimal polynomial of $\sqrt[4]{\alpha}$ over $\mathbb{Q}$ is $\varphi(X) = (X^4 - \alpha)(X^4 - \bar{\alpha}) = X^8 - \mathrm{Tr}(\alpha)X^4 + \mathrm{Nr}(\alpha)$, since it has rational coefficients and the degree 8 is equal to the degree $[L : \mathbb{Q}] = [L : E][E : \mathbb{Q}] = 8$. Thus if L over $\mathbb{Q}$ is Galois, we have $\sqrt[4]{\bar{\alpha}} \in L$, since it is a zero of the same irreducible polynomial, which has $\sqrt[4]{\alpha}$ as its zero.

Hence, we have $L = E(\sqrt[4]{\alpha}) = E(\sqrt[4]{\bar{\alpha}})$. Conversely, if these equalities hold, then the field L is a splitting field of the polynomial $\varphi(X)$, so it is a Galois extension of $\mathbb{Q}$.
The fields $E(\sqrt[4]{\alpha})$ and $E(\sqrt[4]{\bar{\alpha}})$ are Kummer extensions of E of exponent 4 and by **T. 11.4**(c), the equality $E(\sqrt[4]{\alpha}) = E(\sqrt[4]{\bar{\alpha}})$ says that α and $\bar{\alpha}$ represent the same subgroup of E^* modulo E^{*4}. This is equivalent to $\alpha^3 E^{*4} = \bar{\alpha}E^{*4}$ or $\alpha^2 E^{*4} = \bar{\alpha}E^{*4}$ or $\alpha E^{*4} = \bar{\alpha}E^{*4}$, that is, $\alpha\bar{\alpha} = \mathrm{Nr}(\alpha) \in E^{*4}$ or $\alpha^2\bar{\alpha} = \alpha\mathrm{Nr}(\alpha) \in E^{*4}$ or $\alpha^3\bar{\alpha} = \alpha^2\mathrm{Nr}(\alpha) \in E^{*4}$.
Assume that the first case occurs. As a splitting field of $\varphi(X)$, the field L has eight automorphisms mapping its zero $\sqrt[4]{\alpha}$ to one of the numbers: $i^r\sqrt[4]{\alpha}$, $i^r\sqrt[4]{\bar{\alpha}}$ for $r = 0, 1, 2, 3$. Since $\mathrm{Nr}(\alpha)$ is a fourth power in E^*, it is easy to see that $\mathrm{Nr}(\alpha) = r^4$, where $r > 0$ is a rational number. In fact, if $\mathrm{Nr}(\alpha) = (a + bi)^4 = (A + Bi)^2$ for suitable rational numbers a, b, A, B, then we first see that $AB = 0$ and, using this, that $ab = 0$ (since $\mathrm{Nr}(\alpha)$ is a rational number). Hence, we can choose r as the absolute value of a (or b). Let us fix the values of $\sqrt[4]{\alpha}$ and $\sqrt[4]{\bar{\alpha}}$ in such a way that $\sqrt[4]{\alpha}\sqrt[4]{\bar{\alpha}} = r$ and define $\sigma(\sqrt[4]{\alpha}) = i\sqrt[4]{\bar{\alpha}}$. Taking fourth powers on both sides, we get $\sigma(\alpha) = \bar{\alpha}$, so σ is nontrivial on E, which shows that $\sigma(i) = -i$. Using this and the equality $\sigma(\sqrt[4]{\bar{\alpha}}) = -i\sqrt[4]{\alpha}$, which follows from the equality $\sqrt[4]{\alpha}\sqrt[4]{\bar{\alpha}} = r$ and the definition of σ, we check that σ has order 4. The automorphism $\tau(\sqrt[4]{\alpha}) = -\sqrt[4]{\alpha}$ has order 2, which is also a consequence of our choice of $\sqrt[4]{\alpha}$ and $\sqrt[4]{\bar{\alpha}}$. We check that $\sigma\tau = \tau\sigma$, which implies that the group generated by these two automorphisms is isomorphic to $\mathbb{Z}_2 \times \mathbb{Z}_4$. Thus, it is the whole Galois group $G(L/\mathbb{Q})$.
In the second case, we have $\alpha^2\bar{\alpha} = \beta^4$, where $\beta \in E^*$. This equality says that $\bar{\alpha} = \gamma^2$ for some $\gamma \in E^*$, so $\alpha = \bar{\gamma}^2$ and we get $\bar{\gamma}^4\gamma^2 = \beta^4$. Hence, once again, we see that γ is a square in E^* and, consequently, α is a fourth power of an element in E^*. This is impossible, since L has degree 4 over E.
In the third case, we have $\alpha^3\bar{\alpha} = \beta^4$, where $\beta \in E^*$. Let us fix $\sqrt[4]{\alpha}$ and $\sqrt[4]{\bar{\alpha}}$ in such a way that $\sqrt[4]{\alpha^3}\sqrt[4]{\bar{\alpha}} = \beta$. Consider two automorphisms σ, τ of the field L such that $\sigma(\sqrt[4]{\alpha}) = i\sqrt[4]{\alpha}$ and $\tau(\sqrt[4]{\alpha}) = i\sqrt[4]{\bar{\alpha}}$. We check that σ has order 4, τ has order 2 and $\tau\sigma\tau = \sigma^{-1}$. In fact, taking the fourth powers in the equalities defining σ and τ, we get $\sigma(\alpha) = \alpha$ and $\tau(\alpha) = \bar{\alpha}$. Thus σ is the identity on E, and τ is the complex conjugation on E. Hence, we have $\sigma(\beta) = \beta$ and $\tau(\beta) = \bar{\beta}$. Using this and the equality $\sqrt[4]{\alpha^3}\sqrt[4]{\bar{\alpha}} = \beta$, we find that $\sigma(\sqrt[4]{\bar{\alpha}}) = i\sqrt[4]{\bar{\alpha}}$ and $\tau(\sqrt[4]{\bar{\alpha}}) = -i\sqrt[4]{\alpha}$. Now, we have $\sqrt[4]{\alpha} \overset{\sigma}{\mapsto} i\sqrt[4]{\alpha} \overset{\sigma}{\mapsto} -\sqrt[4]{\alpha} \overset{\sigma}{\mapsto} -i\sqrt[4]{\alpha} \overset{\sigma}{\mapsto} \sqrt[4]{\alpha}$ and $\sqrt[4]{\alpha} \overset{\tau}{\mapsto} i\sqrt[4]{\bar{\alpha}} \overset{\tau}{\mapsto} \sqrt[4]{\alpha}$, that is, $\sigma^4 = 1$ and $\tau^2 = 1$. Moreover $\sqrt[4]{\alpha} \overset{\sigma}{\mapsto} i\sqrt[4]{\alpha} \overset{\tau}{\mapsto} \sqrt[4]{\bar{\alpha}} \overset{\sigma}{\mapsto} i\sqrt[4]{\bar{\alpha}}$, so $\sigma\tau\sigma = \tau$, that is, $\tau\sigma = \sigma^{-1}\tau$. These relations say that $G(L/\mathbb{Q})$ consists of automorphisms $\sigma^i\tau^j$, where $i = 0, 1, 2, 3, j = 0, 1$. These eight elements form a group isomorphic to the square group D_4 (see p. 239 on the definition of D_4 by generators and relations).
(d) We fix a value of $\sqrt[4]{\alpha}$ and choose $\sqrt[4]{\bar{\alpha}}$ so that $\sqrt[4]{\alpha}\sqrt[4]{\bar{\alpha}} = \mathrm{Nr}(f)|g|$. Then one may show as in (c) that the automorphism $\sigma : \sqrt[4]{\alpha} \mapsto \sqrt[4]{\bar{\alpha}}$ has order 2 and that the quotient of the Galois group $G(\mathbb{Q}(\sqrt[4]{\alpha})/\mathbb{Q})$ by $\langle\sigma\rangle$ is a cyclic group (of course, of order 4). But it is more instructive to find the Galois group of the field $\mathbb{Q}(\sqrt[4]{\alpha})^{\langle\sigma\rangle} = \mathbb{Q}(\gamma)$, $\gamma = \sqrt[4]{\alpha} + \sqrt[4]{\bar{\alpha}}$, which has degree 4 over $\mathbb{Q}$ by **T. 6.2**, since $[\mathbb{Q}(\sqrt[4]{\alpha}) : \mathbb{Q}] = 8$.

In fact, we have

$$\gamma^4 = \alpha + \bar{\alpha} + 4\sqrt[4]{\alpha}\sqrt[4]{\bar{\alpha}}[(\sqrt[4]{\alpha})^2 + (\sqrt[4]{\bar{\alpha}})^2] + 6(\sqrt[4]{\alpha})^2(\sqrt[4]{\bar{\alpha}})^2$$
$$= \mathrm{Nr}(f)g^2[\mathrm{Tr}(f)^2 - 2\mathrm{Nr}(f)] + 4\mathrm{Nr}(f)g[\gamma^2 - 2\mathrm{Nr}(f)g] + 6\mathrm{Nr}(f)^2g^2$$
$$= 4\mathrm{Nr}(f)g\gamma^2 + \mathrm{Nr}(f)g^2[4\mathrm{Nr}(f) - \mathrm{Tr}(f)^2].$$

Thus γ is a zero of the polynomial $X^4 - 4\mathrm{Nr}(f)gX^2 + \mathrm{Nr}(f)g^2[4\mathrm{Nr}(f) - \mathrm{Tr}(f)^2]$ whose zeros are $\sqrt{2g\left(\mathrm{Nr}(f) \pm \frac{\mathrm{Tr}(f)}{2}\sqrt{\mathrm{Nr}(f)}\right)}$.

(e) Taking $f = k + li$, we get $\mathrm{Tr}(f) = 2k$ and $\Delta = \mathrm{Nr}(f) = k^2 + l^2$. Thus, we may choose $\gamma = \sqrt{m(\Delta + k\sqrt{\Delta})}$, where $m = 2g$. We may assume that m is a square-free integer (of course, $m = 1$ is a possibility). For example, when $f = 1 + 2i, m = 1$, we get the extension $\mathbb{Q}(\sqrt{5+\sqrt{5}})$ of $\mathbb{Q}$.

11.8 We have $\mathrm{Tr}(1) = p \cdot 1 = 0$. According to the additive version of Hilbert's Theorem 90 (see **T. 11.2**(b)), there is an $\alpha \in L$ such that $\alpha - \sigma(\alpha) = 1$, where σ is a generator of the Galois group $G(L/K)$. The images of α by the automorphisms of L over K are $\alpha, \sigma(\alpha) = \alpha - 1, \sigma^2(\alpha) = \alpha - 2, \ldots, \sigma^{p-1}(\alpha) = \alpha - (p-1)$. Since α has p different images, we have $L = K(\alpha)$ by Exercise 9.24. Now we find the polynomial of degree p in $K[X]$ having these zeros: $f(X) = \prod_{i=0}^{p-1}(X - (\alpha - i))$. We have, in fact, $f(X) = X^p - X - a$, where

$$a = \alpha(\alpha - 1)\cdots(\alpha - (p-1)).$$

In order to prove this, we check that $f(0) = -a$ (observe that for $p = 2$, we have $-1 = 1$ and for $p > 2$, the number of factors defining $f(X)$ is odd) and $X^p - X$ has the same value $\alpha^p - \alpha = a'$ for all $\alpha - i, i = 0, 1, \ldots, p-1$ thanks to Fermat's Little Theorem (see Exercise 6.3(c)). We have $a' \in K$, since a' is invariant with respect to all automorphisms of L (that is, $a' \in L^{G(L/K)} = K$). Thus the zeros of $X^p - X - a'$ are the same as the zeros of $f(X)$, which gives the required equality and, of course, $a' = a$.

11.9 Since L is a Galois extension of K, we can choose a prime number q dividing $[L : K]$ and a subgroup of order q in $G(L/K)$ (see A.9.3). The degree of L over the fixed field of such a subgroup is q. Therefore, we assume that we have an extension $K \subset L$ such that $[L : K] = q$ is a prime number. Consider now two cases I and II.

Case I: q is not equal to the characteristic of K (which may be zero). Notice that all q-th roots of unity are in K. This is clear when $q = 2$. If $q > 2$ and the polynomial $X^q - 1$ has an irreducible factor of some degree $d > 1$, then $d < q$ and $d \mid q$, since each zero of this factor in L generates an extension of degree d over K. But this is of course impossible, since q is a prime number. Thus, all irreducible factors of $X^q - 1$ have degree 1. Let ε generate the group of q-th roots of unity in K. Since L is a cyclic Galois extension of degree q, we have $L = K(\alpha)$, where $\alpha^q = a \in K$.

Let $\beta \in L$ be such that $\beta^q = \alpha$. If q is odd, then taking the norms, we get $\mathrm{Nr}(\beta)^q = \mathrm{Nr}(\alpha) = a = \alpha^q$. Hence, we have $\alpha = \varepsilon^i \mathrm{Nr}(\beta) \in K$ for some i, which is a contradiction. (Notice that for $q = 2$, we have $\mathrm{Nr}(\alpha) = -a$, so the argument fails.)

Case II: q is equal to the characteristic p of K. By the "additive" Hilbert 90 (or the description of Artin–Schreier extensions), we have $L = K(\alpha)$, where α is a zero of an irreducible polynomial $X^p - X - a$, $a \in K$ (see Exercise 11.8). We want to show that this is impossible, since each such polynomial has a zero in K. The elements $1, \alpha, \ldots, \alpha^{p-1}$ form a basis of L over K. If $x = a_0 + a_1\alpha + \cdots + a_{p-1}\alpha^{p-1}$, $a_i \in K$, then $x^p - x = b_0 + b_1\alpha + \cdots + b_{p-1}\alpha^{p-1}$ for some $b_i \in K$. We want to find b_{p-1} expressed by a_i. In the computation below we denote by $\cdots$ all terms which are expressed as a linear combination of the powers $1, \alpha, \ldots, \alpha^{p-2}$. We have:

$$x^p - x = (\sum_{i=0}^{p-1} a_i\alpha^i)^p - \sum_{i=1}^{p-1} a_i\alpha^i = \sum_{i=0}^{p-1} a_i^p\alpha^{ip} - \sum_{i=1}^{p-1} a_i\alpha^i$$

$$= \sum_{i=0}^{p-1} a_i^p(\alpha + a)^i - \sum_{i=1}^{p-1} a_i\alpha^i = (a_i^p - a_i)\alpha^{p-1} + \cdots$$

Now choose $x \in L$ to be an arbitrary zero of the equation $X^p - X - a\alpha^{p-1}$ in the field L—such a zero exists since L is algebraically closed. Then the left-hand side above is $a\alpha^{p-1}$, so the corresponding a_i satisfies $a_i^p - a_i = a$, which gives a zero a_i of $X^p - X - a$.

Thus the second case $q = p$ is also excluded and the only possibility that remains is that L is an algebraically closed field of finite dimension over its subfield K and $[L : K]$ is a power of 2.

Assume that $[L : K] = 2$. Denote by i a zero of $X^2 + 1$ in L. We claim that $L = K(i)$. Assume that $i \in K$ and let $L = K(\alpha)$, where $\alpha^2 = a \in K$. Take $\beta \in L$ such that $\beta^2 = \alpha$. Then $\mathrm{Nr}(\beta)^2 = \mathrm{Nr}(\alpha) = -a = -\alpha^2$, which implies that $\alpha = \pm i\mathrm{Nr}(\beta) \in K$, which is impossible. Thus, we have $i \notin K$, that is, $L = K(i)$.

If now $[L : K]$ is a power of 2 bigger than 2, then the Galois group $G(L/K)$ has a subgroup of order 4. Assume that K is its fixed field, so $[L : K] = 4$. A group of order 4 contains a subgroup of order 2, so there is a subfield M of L containing K such that $[L : M] = 2$. Let M be any such subfield of L. Then $L = M(i)$ as we already know. But $[K(i) : K] = 2$ ($i \notin K$), so we can take $M = K(i)$ and we get a contradiction—for such a choice of M, we have $i \in M$ and $i \notin M$.

Problems of Chap. 12

12.2 (a) Since G/N is abelian, we have $g_1g_2g_1^{-1}g_2^{-1} \in N$ when $g_1, g_2 \in G$ (since $Ng_1g_2 = Ng_1Ng_2 = Ng_2Ng_1 = Ng_2g_1$, we have $Ng_1g_2g_1^{-1}g_2^{-1} = N$). Let (a, b, c) be an arbitrary cycle of length 3 and let $g_1 = (a, b, d)$, $g_2 = (c, e, a)$, where

$a, b, c, d, e \in \{1, 2, \ldots, n\}$ for some $n \geq 5$. By the assumption, we have $g_1, g_2 \in G$. Thus

$$g_1 g_2 g_1^{-1} g_2^{-1} = (a, b, d)(c, e, a)(d, b, a)(a, e, c) = (a, b, c) \in N,$$

which shows that any cycle of length 3 is in N.
(b) Assume that A_n is solvable for some $n \geq 5$, that is, there is a sequence $G_0 = A_n \supset G_1 \supset \cdots \supset G_{m-1} \supset G_m = \langle(1)\rangle$ such that G_{i+1} is normal in G_i and G_i/G_{i+1} is abelian for $i = 0, \ldots, m-1$. Thus $G_0 = A_n$ has a normal subgroup G_1 such that G_0/G_1 is abelian. But every cycle (a, b, c), where $a, b, c \in \{1, 2, \ldots, n\}$, is in $G_0 = A_n$ (since $(a, b, c) = (a, b)(b, c)$ is even), so by (a), the group G_1 also contains every cycle of length 3. In the same way, the group G_2 contains all cycles of length 3. Using induction, we see that all subgroups G_i must contain all cycles of length 3. This is a contradiction, since G_m has only one element. Thus, the group A_n is not solvable when $n \geq 5$.

12.3 Among all possible chains $G = G_0 \supset G_1 \supset \ldots \supset G_n = \{e\}$ choose a chain with maximal possible value of n. If for some $i < n$ the quotient G_i/G_{i+1} is not cyclic of prime order, then it contains a nontrivial subgroup H/G_{i+1}, where $G_i \supset H \supset G_{i+1}$ (see A.2.15). This contradicts our choice of n. In fact, G_{i+1} is a normal subgroup of H, since it is normal in G_i, and H/G_{i+1} is an abelian group as a subgroup of G_i/G_{i+1}. The group H is normal in G_i as an inverse image of a normal subgroup H/G_{i+1} in G_i/G_{i+1}, by the natural surjection of G_i onto G_i/G_{i+1} (see A.2.13). Finally, the quotient G_i/H is abelian as the image of a natural surjection of G_i/G_{i+1} onto it (see A.2.12).

12.4 (e) We have $\varphi_{a,b}^n(x) = \varphi_{a^n, b(a^{n-1}+\cdots+a+1)}$ (see Exercise 12.4(b) on p. 171), so if the order of $\varphi_{a,b}$ is p, then $a^p = 1$. Since, according to Fermat's Little Theorem, $a^p = a$, we get $a = 1$. Thus all elements of order p are in $\mathcal{T}_p$.

12.5 First, we prove that (a) and (c) are equivalent. Of course, (c) implies (a), since the group $\mathcal{G}_p$ is solvable, so if G is isomorphic to its subgroup, then it is also solvable by **T. 12.1**(a).

Assume that G is a solvable group. We choose $X = \{0, 1, \ldots, p-1\}$, so G acts transitively on X (see section "Group Actions on Sets" in Appendix). We prove the claim by induction on the order of G. The order of G is at least p by (d). If it is p, then G is a cyclic group with a generator σ of order p. This generator σ must be a cycle of length p, since every permutation is a composition of disjoint cycles whose lengths divide the order of the group element. So σ must be one cycle of length p, since the order of σ is a prime number p. Let $\sigma = (a_0, \ldots, a_{p-1})$, $a_i \in X$. Let τ be the permutation $\tau(a_i) = i$. Then $\tau\sigma\tau^{-1} = (0, 1, \ldots, p-1)$ (see A.8.1(c)). Thus G is conjugate to $\mathcal{T}_p$.

Assume now that the claim is true for groups whose orders are less than the order of G and that G is not of order p. As a solvable group whose order is not a prime number, G has a nontrivial normal subgroup N (see Exercise 12.3). Since the order of N is smaller than the order of G and $N \subset S_p$ acts transitively on X by (e), we can use the inductive assumption, which says that N is conjugate to a subgroup of $\mathcal{G}_p$ containing $\mathcal{T}_p$. We may assume that $\mathcal{T}_p \subseteq N \subseteq \mathcal{G}_p$ (and take instead of G a conjugate group containing N, which we still denote by G). We want to show that $G \subseteq \mathcal{G}_p$. Let $\sigma \in G$. Since the order of $\varphi_{1,1}$ is p, the order of $\sigma\varphi_{1,1}\sigma^{-1}$ is also p and this element belongs to

$$\sigma\mathcal{T}_p\sigma^{-1} \subseteq \sigma N\sigma^{-1} \subseteq N \subseteq \mathcal{G}_p.$$

As we know (see Exercise 12.4(e)), the group $\mathcal{G}_p$ contains exactly one subgroup consisting of all elements of order p—the subgroup $\mathcal{T}_p$ generated by $\varphi_{1,1}$. Hence,

$$\sigma\varphi_{1,1}\sigma^{-1} = \varphi_{1,1}^a = \varphi_{1,a}$$

for some a, $0 < a < p$. If $r \in \mathbb{F}_p$, then $\sigma\varphi_{1,1}(r) = \varphi_{1,1}^a\sigma(r)$, that is, $\sigma(r+1) = \sigma(r) + a$. Hence $\sigma(r+2) = \sigma(r) + 2a$, and repeating the argument, $\sigma(r+x) = \sigma(r) + ax$. Putting $r = 0$ and $\sigma(0) = b$, we get $\sigma(x) = ax + b$. Thus $\sigma \in \mathcal{G}_p$ and our proof of the inclusion $G \subseteq \mathcal{G}_p$ is completed.

Now we prove that (b) and (c) are equivalent. Of course, (c) implies (b) by Exercise 12.4(d). Conversely, assume that every element of G has at most one fixed point on X. Let f_σ be the number of fixed points of $\sigma \in G$ acting on X. According to Bernside's Lemma (see A.9.2(c)),

$$\frac{1}{|G|}\sum f_\sigma$$

is the number of orbits of G in its action on X. Since the action is transitive, this sum is 1. Let r_i be the number of elements of G with $f_\sigma = i$. The number f_σ may assume three values: 0, 1 or p (the last one for the identity). Hence, the expression above gives the equality $|G| = p + r_1$. On the other hand, we have $|G| = 1 + r_0 + r_1$ (there is only one element in G with $f_\sigma = p$, namely, the identity of G). Thus, we get that $r_0 = p - 1$ is the number of elements of G without fixed points. As permutations of X, these elements are cycles of length p. The subgroup N generated by one of them must contain all the others (it contains p elements, which with the exception of the identity have order p). Moreover, the subgroup N is normal in G, since an element of N conjugated by an element of G gives again an element of the same order, so it must be an element of N. Exactly as in the proof $(a) \Rightarrow (c)$, we show that N is conjugate to $\mathcal{T}_p$. Denoting $\mathcal{T}_p$ by N (the only subgroup of G and $\mathcal{G}_p$ containing all elements of order p), the same arguments as in the proof $(a) \Rightarrow (c)$ show that $G \subseteq \mathcal{G}_p$.

(d) Let $x \in X$ and let $G_x = \{g \in G | gx = x\}$ be its stabilizer (for the definition, see p. 267). By A.9.2(a), we know that the number of elements in the orbit Gx is equal to the number of cosets of G_x in G, that is, it is equal to $|G|/|G_x|$. Since G is transitive, we have $|Gx| = |X|$, so $|X||G_x| = |G|$. Hence the order of G is divisible by the number of elements in X.

(e) As above (i.e. by A.9.2(a)), the number of elements in an orbit Hx of a group H acting on X is equal to $|H|/|H_x|$, where $H_x = \{h \in |hx = x\}$ is the fixed group of x in H. Assume that there is an element x' which is not in this orbit. Since the group G acts transitively on X, we have $g \in G$ such that $gx = x'$. The number of elements in the orbit Hx' is equal to $|H|/|H_{x'}|$. But $H_{x'} = gH_xg^{-1}$ (see A.9.2(a)). Thus H_x and $H_{x'}$ have the same number of elements, which shows that the orbits of x and x' have the same number of elements, say, m. Let k be the number of orbits of H on X. Of course, we have $km = |X|$. Since $|X| = p$ is a prime, we get $m = 1$ or $m = p$. In the first case, each orbit has only one element, so every element of H acts as the identity on X. In the second case, there is only one orbit, which consists of p elements, so H acts transitively on X.

12.6 Let G act on itself by conjugation (see section "Group Actions on Sets" in Appendix). This means that we take $X = G$ and we define an action of $g \in G$ on $x \in X$ as gxg^{-1}. Let as usual G_x denote the stabilizer of x, that is, $G_x = \{g \in G|gxg^{-1} = x\}$ (G_x is the subgroup of elements which commute with x). Notice that $G_x = G$ if and only if the element x belongs to the center $C(G)$ of G (see p. 245). Let c be the order of the center, that is, the number of elements for which the orbit of the action of G on X has only one element. If there are other orbits Gx having more than one element, then every such orbit has $|G|/|G_x| = p^{n_x}$ elements, where $n_x > 0$ (see A.9.2(a)), since the orders of G and G_x are different powers of p. By A.9.2(a), we have

$$|G| = \sum_x \frac{|G|}{|G_x|} = c + \sum_{x, n_x > 0} \frac{|G|}{|G_x|},$$

where x represent different orbits of the action of G on X. Since both $|G|$ and the last sum to the right in the formula above are divisible by p, we get that c must also be divisible by p. Thus the center is not trivial (has more than one element).

A proof that G is solvable can be given by induction with respect to the order of the group. If G has order p, then it is cyclic, so as an abelian group it is solvable (see Exercise 12.1(a)). Let $|G| = p^n$ and assume that p-groups of order less than p^n are solvable. Consider the center $C(G)$ (see p. 245). If $C(G) = G$, then G is abelian and therefore a solvable group. If $C(G) \neq G$, both $C(G)$ and $G/C(G)$ are solvable groups (the first one is even abelian, but the second definitely has order less than the order of G). Thus G is solvable by **T. 12.1**(c). Notice that the center $C(G)$ is normal in G (see p. 245).

Problems of Chap. 13

13.1 Example. We show that $L = \mathbb{Q}\left(\sqrt{\sqrt[3]{2} + \sqrt[4]{2}}, \sqrt[5]{2}\right)$ is a radical extension of $\mathbb{Q}$. We construct the following chain of extensions in which each bigger field is an extension of the preceding one by a root from an element belonging to this smaller field:

$$K_0 = \mathbb{Q} \subset K_1 = K_0(\sqrt[3]{2}) \subset K_2 = K_1(\sqrt[4]{2}) \subset K_3 = K_2\left(\sqrt{\sqrt[3]{2}+\sqrt[4]{2}}\right) \subset K_4 = K_3(\sqrt[5]{2}) = L.$$

13.2 Example. We prove that the equation $X^{12}-3X^4-3=0$ is solvable by radicals. Denote X^4 by T. Then the equation reads $T^3-3T-3=0$. This equation is solvable by radicals, which is shown in Chap. 1. If α is any solution of this equation, then in order to find a solution of the starting equation, we have to solve $X^4=\alpha$. Each such equation is of course solvable by radicals, so we can express all solutions of the equation $X^{12}-3X^4-3=0$ by radicals.

13.3 Let $K \subseteq L$ be a radical extension as in (13.1) and consider a longest possible chain of fields

$$K = K_0 \subset K_1 \subset \ldots \subset K_n = L$$

such that $K_i = K_{i-1}(\alpha_i)$, where $\alpha_i^{r_i} \in K_{i-1}$ and r_i are positive integers for $i = 1, \ldots, n-1$ (observe that we assume that the next field is strictly bigger than the preceding one). We claim that all r_i must be prime numbers. Assume that some r_i is not a prime. Then $r_i = pr_i'$ where p is a prime number and $r_i' > 1$ is an integer. Instead of $K_{i-1} \subset K_i = K_{i-1}(\alpha_i)$, we can take $K_{i-1} \subseteq K_i' = K_{i-1}(\beta_i) \subseteq K_i = K_{i-1}(\alpha_i)$, where $\beta_i = \alpha_i^{r_i'}$. Thus, we have $\beta_i^p = \alpha_i^{r_i} \in K_{i-1}$. This chain may extend the chain above and then we get a contradiction. But it may happen that we have an equality $K_{i-1} = K_i'$ or $K_i' = K_i$ (but not both!) and then either $K_{i-1} = K_i'$, so we can replace r_i by a smaller number r_i', or $K_i' = K_i$, which makes it possible to replace r_i by a prime number p. In the first case, we have to repeat the argument with r_i' instead of r_i if r_i' is not a prime number. Continuing, we get that all r_i can be chosen as prime numbers.

13.4 (a) Denote by L a splitting field of $f(X)$ over K. The complex conjugation shifts two nonreal zeros and fixes all the remaining real $p-2$ zeros. We order the zeros so that the nonreal zeros have indices $1, 2$. Thus, the complex conjugation is the transposition $\tau = (1, 2)$. The action of the Galois group $G(L/K)$ on the zeros is transitive, since the polynomial is irreducible (see Exercise 9.4(b)). The set of zeros has p elements, so the order of the Galois group is divisible by p by Exercise 12.5(d). By Cauchy's Theorem (see A.9.3), the group $G(L/K)$ contains an element of order p. Since such an element is a permutation of the set of p zeros of $f(X)$, it must be a cycle of length p. Some power of this cycle maps 1 to 2, so we can choose this power as σ and, if necessary, we can change the numbering of the remaining zeros so that $\sigma = (1, 2, \ldots, p)$. Now according to A.8.4, the Galois group as a group generated by σ and τ is equal to the whole group S_p.
(b) The derivative $5X^4 - p^2$ has 2 real zeros $x = \pm\frac{\sqrt{p}}{\sqrt[4]{5}}$. The values of the function $f(X) = X^5 - p^2X - p$ at these points are $f\left(\frac{\sqrt{p}}{\sqrt[4]{5}}\right) = \frac{p\sqrt{p}}{5\sqrt[4]{5}} - p^2\frac{\sqrt{p}}{\sqrt[4]{5}} - p = \frac{p\sqrt{p}}{\sqrt[4]{5}}\left(\frac{1}{5} - p - \frac{\sqrt[4]{5}}{\sqrt{p}}\right) < 0, f\left(-\frac{\sqrt{p}}{\sqrt[4]{5}}\right) = -\frac{p\sqrt{p}}{5\sqrt[4]{5}} + p^2\frac{\sqrt{p}}{\sqrt[4]{5}} - p = \frac{p\sqrt{p}}{\sqrt[4]{5}}\left(p - \frac{1}{5} - \frac{\sqrt[4]{5}}{\sqrt{p}}\right) > 0$

for every prime p. Thus the polynomial $f(X)$ has exactly 3 real zeros (and 2 nonreal). It is irreducible by Eisenstein's criterion. According to (a) its Galois group is S_5.

13.6 Let $f(X) = 0$, where $f(X) \in \mathbb{Q}[X]$, be irreducible and assume that it has 3 real and 2 nonreal (conjugate) complex roots. It follows from Exercise 13.4(a) that the degree of the splitting field of such a polynomial over any number field over which the polynomial is irreducible equals 120.

Let, as in the text of the exercise,

$$\mathbb{Q} = K_0 \subseteq K_1 \subseteq \ldots \subseteq K_n = L$$

be a chain of fields such that $K_i = K_{i-1}(\alpha_i)$, where $\alpha_i^{r_i} \in K_{i-1}$, the exponents r_i are prime numbers (see Exercise 13.3) for $i = 1, \ldots, n-1$ and the splitting field of $f(X)$ over $\mathbb{Q}$ is contained in L. Let $i < n$ be the least index such that $f(X)$ is irreducible in K_{i-1} but reducible in K_i (i must exist, since $f(X)$ is irreducible in $\mathbb{Q}$ but reducible in L). We show that $f(X)$ must still be irreducible in K_i, which gives a contradiction showing that $f(X) = 0$ cannot be solvable by radicals.

Denote by β_j, $j = 1, 2, 3, 4, 5$, the zeros of $f(X)$ in L and let $M = K_{i-1}(\beta_1, \ldots, \beta_5)$ be a splitting field of $f(X)$ over K_{i-1}.
(a) We prove that $[K_i : K_{i-1}] = 5$, where $K_i = K_{i-1}(\alpha_i)$, $\alpha_i^5 = a_i \in K_{i-1}$.
According to the assumption, the degree $[K_i : K_{i-1}]$ is a prime number, but since the polynomial $f(X)$ of degree 5 is irreducible in K_{i-1} and reducible in K_i, Nagell's lemma says that $[K_i : K_{i-1}] = 5$.
(b) Assume that the polynomial $f(X)$ is reducible over K_i. We show that this cannot happen considering two possibilities: $f(X) = g(X)h(X)$ over K_i, where $\deg(g) = 2$, $\deg(h) = 3$ or $\deg(g) = 1$, $\deg(h) = 4$.
We start with the first case ($\deg(g) = 2$, $\deg(h) = 3$). As we know, the degree $[M : K_{i-1}] = 120$ (see Exercise 13.4(a)). But we have the tower $K_{i-1} \subset K_i \subset K_i(\beta_1, \ldots, \beta_5)$ in which $[K_i : K_{i-1}] = 5$ and $[K_i(\beta_1, \ldots, \beta_5) : K_i] \leq 2 \cdot 6 = 12$, since we extend by zeros of a quadratic polynomial $g(X)$ (at most a quadratic extension) and by zeros of a cubic polynomial $h(X)$ (an extension of degree at most equal 6). Thus $[K_i(\beta_1, \ldots, \beta_5) : K_{i-1}] \leq 5 \cdot 2 \cdot 6 = 60$. But M is a subfield of $K_i(\beta_1, \ldots, \beta_5)$ containing K_{i-1} and has degree 120 over K_{i-1}, which gives a contradiction. Hence $f(X)$ cannot be factorized as a product of $g(X)$ and $h(X)$.
Now, we consider the second case ($\deg(g) = 1$, $\deg(h) = 4$). Let $N = K_{i-1}(\alpha_i, \varepsilon)$, where $\varepsilon^5 = 1, \varepsilon \neq 1$, be a splitting field of $X^{r_i} - a_i$ over K_{i-1} (recall that the zeros of $X^{r_i} - a_i$ are $\varepsilon^j \alpha_i$ for $j = 0, 1, 2, 3, 4$—see Exercise 5.11(c)). We have $K_{i-1} \subset K_i \subset N = K_{i-1}(\alpha_i, \varepsilon)$ and the degree $[N : K_{i-1}] \leq 5 \cdot 4 = 20$, since ε is a zero of $X^4 + X^3 + X^2 + X + 1$. But $f(X)$ is irreducible over K_{i-1} and having a factor of degree 1 over K_i, it has a zero in N. Since N is normal over K_{i-1}, all the zeros of $f(X)$ are in N. Hence N contains M, whose degree over K_{i-1} is 120. This is a contradiction, which means that factorization of $f(X)$ over K_i of the given type is impossible.

13.7 There are at least two possibilities. Using **T. 13.1** and Exercise 12.1(a),(b),(c) together with **T. 12.1**(a), we get that the Galois groups of all equations of degree ≤ 4 are solvable (see also the solution of Exercise 13.12).

A more direct way is to use the fact that the equations of degree ≤ 4 really are solvable by radicals, which was shown in Chap. 1. Of course, in each case it is possible to give an explicit chain of fields like (13.1) such that K is the field containing the coefficients of an equation, and L is a field containing one or more of its solutions. For example, for a cubic equation $X^3+pX+q=0$, Cardano's formula (1.5) gives the following chain:

$$K = K_0 = \mathbb{Q}(p,q) \subseteq K_1 = K_0\left(\sqrt{\frac{q^2}{4}+\frac{p^3}{27}}\right) \subseteq L = K_2 = K_1\left(\sqrt[3]{-\frac{q}{2}+\sqrt{\frac{q^2}{4}+\frac{p^3}{27}}}\right).$$

Notice that $\beta = \sqrt[3]{-\frac{q}{2}-\sqrt{\frac{q^2}{4}+\frac{p^3}{27}}} \in L$ as $\alpha\beta = \pm p/3$, where $\alpha = \sqrt[3]{-\frac{q}{2}+\sqrt{\frac{q^2}{4}+\frac{p^3}{27}}}$.

13.9 (a) Let L be a splitting field of an irreducible polynomial $f(X)$ over a field K. Assume that L is generated over K by any two zeros of $f(X)$. If a nontrivial automorphism σ in $G(L/K)$ has more than one fixed point as a permutation of the zeros $\alpha_1,\dots,\alpha_p$ of $f(X)$ and, say, α_1,α_2 are such zeros, then $\sigma(\alpha_i)=\alpha_i$ for $i=1,2$. But $L = K(\alpha_1,\alpha_2)$, so σ is the identity on L, which contradicts its choice. Thus, every nontrivial automorphism in $G(L/K)$ has at most one fixed point
Conversely, assume that every $\sigma \in G(L/K)$, $\sigma \neq 1$, has at most one fixed point among the zeros of $f(X)$. If there are two zeros α_i,α_j of $f(X)$ which do not generate L, then there is a zero α_k of $f(X)$ such that $\alpha_k \in L \setminus K(\alpha_i,\alpha_j)$. Take a nontrivial automorphism $\sigma \in G(L/K(\alpha_i,\alpha_j))$ (see **T. 9.1** and **T. 9.3**). Then σ has two fixed points in the set of zeros of $f(X)$, which is a contradiction.
(b) If $f(X) \in K[X]$ has prime degree p and is solvable by radicals, then its Galois group $G(K_f/K)$ is solvable (see **T. 13.1**). By Exercise 12.5 every nontrivial element of this group has at most one fixed element (as a permutation of the zeros of $f(X)$). By (a), we get that K_f is generated over K by any two zeros of $f(X)$.
Conversely, assume that K_f is generated over K by any two zeros of $f(X)$. Then the Galois group $G(K_f/K)$ is transitive as a permutation group on the set of zeros of $f(X)$ (see Exercise 7.4) and every nontrivial automorphism has at most one zero of $f(X)$ as a fixed element (by (a)). According to Exercise 12.5, the group $G(K_f/K)$ is solvable. Hence the equation $f(X)=0$ is solvable by radicals by **T. 13.1**.

13.11 (a) Let α_i, $i = 1,2,3,4,5$, denote the zeros of $f(X)$, so $\Delta(f) = \prod_{1\leq i<j\leq 5}(\alpha_i-\alpha_j)^2$. The polynomial has 1, 3 or 5 real zeros. If all zeros are real, then of course, $\Delta(f)>0$. If only one zero is real, say, $\alpha_1 = a$ and the remaining $\alpha_{2,3} = b \pm ci$, $\alpha_{4,5} = d \pm ei$ are nonreal, where $a,b,c,d,e \in \mathbb{R}$, $ce \neq 0$, then $\Delta(f)$ is a product of ten differences $\alpha_i-\alpha_j$ and it is easy to see that only $2ci$ and $2ei$ contribute with nonreal factors, while the remaining eight factors form four pairs of conjugate complex numbers. Thus the product of squares is positive, that is, we also have $\Delta(f)>0$ in this case. If exactly 3 zeros are real, say, $\alpha_1 = a$, $\alpha_2 = b$, $\alpha_3 = c$, $\alpha_{4,5} = d+ei$, where $a,b,c,d,e \in \mathbb{R}$, $e \neq 0$, then in a similar way, we see that in the product, there is only one nonreal factor $\pm 2ei$, whose square gives a negative real number. Thus $\Delta(f) < 0$ in this case. Hence $\Delta(f) < 0$ if and only if the polynomial has exactly 3 real zeros and by Exercise 13.10, this implies that the equation $f(X) = 0$ is not solvable by radicals.

13.12 (a) The Galois group $G(L/K)$, where $K = \mathbb{Q}(\varepsilon, s_1, s_2, s_3)$ and $L = K(X_1, X_2, X_3) = \mathbb{Q}(\varepsilon, X_1, X_2, X_3)$, is S_3 and consists of all permutations of the variables X_1, X_2, X_3 (or the indices $1, 2, 3$)—see the proof of **T. 13.2**. The field L is a splitting field of the general polynomial $f(X) = X^3 - s_1X^2 + s_2X - s_3$ over the field K.

The group S_3 is solvable according to Exercise 12.1(b). In fact, we have the sequence $S_3 \supset A_3 \supset (1)$, where A_3 is the group of even permutations of $1, 2, 3$ generated by the cycle $(1, 2, 3)$. In this sequence, the group A_3 is normal in S_3 and S_3/A_3 is cyclic of order 2, and A_3 is cyclic of order 3. The fixed field corresponding to S_3 is, of course, K, and the fixed field corresponding to A_3 is, as usual (see Exercise 9.24) the field generated by $\sqrt{\Delta} = (X_1 - X_2)(X_2 - X_3)(X_3 - X_1)$, where $\Delta = (X_1 - X_2)^2(X_2 - X_3)^2(X_3 - X_1)^2$ is the discriminant of $f(X)$. Let $K_1 = K_0(\sqrt{\Delta})$. We have $[K(\sqrt{\Delta}) : K] = 2$. By the Tower Law, we have $[L : K(\sqrt{\Delta})] = 3$ and L is a cyclic extension of K_1. By **T. 11.3**, we can find $\alpha \in L$ such that $L = K_1(\alpha)$ and $\alpha^3 \in K_1$. In fact, according to the first proof of this theorem (see p. 131), we can take $\alpha = X_1 + \varepsilon X_2 + \varepsilon^2 X_3$. Thus, we have the following chain

$$K_0 = K \subset K_1 = K_0(\sqrt{\Delta}) \subset K_2 = K_1(\alpha) = L.$$

(b) In order to solve the equation $f(X) = 0$, that is, to express the zeros X_1, X_2, X_3 in terms of the coefficients s_1, s_2, s_3 using the arithmetic operations and roots, we observe now that the nontrivial automorphism of the quadratic field extension $K(\sqrt{\Delta}) \supset K$ is induced by any element of order 2 in S_3. Take $\beta = X_1 + \varepsilon X_3 + \varepsilon^2 X_2$, which is obtained from α by an element of order 2 in S_3 (here $(2, 3)$). We also have $\beta^3 \in K_1$. Moreover $\alpha^3 + \beta^3, \alpha^3\beta^3 \in K$, since these elements are fixed by all automorphisms (see **T. 9.1**(a)). Now we compute $\alpha^3 + \beta^3$ and $\alpha\beta$ (in fact, the easiest way to carry out quick computations is to use e.g. Maple) obtaining:

$$\alpha^3 + \beta^3 = 2s_1^3 - 9s_1s_2 + 27s_3, \qquad \alpha^3\beta^3 = (s_1^2 - 3s_2)^3,$$

so α^3, β^3 are zeros of the quadratic equation $T^2 - (2s_1^3 - 9s_1s_2 + 27s_3)T + (s_1^2 - 3s_2)^3 = 0$. Notice that $\alpha\beta = s_1^2 - 3s_2$. Now we can solve this equation, which gives α^3, β^3 and taking the third roots so that $\alpha\beta = s_1^2 - 3s_2$, we find three possibilities for α, β. Finally, notice that the equation system:

$$\begin{aligned} X_1 + \varepsilon X_2 + \varepsilon^2 X_3 &= \alpha \\ X_1 + \varepsilon^2 X_2 + \varepsilon X_3 &= \beta \\ X_1 + X_2 + X_3 &= s_1 \end{aligned}$$

gives $X_1 = \frac{1}{3}(s_1 + \alpha + \beta), X_2 = \frac{1}{3}(s_1 + \varepsilon^2\alpha + \varepsilon\beta), X_3 = \frac{1}{3}(s_1 + \varepsilon\alpha + \varepsilon^2\beta)$.

Remark. The above solution of the cubic equations seems to result in more complicated formulae than in Cardano's method. But in reality, the formulae are identical when we specialize and choose an equation with $s_1 = 0$. This is always possible when we work with concrete equations and number coefficients. Notice that in the case of the quadratic equation which we used above, $T^2 - (2s_1^3 - 9s_1s_2 + 27s_3)T + (s_1^2 - 3s_2)^3 = 0$, its discriminant is equal $-27\Delta(f)$, which can be checked by direct computations (preferably using e.g. Maple).

13.13 Let L be a radical extension of K such that

$$K_0 = K \subset K_1 \subset \ldots \subset K_n = L \subset \mathbb{R},$$

where $K_i = K_{i-1}(\alpha_i)$, $\alpha_i^{r_i} = a_i \in K_i$, r_i are prime numbers for $i = 1, \ldots, n$ and $K_f = K(x_1, \ldots, x_n) \subset L$, where x_i are the zeros of $f(X)$ in L. If all zeros of $f(X)$ are in K, then $K_f = K$ and the claim is true. Assume that one of the zeros of $f(X)$, say x_1, is not in K. Choose $i \geq 1$ such that $x_1 \notin K_{i-1}$ but $x_1 \in K_i$. Since $K_{i-1} \subset K_{i-1}(x_1) \subseteq K_i$ and $[K_{i-1}(x_1) : K_{i-1}] > 1$ divides the prime number $[K_i : K_{i-1}] = r_i$, we have $K_i = K_{i-1}(\alpha_i) = K_{i-1}(x_1)$ (K_i and $K_{i-1}(x_1)$ have the same prime degree r_i over K_{i-1}). Thus $X^{r_i} - a_i$ is irreducible over K_{i-1} (see Exercise 5.12) and has a zero α_i in $K_i = K_{i-1}(x_1)$. As a consequence, the polynomial $X^{r_i} - a_i$ splits in the splitting field $K_{i-1}(x_1, \ldots, x_n)$ of $f(X)$ over K_{i-1}. But if $r_i > 2$, we get a contradiction, since the field $K_{i-1}(x_1, \ldots, x_n) \subseteq L$ is real, while all the zeros of $X^{r_i} - a_i$ with the exception of α_i are nonreal. Hence, the only possibility is that all $r_i = 2$, so that the degree $[L : K]$ is a power of 2. This shows that the degree $[K_f : K]$ as a divisor of $[L : K]$ is also a power of 2.

13.14 (a) Let K be a finite extension of $\mathbb{R}$ of odd degree. Then $K = \mathbb{R}(\alpha)$ by **T. 8.1**. Let $f(X)$ be the minimal polynomial of α over $\mathbb{R}$. Since $f(X)$ is irreducible over $\mathbb{R}$ and has a real zero as a polynomial over $\mathbb{R}$ of odd degree, this degree must be equal to 1. Hence $f(X) = X - \alpha$ and $[K : \mathbb{R}] = 1$, that is, $K = \mathbb{R}$.
(b) A quadratic extension of $\mathbb{C}$ is generated by a zero of an irreducible quadratic polynomial over $\mathbb{C}$. But any such polynomial has a complex zero by the known formulae for its zeros. Hence any quadratic polynomial over $\mathbb{C}$ is reducible and quadratic extensions of $\mathbb{C}$ do not exist.
If K is a quadratic extension of the real numbers $\mathbb{R}$, then $K = \mathbb{R}(\alpha)$, where α is a zero of a quadratic polynomial. By the known formula, the zeros of this polynomial are complex numbers, so $K = \mathbb{C}$.
(c) Let $f(X) \in \mathbb{C}[X]$ be an irreducible polynomial. We have to prove that its degree is 1. A splitting field K of $f(X)$ over $\mathbb{C}$ is a finite Galois extension of $\mathbb{C}$. We assume that it is also a Galois extension of $\mathbb{R}$. In fact, if $K = \mathbb{C}(\alpha)$, then we can take the minimal polynomial of α over $\mathbb{R}$ and multiply it by $X^2 + 1$. Such a polynomial has a splitting field over $\mathbb{R}$ which contains K. Thus, we assume that we have a Galois extension K of $\mathbb{R}$ containing $\mathbb{C}$. This field K has even degree over $\mathbb{R}$ equal to at least 2. We prove that it equals 2, that is, $K = \mathbb{C}$ by (b) (the only quadratic extension of $\mathbb{R}$).

Assume that K has (even) degree > 2 over $\mathbb{R}$. Let $G = G(K/\mathbb{R})$ be the Galois group of K over $\mathbb{R}$. Let $|G| = 2^k m$, where m is odd and let H be the Sylow 2-subgroup of G (see A.9.4). Then $[K^H : \mathbb{R}]$ has odd degree m, so $K^H = \mathbb{R}$ by (a). Hence $m = 1$ and G is a 2-group.
By Exercises 12.6 and 12.3, there is a chain of subgroups of G:

$$G = G_0 \supset G_1 \supset \ldots \supset G_k = \{e\}$$

such that G_{i+1} is normal in G_i and the quotient group G_i/G_{i+1} is cyclic of order 2 for $i = 0, 1, \ldots, k-1$. By **T. 9.2**, there is a corresponding chain of subfields of K such that

$$\mathbb{R} = K^{G_0} \subset \mathbb{C} = K^{G_1} \subset K^{G_2} \subset \cdots \subset K^{G_k} = K$$

in which K^{G_1} is a quadratic extension of $\mathbb{R}$, so $K^{G_1} = \mathbb{C}$ by (b), and K^{G_2} is a quadratic extension of $\mathbb{C}$ if $k > 1$. However, this is impossible by (b), so $k = 1$. Consequently, $K = \mathbb{C}$, which means that $\mathbb{C}$ does not have any nontrivial algebraic extensions.

Problems of Chap. 14

14.2 It is well-known how to construct a perpendicular projection of a point on a given line or a line perpendicular to a given one in a given point.

If a point (a, b) is constructible from X, then also $(a, 0)$ and $(0, b)$ are constructible as projections of (a, b) onto the lines (the axes) through $(0, 0), (1, 0)$ and $(0, 0), (0, 1)$, respectively. Hence the numbers a, b are constructible since $|a|$ is the distance between $(a, 0)$ and $(0, 0)$, and $|b|$ between $(0, b)$ and $(0, 0)$.

Conversely, if a, b are constructible, then we can find the points $(a, 0), (0, b)$ and constructing the lines perpendicular to the axes at these points, we find (a, b) as their intersection.

14.3 (a) In general, if a, b are given constructible numbers, then the numbers $a \pm b, ab, \sqrt{|a|}$ and a/b for $b \neq 0$ can be constructed (see the proof of **T. 14.1** on p. 141). Thus, the number $\sqrt{a^2 + b^2}$ is also constructible—given $|a|$ and $|b|$, we construct a^2, b^2, then $a^2 + b^2$ and finally $\sqrt{a^2 + b^2}$. Another possibility is to construct a right triangle with legs (catheti) $|a|, |b|$—its hypothenuse is just $\sqrt{a^2 + b^2}$.

14.6 (b) There are integers a and b such that $1 = ak + bl$, that is, $\frac{1}{kl} = a\frac{1}{l} + b\frac{1}{k}$. Thus the angle $\frac{2\pi}{kl}$ is constructible if the angles $\frac{2\pi}{k}$ and $\frac{2\pi}{l}$ are constructible. The remaining part follows from (a).
(d) If the angle $\frac{2\pi}{p^2}$ is constructible, then the point $(\cos\frac{2\pi}{p^2}, \sin\frac{2\pi}{p^2})$ is constructible. This implies that $[L : \mathbb{Q}]$ is a power of 2, where $L = \mathbb{Q}(\cos\frac{2\pi}{p^2}, \sin\frac{2\pi}{p^2}, i) \supseteq \mathbb{Q}(\varepsilon), \varepsilon = \cos\frac{2\pi}{p^2} + i\sin\frac{2\pi}{p^2}$ (apply **T. 14.1** and notice that the degree of

$\mathbb{Q}(\cos\frac{2\pi}{p^2}, \sin\frac{2\pi}{p^2})$ over $\mathbb{Q}$ is a power of 2 if and only if the degree of L is such a power). But $[\mathbb{Q}(\varepsilon) : \mathbb{Q}] = \varphi(p^2) = p(p-1)$ is not a power of 2 if $p > 2$.
(e) The angle $\frac{2\pi}{p}$ is constructible if and only if the point $P = (\cos\frac{2\pi}{p}, \sin\frac{2\pi}{p})$ is constructible. The point P is constructible if and only if $[L : \mathbb{Q}]$ is a power of 2, where $L = \mathbb{Q}(\cos\frac{2\pi}{p}, \sin\frac{2\pi}{p}, i) \supseteq \mathbb{Q}(\varepsilon)$, $\varepsilon = \cos\frac{2\pi}{p} + i\sin\frac{2\pi}{p}$ (apply **T. 14.1** and **T. 14.2**). But $[\mathbb{Q}(\varepsilon) : \mathbb{Q}] = p-1$, that is, P is constructible if and only if $p = 2^T + 1$. If now $T = 2^t u$, where u is odd, then $2^T + 1$ is not a prime when $u > 1$ (in fact, $a^u + b^u = (a+b)(a^{u-1} - a^{u-2}b + \cdots + b^{u-1})$ for $a = 2^t, b = 1$). Hence, $T = 2^t$.

Problems of Chap. 15

15.1 (a) Assume first that $f(X)$ is reducible over K. Then it has a zero $\alpha_1 \in K$. The quadratic polynomial $g(X) = f(X)/(X-\alpha_1) = (X-\alpha_2)(X-\alpha_3)$ has its coefficients in K, so $g(\alpha_1) = (\alpha_1-\alpha_2)(\alpha_1-\alpha_3) \in K$. Hence, we have $\Delta(f) = g(\alpha_1)^2(\alpha_2-\alpha_3)^2$ and $K(\sqrt{\Delta}) = K(\alpha_2-\alpha_3) = K(\alpha_1,\alpha_2,\alpha_3)$, since $\alpha_1, \alpha_2+\alpha_3 \in K$ and $\mathrm{char}(K) \neq 2$. Of course, we have $K(\sqrt{\Delta}, \alpha_1) = K_f$.
Now assume that $f(X)$ is irreducible over K. A splitting field K_f has at most degree 6 (as a splitting field of a polynomial of degree 3) and at least degree 3 over K (since $[K(\alpha_1) : K] = 3$). If $\sqrt{\Delta} \in K$, then the Galois group $G(K_f/K)$ only has even permutations (see Exercise 9.25), so we have $G(K_f/K) = A_3$ and $[K_f : K] = 3$. Of course, we have $K(\sqrt{\Delta}, \alpha_1) = K_f$.
If $\sqrt{\Delta} \notin K$, then the Galois group $G(K_f/K)$ has both odd and even permutations (see Exercise 9.25), so we must have $G(K_f/K) = S_3$ and $[K_f : K] = 6$. Since $[K(\sqrt{\Delta}) : K] = 2$ and $f(X)$ is still irreducible over $K(\sqrt{\Delta})$ (see Exercise 4.14(a)), we have $[K(\sqrt{\Delta}, \alpha_1) : K] = 6$, that is, $K(\sqrt{\Delta}, \alpha_1) = K_f$.

15.3 (a) Depending on the choice of the factorization $f(X) = X^4 + pX^2 + qX + r = (X^2 + aX + b)(X^2 + a'X + b')$, the zeros of the first factor can be taken as α_1, α_4 or α_2, α_4 or α_3, α_4. Hence $a = -(\alpha_i + \alpha_4)$ for $i = 1, 2, 3$ and the corresponding zero a^2 of $r(f)$ is one of the $\beta_i = (\alpha_i + \alpha_4)^2$ for $i = 1, 2, 3$.
(b) Using Exercise 9.25, the notations in (a) and the equality $\alpha_1+\alpha_2+\alpha_3+\alpha_4 = 0$, we have

$$\Delta(r(f)) = (\beta_1 - \beta_2)^2(\beta_2 - \beta_3)^2(\beta_3 - \beta_1)^2$$

$$= (\alpha_1 - \alpha_2)^2(\alpha_1 - \alpha_4)^2(\alpha_2 - \alpha_3)^2(\alpha_2 - \alpha_4)^2(\alpha_3 - \alpha_1)^2(\alpha_3 - \alpha_4)^2 = \Delta(f).$$

Thus $\Delta(r(f)) \neq 0$ if and only if $\Delta(f) \neq 0$, that is, all α_i are different if and only if all β_i have this property.
(c) It is clear that

$$K_{r(f)}(\alpha_i) = K(\beta_1, \beta_2, \beta_3, \alpha_i) \subseteq K(\alpha_1, \alpha_2, \alpha_3, \alpha_4) = K_f.$$

In order to prove the converse inclusion, choose $i = 4$ and consider the equations:

$$f_i(x) = (X + \alpha_4)^2 - \beta_i$$

for $i = 1, 2, 3$. This quadratic polynomial has its coefficients in the field $K_{r(f)}(\alpha_4)$. Assume that $f_i(X)$ is irreducible in this field. By (a), it has a zero $X = \alpha_i$, which is a zero of the polynomial $f(X)$. Hence $f_i(X)$ divides $f(X)$ (as an irreducible polynomial having a common zero). Thus one more of $\alpha_1, \alpha_2, \alpha_3, \alpha_4$ different from α_i is a zero of $f_i(X)$, say, α_j, where $j \neq i$. Then

$$f_i(\alpha_i) = f_i(\alpha_j) = 0,$$

that is,

$$(\alpha_i + \alpha_4)^2 = (\alpha_j + \alpha_4)^2$$

for $j \neq i$, which says that $\beta_i = \beta_j$—a contradiction according to (b). Thus the polynomial is reducible in the field $K_{r(f)}(\alpha_4)$ and its zero α_i belongs to it. This holds for $i = 1, 2, 3$ (and, of course, also for $i = 4$) and shows that also $K_f \subseteq K_{r(f)}(\alpha_4)$. Together with the converse inclusion, we get $K_f = K_{r(f)}(\alpha_4)$.

15.5 We give at the same time a solution of both (a) and (b) as the essential difference is only concerned with the distinction between the cases when the Galois group is D_4 or C_4 (Case 2 below). We have three cases:

Case 1: The resolvent $r(f)$ has no zeros in K. This means that $r(f)$ is an irreducible cubic over K. According to Exercise 15.1(a), the splitting field $K_{r(f)}$ has degree 3 or 6 over K depending on the discriminant $\Delta = \Delta(r(f)) = \Delta(f)$: the first case when $\sqrt{\Delta} \in K$, and the second, when $\sqrt{\Delta} \notin K$.

If $\sqrt{\Delta} \in K$ (so the degree of $K_{r(f)}$ over K is 3), then the polynomial $f(X)$ of degree 4 is still irreducible over $K_{r(f)}$ (see Exercise 4.2). Hence the splitting field $K_f = K(\sqrt{\Delta}, \alpha_1)$, where α_1 is any zero of $f(X)$ (see Exercise 15.3(c)), has degree 4 over $K_{r(f)}$. Thus the degree of K_f over K equals 12, which means that the Galois group $G(K_f/K)$ is a subgroup of order 12 of S_4. There is only one such subgroup (see A.8.2) and it is the alternating group A_4 of all even permutations of $1, 2, 3, 4$.

If $\sqrt{\Delta} \notin K$, then the Galois group $G(K_f/K)$ must contain both even and odd permutations by Exercise 9.25 and its order is divisible by 6, since the splitting field K_f over K contains a subfield $K_{r(f)}$ of degree 6 over K. Moreover, the Galois group is transitive, since $f(X)$ is irreducible over K. There is only one such subgroup of S_4 and it is S_4 itself (see section "Transitive Subgroups of Permutation Groups" in Appendix concerning the claim that S_4 is the only transitive subgroup of S_4 of order divisible by 6).

Case 2: The resolvent $r(f)$ has exactly one zero in K, say $\beta = \beta_1 = (\alpha_1 + \alpha_4)^2 \in K$ (see Exercise 15.3(a)), that is, $[K_{r(f)} : K] = 2$.

First of all, let us notice that in this case, we have $\sqrt{\Delta} \notin K$. In fact, the splitting field of the resolvent $K_{r(f)} = K(\sqrt{\Delta}, \beta) = K(\sqrt{\Delta})$ (see Exercise 15.1) must be bigger than K, since otherwise all the zeros of the resolvent are in K. Thus $\sqrt{\Delta}$ cannot belong K, so the Galois group $G(K_f/K)$ must contain both even and odd permutations by Exercise 9.25.

If f is irreducible over $K_{r(f)}$, then K_f has degree 8 over K ($K_f = K_{r(f)}(\alpha_1)$ has degree 4 over $K_{r(f)}$). Thus $G(K_f/K) = D_4$, since the (transitive) subgroups of order 8 in S_4 are isomorphic to D_4 (see section "Transitive Subgroups of Permutation Groups" in Appendix). We check that in this case $\sqrt{\beta\Delta} \notin K$ if $\beta \neq 0$ and $\sqrt{\delta\Delta} \notin K$ if $\beta = 0$. In fact, if $\beta \neq 0$, then $K(\sqrt{\beta})$ is a quadratic field over K (if β is a square in K, then $f(X)$ is reducible over K according to Exercise 3.9, which is excluded). If $\sqrt{\beta\Delta} \in K$, then $K(\sqrt{\beta}) = K(\sqrt{\Delta}) = K_{r(f)}$ and the polynomial $f(X)$ is reducible over $K_{r(f)}$ according to Exercise 3.9, since β is a nonzero square in $K_{r(f)}$. This contradicts the assumption that $f(X)$ is irreducible over $K_{r(f)}$. Similarly, if $\beta = 0$, then $q = 0$. Since the polynomial $f(X) = X^4 + pX^2 + r$ is irreducible over K, we know that $\delta = p^2 - 4r$ is not a square in K, so $K(\sqrt{\delta})$ is a quadratic field over K. If $\sqrt{\delta\Delta} \in K$, then $K(\sqrt{\delta}) = K(\sqrt{\Delta}) = K_{r(f)}$ and the polynomial $f(X) = X^4+pX^2+r$ is reducible over $K_{r(f)}$ contrary to the assumption. Thus, we have $\sqrt{\delta\Delta} \notin K$.

If f is reducible over $K_{r(f)}$, then its irreducible factors over this field have degree 2, since no zero of this polynomial can belong to the field $K_{r(f)}$ (of degree 2 over K). Hence $K_f = K_{r(f)}(\alpha_1)$ has degree 2 over $K_{r(f)}$. Thus the Galois group $G(K_f/K)$ has order 4, is transitive and contains both even and odd permutations. The only such subgroups of S_4 are cyclic (see section "Transitive Subgroups of Permutation Groups" in Appendix). Since in this case the polynomial $f(X)$ is reducible over $K_{r(f)}$, the same considerations as above show that $\sqrt{\beta\Delta} \in K$ if $\beta \neq 0$ and $\sqrt{\delta\Delta} \in K$ if $\beta = 0$. Since the cases of $f(X)$ reducible and irreducible over K exclude each other, we get that the conditions on $\sqrt{\beta\Delta}$ and $\sqrt{\delta\Delta}$ in the cases D_4 and C_4 are necessary and sufficient.

Case 3: The resolvent has all three zeros in K. Since now $K_{r(f)} = K$, the splitting field $K_f = K(\alpha)$, where α is any zero of $f(X)$ by Exercise 15.3(c). Since $f(X)$ is irreducible, this field has degree 4 over K. Since all $\beta_i = (\alpha_i + \alpha_4)^2$ are in K, they are fixed by the four permutations in V_4 (see Exercise 15.4(a) for the notation), which is the Galois group $G(K_f/K)$ in this case.

15.6 Assume that

$$f(X) = X^4 + pX^2 + qX + r = (X^2 - aX + b)(X^2 + aX + b'),$$

where $a, b, b' \in K$, $\delta = a^2 - 4b$ and $\delta' = a^2 - 4b'$ are not squares in K (since otherwise, the polynomial $f(X)$ has zeros in K). The zeros of $f(X)$ are $a/2 \pm \sqrt{\delta}$ and $-a/2 \pm \sqrt{\delta'}$. The field $K_f = K(\sqrt{\delta}, \sqrt{\delta'})$ is a splitting field of $f(X)$ over K.

The zeros of the resolvent $r(f)$ are according to Exercise 15.3(a): $a^2, (\sqrt{\delta} \pm \sqrt{\delta'})^2 = \delta + \delta' \pm 2\sqrt{\delta\delta'}$. The resolvent $r(f)$ has only one zero in K if and only if $\sqrt{\delta\delta'} \notin K$, which is equivalent to $K(\sqrt{\delta}) \neq K(\sqrt{\delta'})$, that is, $K_f = K(\sqrt{\delta}, \sqrt{\delta'})$ has degree 4 over K. The Galois group $G(K_f/K)$, in this case, is of course isomorphic to V—the Klein four-group, since all the automorphisms are given by $\sqrt{\delta} \mapsto \pm\sqrt{\delta}$ and $\sqrt{\delta'} \mapsto \pm\sqrt{\delta'}$ (there are four automorphisms and the nontrivial has order 2).

The resolvent $r(f)$ has 3 zeros in K if and only if $\sqrt{\delta\delta'} \in K$, which is equivalent to $K(\sqrt{\delta}) = K(\sqrt{\delta'})$, that is, $K_f = K(\sqrt{\delta})$. Thus in this case the splitting field K_f is quadratic over K and its Galois group is C_2—the cyclic group of order 2.

15.9 (a) Let $k' = k(\theta)$, where θ is a zero of its minimal polynomial $g(X) \in k[X]$ (see **T. 8.2**) of some degree m. The elements $1, \theta, \ldots, \theta^{m-1}$ form a basis of k' over k. Of course, we have $K' = K(\theta)$ as $K' = k'(Y_1, \ldots, Y_n) = k(\theta, Y_1, \ldots, Y_n) = K(\theta)$. We want to show that the elements $1, \theta, \ldots, \theta^{m-1}$ are still linearly independent over K, that is, $[K' : K] = m$. In fact, if they are linearly dependent over K, then there is a linear combination $\sum_{i=0}^{m-1} \varphi_i(Y_1, \ldots, Y_n)\theta^i = 0$, where $\varphi_i(Y_1, \ldots, Y_n)$ are polynomials of $Y_1, \ldots, Y_n$ (we can always multiply all rational functions by their common denominator to get polynomials as coefficients in such a sum). Now we gather all different monomials of the type $Y_1^{i_1} \cdots Y_n^{i_n}$. These monomials are linearly independent over the field k'. The coefficients of these monomials in the linear combination above are of the type $\sum_{i=0}^{m-1} a_i\theta^i$, where $a_i \in k$. Linear independence of the monomials implies that all these coefficients must be equal to 0. Hence, all a_i are equal to 0, since $1, \theta, \ldots, \theta^{m-1}$ are linearly independent over k. Thus, all the polynomials $\varphi_i(Y_1, \ldots, Y_n)$ are equal to 0. This shows that $1, \theta, \ldots, \theta^{m-1}$ also form a basis of K' over K. Hence, the polynomial $g(X)$ is irreducible over K as its degree m is equal to the degree $[K' : K] = m$ (see **T. 4.2**).
The field $k' = k(\theta)$ is a splitting field of $g(X)$ over k and, of course, the field $K' = K(\theta)$ is a splitting field of the same polynomial over K. If $\theta_1 = \theta, \ldots, \theta_m$ are all the zeros of $g(X)$ in k', as well as in K', then by Exercise 6.1, each automorphism in $G(k'/k)$, as well as in $G(K'/K)$, is defined by the image $\sigma_i(\theta) = \theta_i$, so we have an isomorphism between these two groups mapping each element of the first to the corresponding element of the second defined by the same action on θ (and as identity on the elements of K).

15.10 Example 1. We show that the Galois group of $f(X) = X^5 + X^2 + 1$ is S_5. This polynomial is irreducible modulo 2, $X^5 + X^2 + 1 \equiv (X^3 + X^2 + 4X + 3)(X^2 + 4X + 2) \pmod 5$ and $X^5 + X^2 + 1 \equiv (X + 2)(X^4 + X^3 + X^2 + 2X + 2) \pmod 3$.

Thus, the Galois group contains a cycle of length 5, a cycle of length 4, and a permutation which is a product of cycles of length 2 and 3. A square of the last permutation is a cycle of length 3, and its cube is a cycle of length 2. Hence the Galois group $G(\mathbb{Q}_f/\mathbb{Q})$ contains elements of orders $2, 3, 4, 5$ (use **T. 15.4**), so its order must be divisible by $3 \cdot 4 \cdot 5 = 60$ (see A.2.6). Moreover, it contains odd permutations (for example, a cycle of length 2). Thus, the Galois group cannot be A_5 (the group of all even permutations) and consequently, it must contain 120 elements (the maximal possible order of the Galois group for a polynomial of degree 5—see Exercise 9.3). Here we use the fact that there is only one subgroup in S_5 of index 2, namely, the group A_5 (see A.8.2).
Example 2. We show that the Galois group of $f(X) = X^5 + 20X + 16$ is A_5. First we find that the polynomial is irreducible modulo 3, $X^5 + 20X + 16 \equiv (X^3 + 2X^2 + 5X + 5)(X-4)(X-5) \pmod 7$ and $X^5 + 20X + 16 \equiv (X-5)(X^2 + 12X + 14)(X^2 + 17X + 2) \pmod{23}$. Thus by **T. 15.4**, the Galois group contains elements of orders 5, 3 and 2.

Since the polynomial is irreducible over $\mathbb{Q}$, the Galois group is transitive on its zeros and its order must be at least equal to $2\cdot 3\cdot 5 = 30$. As we know (see p. 276) the order of such a transitive subgroup is at least 60. Now we look at the discriminant of this polynomial, finding that $\Delta(f) = 2^{16}5^6$, so it is a square. Hence by Exercise 9.25, the Galois group $G(\mathbb{Q}_f/\mathbb{Q})$ consists of even permutations of the zeros, so it is A_5.

15.11 (a) We assume that $n > 3$ (it is easy to handle $n = 1, 2, 3$). Choose a prime number $p > 3$ such that $p > n - 2$. Choose a monic irreducible polynomial $f_2(X)$ of degree n in $\mathbb{F}_2[X]$, an irreducible polynomial $g(X)$ of degree $n - 1$ in $\mathbb{F}_3[X]$ and an irreducible quadratic polynomial $h(X)$ in $\mathbb{F}_p[X]$. Define $f_3(X) = Xg(X)$ and $f_p(X) = \prod_{i=0}^{n-2}(X+i)g(X)$, which are polynomials of degree n. Using the Chinese Remainder Theorem (see A.7.1 and A.7.2), choose a polynomial $f(X) \in \mathbb{Z}[X]$ of degree n such that the reduction of $f(X)$ modulo 2 equals $f_2(X)$, the reduction modulo 3 equals $f_3(X)$ and the reduction modulo p equals $f_p(X)$. Then the Galois group of $f(X)$ over $\mathbb{Q}$ is S_n. In fact, the polynomial $f(X)$ is irreducible over $\mathbb{Q}$, since its reduction modulo 2 is irreducible over $\mathbb{F}_2$, so its Galois group is transitive (on its zeros, which we number by $1, 2, \ldots, n$ and identify the Galois group with a subgroup of S_n). According to **T. 15.4** applied to the reduction modulo 3, the Galois group $G(\mathbb{Q}_f/\mathbb{Q})$ contains a cycle of length $n - 1$ (which we may identify with the cycle $\sigma = (1, \ldots, n-1)$). Taking the reduction of $f(X)$ modulo p, the same theorem implies that $G(\mathbb{Q}_f/\mathbb{Q})$ contains a transposition. It follows from A.8.5 that the Galois group $G(\mathbb{Q}_f/\mathbb{Q})$ is the group S_n.

(b) **Example 1**. We construct a polynomial of degree 6 with Galois group S_6 over $\mathbb{Q}$. Let $f_2(X) = X^6+X+1, f_3(X) = X(X^5+2X+2)$ and $f_7(X) = (X^2-1)(X^2-4)(X^2+1) = X^6+3X^4+6X^2+4$. Then the polynomial $f(X) = X^6+24X^4+20X^2+35X+39$ has Galois group S_6 over $\mathbb{Q}$ according to our construction in (a) (see Example A.7.2 how to find the coefficients of the polynomial $f(X)$).

Example 2. We construct a polynomial of degree 8 with Galois group S_8 over $\mathbb{Q}$. Let $f_2(X) = X^8 + X^4 + X^3 + X + 1$, $f_3(X) = X(X^7 + X^2 + 2)$ and $f_{11}(X) = (X^2 - 1)(X^2 - 4)(X^2 - 9)(X^2 + 1) = X^8 + 9X^6 + 2X^4 + 2X^2 + 8$. Then the polynomial $f(X) = X^8+42X^6+57X^4+55X^3+24X^2+11X+63$ has Galois group S_8 over $\mathbb{Q}$ according to our construction in (a).

Appendix: Groups, Rings and Fields

Equivalence Relations

A **relation** on a set X is any subset $\mathcal{R}$ of the product $X \times X = \{(x, y)|x, y \in X\}$. For example, the relation "less than or equal" on the set of real numbers is the subset $\mathcal{R} \subset \mathbb{R} \times \mathbb{R}$ consisting of all pairs (x, y) such that $x \leq y$. If we don't have any special reason to use a specific notation for a relation, we shall write $x \sim y$ when $x, y \in X$ are in a relation $\mathcal{R}$ on X, that is, $(x, y) \in \mathcal{R}$.

We say that a relation $\mathcal{R}$ on X is an **equivalence relation** if it is **reflexive, symmetric** and **transitive**, that is, if for $x, y, z \in X$, we have

(a) $x \sim x$ (**reflexivity**);
(b) if $x \sim y$, then $y \sim x$ (**symmetry**);
(c) if $x \sim y$ and $y \sim z$, then $x \sim z$ (**transitivity**).

An **equivalence class** $[x]$ of $x \in X$ with respect to an equivalence relation $\mathcal{R}$ is the set of all $x' \in X$, which are related to x with respect to this relation, that is, $[x] = \{x' \in X|x' \sim x\}$ (in terms of pairs, $[x] = \{x' \in X|(x', x) \in \mathcal{R}\}$). Every element belonging to an equivalence class is called its **representative**.

The equivalence relations play a very important role, since they split all the elements of X into equivalence classes so that every class contains elements with some special property ("relatives" to a fixed $x \in X$):

Proposition A.1.1. *The equivalence classes of an equivalence relation on X form a partition of X, that is, every element of X belongs to an equivalence class and different equivalence classes are disjoint.*

Proof. Every element $x \in X$ belongs to an equivalence class—its own class $[x]$. If $[x_1]$ and $[x_2]$ are two equivalence classes with a common element x, then $x_1 \sim x$, $x_2 \sim x$, which implies that $x_1 \sim x_2$ by symmetry and transitivity. If now $y \in [x_1]$, then $y \sim x_1$, so $x_1 \sim x_2$ implies that $y \sim x_2$, that is, $y \in [x_2]$. In the same way, if $y \in [x_2]$, then $y \in [x_1]$. Thus, $[x_1] = [x_2]$, which means that two equivalence classes

J. Brzeziński, *Galois Theory Through Exercises*, Springer Undergraduate Mathematics Series, https://doi.org/10.1007/978-3-319-72326-6

with at least one common element already coincide. Hence, different equivalence classes must be disjoint. □

In general, if X is any set, then its **partition** is a family X_i of subsets of X (for i in an index set) which cover X (that is, $X = \bigcup X_i$) and are disjoint (that is, $X_i \cap X_j = \emptyset$ if $i \neq j$). It is clear that any partition defines an equivalence relation on X when we declare that two elements are equivalent when they are in the same subset X_i. Of course, the equivalence classes are just the sets X_i. Thus essentially, the notion of equivalence relation is exactly the same as the notion of partition.

In this book, we meet many equivalence relations. The following example is one of the most important:

Example A.1.2. Let $X = \mathbb{Z}$ be the set of integers and let n be a fixed positive integer. Define $\mathcal{R}$ as the set of all pairs of integers (x, y) such that x, y give the same residue when divided by n. We check immediately that it is an equivalence relation on $\mathbb{Z}$. An equivalence class $[x]$ consists of all integers which give the same residue when divided by n. Since $x = nq + r$, where r is the unique residue of x such that $0 \leq r < n$, we have exactly n possibilities for r: $0, 1, \ldots, n-1$. Thus, we get that there are exactly n different equivalence classes represented by these residues. This set of equivalence classes $[0], [1], \ldots, [n-1]$ is often denoted by $\mathbb{Z}_n$ and is called the set of **residues modulo** n. Very often the notation $[x]$ is simplified to x with no risk of confusion, and we write $\mathbb{Z}_n = \{0, 1, \ldots, n-1\}$. The set $\mathbb{Z}_n$ also inherits addition and multiplication (modulo n) from $\mathbb{Z}$, which turns it into a ring (see below A.2.9 and A.3.6).

Groups

A **group** is a set G with a binary operation which maps every pair $g_1, g_2 \in G$ onto $g_1g_2 \in G$ in such a way that

(a) $(g_1g_2)g_2 = g_1(g_2g_3)$ (**associativity**),
(b) there is an $e \in G$ such that $eg = ge = g$ for each $g \in G$ (existence of an **identity**),
(c) for each $g \in G$ there exists a $g' \in G$ such that $gg' = g'g = e$ (existence of an **inverse**).

One checks easily that in any group, there is only one identity e (sometimes called the **neutral element**) and for each g there is only one inverse g'. The inverse of an element $g \in G$ is denoted by g^{-1}. A group is called **abelian** if it is commutative, that is, $g_1g_2 = g_2g_1$ for every pair $g_1, g_2 \in G$.

The multiplicative notation of the group operation is more convenient (more compact) and it is usually used when the general theory of groups is presented, but sometimes the group operation is denoted as addition and then the binary operation maps a pair $g_1, g_2 \in G$ onto $g_1 + g_2 \in G$. The associativity then means that $(g_1 + g_2) + g_3 = g_1 + (g_2 + g_3)$, the identity (now called zero) is usually denoted by 0, and the inverse of $g \in G$ is then called the **opposite element** and is denoted by $-g$.

Most of the groups considered in this text are finite, that is, G has finitely many elements. If G is finite, then its number of elements is denoted by $|G|$ and called the **order** of the group G. If G has infinitely many elements, we often write $|G| = \infty$.

A **subgroup** H of G is a nonempty subset of G which is also a group when the binary operation is restricted to it. A subgroup H which is not equal to G is called **proper**. A nonempty subset H of G is a subgroup if and only if $h_1h_2 \in H$ for each pair $h_1, h_2 \in H$ and $h^{-1} \in H$ whenever $h \in H$. In fact, the associativity holds as it holds in G and the identity element is in H as $e = hh^{-1}$ for any $h \in H$. For a finite subset of a group, the situation is a little better:

Proposition A.2.1. *If H is a finite subset of a group G, then it is a subgroup if and only if $h_1h_2 \in H$ whenever $h_1, h_2 \in H$.*

Proof. We have to show that $h^{-1} \in H$ when $h \in H$. All powers h^n when $n = 1, 2, \ldots$ are in H and since H is finite, they cannot all be different. Thus, there exists $m < n$ such that $h^m = h^n$, which gives $h^{n-m} = e$ in G. But this equality says that $h^{-1} = h^{n-m-1} \in H$. □

If $g \in G$, then the powers of g are defined in the usual manner, that is, g^n is a product of n factors g, when $n > 0$ and of $-n$ factors g^{-1} when $n < 0$. We define $g^0 = e$. It is clear that all these integer powers of $g \in G$ form a subgroup of G. It is denoted by $\langle g \rangle$ and is called the **cyclic** group generated by g. Such an element g is called its **generator**. The **order** of g is defined as the order of the group generated by it and is denoted by $o(g)$. In the additive notation, that is, when the group operation is addition, then instead of powers g^n, we have multiples ng. For example, in the group of integers $\mathbb{Z}$ (with addition), the multiples of $g = 1$ are all integers, so $\mathbb{Z} = \langle 1 \rangle$ is a cyclic group with respect to addition and 1 is its generator (another generator is -1). In the group $\mathbb{Q}^*$ of rational numbers with multiplication, the cyclic group $\langle 2 \rangle$ generated by 2 consists of all integer powers 2^n, $n \in \mathbb{Z}$ and is infinite, but the cyclic group $\langle -1 \rangle$ has only two elements. In the field of complex numbers, all solutions of the equation $X^n = 1$, that is, all nth roots of unity, form a cyclic group of order n. This group, often denoted by U_n, consists of the numbers $\varepsilon_k = e^{\frac{2\pi ik}{n}} = \cos\frac{2\pi k}{n} + i\sin\frac{2\pi k}{n}$ for $k = 0, 1, \ldots, n-1$. A generator is ε_1, for which we have $\varepsilon_k = \varepsilon_1^k$, that is, $U_n = \langle \varepsilon_1 \rangle$.

If all integer powers g^n give different elements of the group G, then the cyclic subgroup $\langle g \rangle$ has infinite order. In the opposite case, there are different powers which give the same element, that is, $g^k = g^l$ for some integers $k < l$. Then, we have $g^{l-k} = e$, where $n = l - k > 0$. We can choose the least positive integer n such that $g^n = e$ and then it is easy to check that $\langle g \rangle = \{e, g, \ldots, g^{n-1}\}$. In fact, since for every integer N we have $N = qn + r$, where $0 \le r < n$, it follows that $g^N = g^{qn+r} = g^r$. It is clear that the powers g^r for $r = 0, 1, \ldots, n-1$ are different, since n is the least positive exponent such that $g^n = e$. Thus, such a least positive exponent is just the order of g. Notice that if $g^N = e$ and $o(g) = n$, then $n \mid N$. In fact, if $N = qn + r$, where $0 \le r < n$, then $e = g^N = g^{qn+r} = g^r$, which implies that $r = 0$, since $0 \le r < n$.

Proposition A.2.2. *Let $G = \langle g \rangle = \{e, g, \ldots, g^{n-1}\}$, $g^n = e$, be a cyclic group of order n.*
(a) The element g^k for $0 < k < n$ is a generator of G if and only if $\gcd(k, n) = 1$, so G has $\varphi(n)$ generators, where φ is Euler's function (for the definition of φ, see p. 279).
(b) Every subgroup of G is cyclic and for every divisor d of n, the element $g^{\frac{n}{d}}$ generates a subgroup of order d in G and such a subgroup is unique.

Proof. (a) If g^k generates G, then there is an integer l such $(g^k)^l = g$, that is, $g^{kl-1} = e$. Since the order of g is n, we have $n \mid kl - 1$, which shows that n and k are relatively prime, that is, $\gcd(k, n) = 1$ (if d is a common divisor of n and k, then d must divide 1).
Conversely, if k and n are relatively prime, then $\gcd(k, n) = 1 = kl + nq$ for some integers l, q (see A.3.4(b)). Hence $(g^k)^l = g^{1-nq} = g$ as $g^n = e$. This shows that a power of g^k equals g, so the powers of g^k generate G.
(b) If H is any subgroup of G, we choose the least positive power m of g such that $h = g^m \in H$. If $g^M \in H$ for some integer M, then $m \mid M$, since otherwise $M = mq + r$, where $0 < r < m$ and $g^M = g^{mq+r} = g^r \in H$, which contradicts the choice of m. Thus, $g^n = e \in H$ implies that m divides n. Let $d = n/m$. Thus, we have $h = g^{\frac{n}{d}} \in H$ and H consists of exactly those powers of g whose exponents are divisible by m, that is, $H = \langle h \rangle = \{e, h, \ldots, h^{d-1}\}$ and $h^d = g^{md} = g^n = e$. Thus $H \subseteq G$ uniquely defines $m = n/d$ and, consequently, the subgroup H of G is uniquely defined by its order d (which, as we found, divides n). □

Formally, we say that two groups G and G' are **isomorphic** (denoted by $G \cong G'$) if there is a bijective function $\varphi : G \to G'$ such that $\varphi(g_1 g_2) = \varphi(g_1)\varphi(g_2)$ for all $g_1, g_2 \in G$. Such a function is called an **isomorphism** of groups G and G'. An isomorphism $\varphi : G \to G$ is called an **automorphism** of G.

Example A.2.3. If $G = \langle g \rangle$ is an infinite cyclic group, then G is isomorphic to the group of integers $\mathbb{Z}$ (with addition), and when G is finite of order n, then it is isomorphic to U_n. In the first case, the function $\varphi : G \to \mathbb{Z}$ such that $\varphi(g^n) = n$ defines an isomorphism—different powers of g map to different integers and $\varphi(g^k g^l) = \varphi(g^{k+l}) = k + l = \varphi(g^k) + \varphi(g^l)$. If G is cyclic of order n, then by A.2.2, we define a bijective function $\varphi : G \to U_n$ such that $\varphi(g^r) = \varepsilon_1^r$ for $r = 0, 1, \ldots, n-1$. We then have $\varphi(g^k g^l) = \varphi(g^{k \oplus l}) = \varepsilon_1^{k \oplus l} = \varphi(g^k)\varphi(g^l)$, where $\oplus$ denotes addition modulo n. Notice that $g^n = e$ and $\varepsilon_1^n = 1$ turn addition of exponents into addition modulo n. □

Cyclic subgroups are generated by one element of the group. If A is any subset of G, then there is a smallest subgroup of G containing A—the intersection of all subgroups of G containing A. Such a subgroup consists of all finite products $g_1 \cdots g_k$, where each g_i or its inverse is in A. In fact, a product of two such products and the inverse of such a product keep the required type. If $A = \{g_1, \ldots, g_k\}, g_i \in G$, is a finite set, then the smallest subgroup of G containing A is denoted by $\langle g_1, \ldots, g_k \rangle$. If $G = \langle g_1, \ldots, g_k \rangle$, then we say that the group G is generated by $g_1, \ldots, g_k$ and these elements are called **generators** of G. Notice that cyclic groups are the groups with one generator. Every equality such that a product of the generators or their inverses is equal to the identity is called a **relation** among the generators. For example, if $G = \langle g \rangle$ is a cyclic group of order n, then $g^n = e$ is a relation for the generator g. If a group is not cyclic, for example $G = S_3$, then the number of generators must be at least 2. In the case of $G = S_3$, we have $G = \langle g, h \rangle$, where $g^3 = 1$, $h^2 = 1$ and $hg = g^2h$ (that is, $g^2hg^{-1}h^{-1} = 1$). For example, take $g = (1, 2, 3)$ and $h = (1, 2)$ (as g, h one can choose any element of order 3 and any element of order 2 while (1) is denoted by 1). Generators and relations among them give a very convenient way to describe groups, which is often used in different computer packages. A group G is defined by a set of its generators $g_1, \ldots, g_k$ and relations among them if every group G' with some generators $h_1, \ldots, h_k$ satisfying the same relations (with g_i replaced by h_i) is isomorphic to G (which means "the same").

Example A.2.4. Consider the symmetry group of a regular polygon with n sides. It is usually denoted by D_n $(n \geq 3)$ and called the **dihedral group**. The order of D_n is $2n$. It consists of n rotations by angles $\frac{2\pi k}{n}$ for $k = 0, 1, \ldots, n-1$ and n symmetries in n symmetry axes. As for S_3 (which is D_3), we can choose two elements g, h, where the first g is the rotation by the angle $\frac{2\pi}{n}$ and h is any symmetry of the polygon. Since the rotation g has order n, it generates a cyclic subgroup C_n of D_n of order n. A symmetry h has order 2 and is not in C_n, so the products g^kh for $k = 1, \ldots, n$ give n different elements (symmetries) in the group D_n. Notice that the relations defining D_n generated by g, h are $g^n = 1, h^2 = 1$ and $hgh = g^k$ for some k (depending on the choice of the symmetry h—the simplest choice is $k = -1$, when the relation is $ghgh = 1$).

If A, B are two subsets of a group G, then AB denotes the set of all products ab, where $a \in A$ and $b \in B$. Such a product is of course associative, that is, $A(BC) = (AB)C$ for any subsets A, B, C of G. Notice that if H is a subgroup of G, then $HH = H$. If, in particular, we take $A = \{g\}$ and $B = H$ is a subgroup of G, then the product gH is called a **left coset** of H in G. Similarly, the product Hg is a **right coset** of H in G. Below are some simple, but important properties of cosets:

Proposition A.2.5. *Let H be a subgroup of a group G.*
(a) $gH = H$ if and only if $g \in H$.
(b) If $g' \in gH$, then $g'H = gH$ (every element in a coset represents it).
(c) $g'H = gH$ if and only if $g^{-1}g' \in H$ (additively: $g + H = g' + H$ if and only if $g' - g \in H$).
(d) $g \in gH$ (every element in G belongs to a coset).
(e) If $g \in g_1H \cap g_2H$, then $g_1H = g_2H$ (two different cosets are disjoint).
(f) If H is finite, then $|gH| = |H|$ for any $g \in G$.
(g) The left (or right) cosets of H in G form a partition of the set G.

Proof. (a) If $gH = H$, then $ge = g$ belongs to H. If $g \in H$, then $gH \subseteq H$ and, conversely, $H \subseteq gH$, since $h = g(g^{-1}h) \in gH$ for any $h \in H$ $(g^{-1}h \in H)$.
(b) If $g' = gh, h \in H$, then $g'H = ghH = gH$ by (a).
(c) We have $g'H = gH$ if and only if $g^{-1}g'H = H$ if and only if $g^{-1}g' \in H$ by (a).

(d) We have $g = ge$, $e \in H$.
(e) If $g \in g_1H \cap g_2H$, then $g_1H = g_2H = gH$ by (b).
(f) If H is finite and $H = \{h_1, \dots, h_m\}$, then $gH = \{gh_1, \dots, gh_m\}$ and all products gh_i are different, since $gh_i = gh_j$ implies $h_i = h_j$ (multiply on the left by g^{-1}).
(g) This is clear from (d) and (e) and means that any subgroup defines an equivalence relation on the set G by declaring two elements equivalent if and only if they belong to the same coset. □

Notice that (left) cosets are equivalence classes of the equivalence relation on G such that $g \sim g'$ if and only if $g^{-1}g' \in H$ by A.2.5(c),(d),(e).

A.2.6 Lagrange's Theorem. *The order of a subgroup of a finite group divides its order. In particular, the order of an element in a finite group divides its order and if the group has order n, then $g^n = e$ for all elements g in the group.*

Proof. Let $|G| = n$ and $|H| = m$. The group G is a union of the left cosets of H in G by A.2.5(d). These cosets are disjoint by A.2.5(e) and each of them consists of m elements by A.2.5(e). If i denotes the number of cosets, we have $n = mi$, so m divides n.

If $g \in G$ has order m, then $g^m = e$ and the subgroup $\langle g \rangle$ has order m dividing n. Hence, if $n = mi$, we have $g^n = g^{mi} = (g^m)^i = e$. □

The number of cosets of a subgroup H in a group G is called the **index** of H in G and is denoted by $[G : H]$. This number is finite when G is finite, but it may also be finite when G is infinite. For a finite group G, we have $[G : H] = |G|/|H|$ according to the argument given in the proof of Lagrange's theorem. Notice that in a finite group, the number of left cosets of a subgroup is equal to the number of its right cosets, which easily follows from Lagrange's theorem by noting that the argument in the proof works equally well for right as for left cosets. In general, it is easy to construct a bijection between the sets of right and left cosets (but gH need not be equal to Hg!). Sometimes, we write $[G : H] = \infty$ when there are infinitely many cosets of H in G. Usually what can be said about right cosets can also be said about the left. Sometimes, we simply say cosets when we mean just a fixed type of them—left or right, but we prefer to use left cosets when we give proofs of different statements.

Sometimes the cosets of a subgroup H in a group G form a group with respect to the multiplication of subsets of G. For this to hold, then in particular the product of any two cosets $gHg'H$ must again be a coset. If this is the case, then this coset must be $gg'H$ (of course, $gg' \in gHg'H$, so if the product $gHg'H$ is a coset, it must be $gg'H$ by A.2.5(b)). When a product of any (two) cosets is a coset, then these cosets form a group, since multiplication of cosets is associative, the identity is the coset H and the inverse of gH is $g^{-1}H$. This group is then called the **quotient group** of G modulo H and is denoted by G/H. If G is finite, then the order of G/H is the number of cosets of H in G, that is, the index of H in G. Thus for the orders, we have $|G/H| = |G|/|H|$, which in some way motivates the terminology. The subgroups for which the above construction of the quotient group is possible are called **normal subgroups**. More precisely, we have the following characterization of normal subgroups:

Proposition A.2.7. *Each of the following equivalent conditions defines a normal subgroup H of G:*
(a) The product of any two cosets of H in G is a coset;
(b) For each $g \in G$ the left and the right cosets of g coincide, that is, $gH = Hg$;
(c) For each $g \in G$, we have $g^{-1}Hg \subseteq H$.

Proof. First, we prove that (a) implies both (b) and (c), which are equivalent. Since the product of cosets is a coset, we have $gHgH = g^2H$ (as we know, the product is the coset containing g^2, so it must be g^2H). Thus for every $h \in H$, there exists an $h' \in H$ such that $ghg = g^2h'$, that is, $hg = gh'$. Hence for every $g \in H$, we have $Hg \subseteq gH$, that is, $g^{-1}Hg \subseteq H$. Hence, (a) implies (c). Since (c) holds for any $g \in G$, we can replace g by g^{-1} and we get $gHg^{-1} \subseteq H$. Now notice that this inclusion is equivalent to $gH \subseteq Hg$. Thus, (c) gives both $gH \subseteq Hg$ and $Hg \subseteq gH$, which is (b). Of course, (c) follows immediately from (b). And finally, (b) implies that $gHg'H = gg'HH = gg'H$, so we get (a). □

Notice that the conditions in A.2.7 are satisfied for every subgroup H of any abelian group G. Thus, all subgroups of any abelian group are normal.

If $g \in G$, then the mapping $x \mapsto gxg^{-1}$ for $x \in G$ is called **conjugation** by g. Such a function $\varphi_g(x) = gxg^{-1}$ for $x \in G$ is an automorphism of the group G, which is called an **inner automorphism**. Using this terminology, A.2.7(c) says that H is a normal subgroup of G if and only if it is invariant with respect to inner automorphisms. The fact that H is a normal subgroup of G is often denoted by $H \lhd G$.

If H is any subgroup of a group G, then it is easy to check that $\mathcal{N}(H) = \{g \in G | gHg^{-1} \subseteq H\}$ is a group containing H in which H is, of course, normal by A.2.7(c). The group $\mathcal{N}(H)$ is called the **normaliser** of H in G and is the biggest subgroup of G in which H is normal. Of course, it may happen that $\mathcal{N}(H) = H$ or $\mathcal{N}(H) = G$ and the last case happens precisely when the subgroup H is normal in G.

We give two examples of normal subgroups and of quotient groups.

Example A.2.8. If H is a subgroup of index 2 in a group G, then H is normal in G. In fact, there are two left and two right cosets. One (both left and right) is H. If $g \in G$ is any element not in H, then $gH = G \setminus H$ is the second left coset, and $Hg = G \setminus H$ is the second right coset, so $gH = Hg$ for any $g \in H$. Hence, H is normal in G.

Example A.2.9. In the group of integers $\mathbb{Z}$ with respect to addition, any subgroup is normal, since the group is abelian. Every subgroup is the set of all multiples $\langle n \rangle$ of a fixed integer n. The quotient group $\mathbb{Z}/\langle n \rangle$ consists of the cosets $a + \langle n \rangle$ and each such coset consists of all $b = a + qn$ for integers q, that is, $b - a = qn$. Hence a coset consists of all integers giving the same residue as a when divided by n. It is possible to choose the least nonnegative residue $0 \leq r < n$, that is, each coset has a representative r satisfying this restriction. Hence all cosets are $r + \langle n \rangle (= r + \mathbb{Z}n)$, where $r = 0, 1, \ldots, n-1$. In fact, the elements of the group $\mathbb{Z}/\langle n\rangle$ are simply the equivalence classes of the equivalence relation defined in A.1.2 (the same as above). Usually, we write $\mathbb{Z}/\langle n\rangle = \mathbb{Z}_n$. The addition of the cosets $[r] = r+\mathbb{Z}n$ is sometimes denoted by $[r] \oplus [r']$ or, as in Maple, $r + r'$ mod n. This means that the sum of the residues r, r' in $\mathbb{Z}_n$ is the residue of the sum of the integers r, r' when divided by n. Notice that the group $\mathbb{Z}_n$ with addition modulo n is cyclic of order n, since 1 is its generator. This group is isomorphic to U_n (see A.2.3). According to A.2.2 this group has $\varphi(n)$ generators and r is one of them if and only if $\gcd(r, n) = 1$.

If H is a normal subgroup of G, then we have a natural function $\iota : G \to G/H$ such that $\iota(g) = gH$ mapping any element $g \in G$ onto its coset. This function has a multiplicative property: $\iota(gg') = \iota(g)\iota(g')$, since $gg'H = gHg'H$. In general, a function $\varphi : G \to G'$ is called a **group homomorphism** if this property holds, that is, $\varphi(gg') = \varphi(g)\varphi(g')$ for any $g, g' \in G$. Notice that $\varphi(e) = e'$ and $\varphi(g^{-1}) = \varphi(g)^{-1}$. In fact, we have $\varphi(e) = \varphi(ee) = \varphi(e)\varphi(e)$, which implies $\varphi(e) = e'$. Thus, we have $\varphi(g)\varphi(g^{-1}) = \varphi(gg^{-1}) = \varphi(e) = e'$, that is, $\varphi(g^{-1})$ is the inverse of $\varphi(g)$.

The homomorphism $\iota : G \to G/H$, where H is normal in G, is called the **natural surjection** (defined by a normal subgroup H). Notice that the elements mapping on the identity of G/H, that is, on H, are those $g \in G$ for which $\iota(g) = gH = H$. The equality $gH = H$ is equivalent to $g \in H$ (see A.2.5(a)).

In general, the **kernel** of a group homomorphism $\varphi : G \to G'$ is the set of elements in G whose image in G' is the identity element e'. The kernel is denoted by $\mathrm{Ker}(\varphi)$, that is, $\mathrm{Ker}(\varphi) = \{g \in G | \varphi(g) = e'\}$. Thus, we have just established that if H is a normal subgroup of G, then the kernel of the natural surjection of G onto G/H is H.

Notice that the kernel of a group homomorphism $\varphi : G \to G'$ is always a normal subgroup of G. In fact, if $g, g' \in \mathrm{Ker}(\varphi)$, then $\varphi(gg') = e'$, so $gg' \in \mathrm{Ker}(\varphi)$. We also have $g^{-1} \in \mathrm{Ker}(\varphi)$ if $g \in \mathrm{Ker}(\varphi)$, since $\varphi(g^{-1}) = \varphi(g)^{-1} = e'$ when $\varphi(g) = e'$. Thus the kernel $H = \mathrm{Ker}(\varphi)$ of a homomorphism $\varphi : G \to G'$ is a subgroup of G. In order to check that H is normal, we have to check that $ghg^{-1} \in H$ whenever $h \in H$. This follows immediately, since $\varphi(g^{-1}hg) = \varphi(g^{-1})\varphi(h)\varphi(g) = \varphi(g^{-1})\varphi(g) = \varphi(g^{-1}g) = \varphi(e) = e'$.

Proposition A.2.10. *A homomorphism $\varphi : G \to G'$ is injective, that is, the images of different elements are different if and only if* $\mathrm{Ker}(\varphi) = \langle e\rangle$.

Proof. We have $\varphi(g) = \varphi(g')$ if and only if $\varphi(g'g^{-1}) = e'$, that is, $g'g^{-1} \in \mathrm{Ker}(\varphi)$. Hence, if $\mathrm{Ker}(\varphi) = \langle e\rangle$, then $g'g \in \mathrm{Ker}(\varphi)$ gives $g'g^{-1} = e$, that is, $g' = g$ and, conversely, if the mapping φ is injective and $g \in \mathrm{Ker}(\varphi)$, then $\varphi(g) = e' = \varphi(e)$ gives $g = e$, that is, $\mathrm{Ker}(\varphi) = \langle e\rangle$. □

The quotient group $G/\mathrm{Ker}(\varphi)$ is closely related to the image of G in G'. In fact, all elements in the coset gH, where $H = \mathrm{Ker}(\varphi)$, have the same image in G', since $\varphi(gh) = \varphi(g)\varphi(h) = \varphi(g)$. Hence, we can define a function $\varphi^* : G/H \to G'$ mapping the whole coset gH onto $\varphi(g)$, that is, $\varphi^*(gH) = \varphi(g)$. It is easy to check that this function is an injection from G/H to G' and its image is, of course, $\varphi(G)$. In fact, since H is normal, we have

$$\varphi^*(gHg'H) = \varphi^*(gg'H) = \varphi(gg') = \varphi(g)\varphi(g') = \varphi^*(gH)\varphi^*(g'H),$$

so φ^* is a homomorphism of G/H into G'. But $\text{Ker}(\varphi^*) = \{gH \in G/H | \varphi^*(gH) = \varphi(g) = e'\} = H$, so the kernel of φ^* is the identity element of G/H. Thus φ^* is injective, so it is an isomorphism of $G/\text{Ker}(\varphi)$ with $\varphi(G)$, which is often denoted by $\text{Im}(\varphi)$. This is an important fact, which is very often called the **fundamental theorem on group homomorphisms**:

A.2.11 Fundamental Theorem on Group Homomorphisms. *If $\varphi : G \to G'$ is a group homomorphism, then the kernel $H = \text{Ker}(\varphi)$ is a normal subgroup of G and the function $\varphi^*(gH) = \varphi(g)$ is an injection $\varphi^* : G/\text{Ker}(\varphi) \to G'$. Thus φ^* is an isomorphism of the groups $G/\text{Ker}(\varphi)$ and $\text{Im}(\varphi)$. In particular, if G is finite, we have $|G| = |\text{Ker}(\varphi)||\text{Im}(\varphi)|$.*

Theorem A.2.11 is a special case of a more general result in which a group homomorphism induces a homomorphism of quotient groups. Let H be a normal subgroup of G, and H' a normal subgroup of G'. We say that a group homomorphism $\varphi : G \to G'$ is a **homomorphism of pairs** if $\varphi(H) \subseteq H'$—the pair $H \subseteq G$ is mapped into the pair $H' \subseteq G'$. This situation is very common and we meet it several times in Chap. 12, where we use the following very important consequence:

A.2.12 Group Homomorphism of Pairs. *Let $\varphi : G \to G'$ be a group homomorphism of the pair $H \lhd G$ into the pair $H' \lhd G'$. Then there exists an induced homomorphism*

$$\varphi^* : G/H \to G'/H'$$

such that $\varphi^*(gH) = \varphi(g)H'$. Moreover, we have $\text{Ker}(\varphi^*) = \varphi^{-1}(H')/H$ and $\varphi^*(G/H) = \varphi(G)H'/H'$.

Proof. We have only to check that the function φ^* is well defined, since it would then follow immediately that

$$\varphi^*(g_1Hg_2H) = \varphi^*(g_1g_2H) = \varphi(g_1g_2)H' = \varphi(g_1)\varphi(g_2)H'$$

$$= \varphi(g_1)H'\varphi(g_2)H' = \varphi^*(g_1H)\varphi^*(g_2H).$$

In order to check the definition of φ^*, we must show that if $g_1H = g_2H$, then $\varphi^*(g_1)H' = \varphi^*(g_2)H'$, that is, the definition is independent of the choice of a representatives of a coset of H. But $g_1H = g_2H$ is equivalent to $g_1^{-1}g_2 \in H$, which implies $\varphi(g_1^{-1}g_2) \in H'$, so $\varphi(g_1)^{-1}\varphi(g_2) \in H'$, which gives $\varphi(g_1)H' = \varphi(g_2)H'$, that is, $\varphi^*(g_1H) = \varphi^*(g_2H)$. □

Proposition A.2.13. *Let $\varphi : G \to G'$ be a group homomorphism with kernel $N = \text{Ker}(\varphi)$.*
(a) The image $\varphi(H)$ of a subgroup H of G is a subgroup of G' and the inverse image $\varphi^{-1}(H')$ of any subgroup H' of G' is a subgroup of G containing N. We have $\varphi(\varphi^{-1}(H')) = H'$ and $\varphi^{-1}(\varphi(H)) = NH$. The inverse image $\varphi^{-1}(H')$ is normal in G if H' is normal in G'.

(b) If φ is surjective (so $G' \cong G/N$), then there is a one-to-one correspondence between the subgroups H' of G' and the subgroups H of G containing N in which H' corresponds to the inverse image $\varphi^{-1}(H')$ and H corresponds to its image $\varphi(H)$. In this correspondence, normal subgroups of G containing N correspond to normal subgroups of G'.

Proof. (a) If H' is a subgroup of G', then one checks easily that $\varphi^{-1}(H')$ is a subgroup of G. Of course, the group $H = \varphi^{-1}(H')$ contains N, which is the inverse image of the identity e' of G'. If H' is normal in G', then for any $g \in G$ and $h \in G$ such that $\varphi(h) \in H'$, we have $\varphi(ghg^{-1}) = \varphi(g)\varphi(h)\varphi(g)^{-1} \in H'$, since H' is normal in G'. Hence $ghg^{-1} \in \varphi^{-1}(H')$, so this subgroup of G is normal.
The equality $\varphi(\varphi^{-1}(H')) = H'$ is evident (for any function between sets G, G' not just groups). In order to check the second equality $\varphi^{-1}(\varphi(H)) = HN$, take $g \in \varphi^{-1}(\varphi(H))$. Then $\varphi(g) = \varphi(h)$ for some $h \in H$, so $\varphi(gh^{-1}) = e'$. Hence, we have $gh^{-1} \in N$, which means that $g \in NH$. Conversely, if $g \in NH$, then $g = nh$ for some $n \in N, h \in H$. Thus $\varphi(g) = \varphi(nh) = \varphi(h) \in \varphi(H)$, so $g \in \varphi^{-1}(\varphi(H))$.
(b) Let $X_N(G)$ denote the set of all subgroups of G containing N, and let $X(G')$ denote the set of all subgroups of G'. We have two functions: $\Phi(H) = \varphi(H)$ mapping the subgroups of G to subgroups of G' (see (a)), and $\Psi(H') = \varphi^{-1}(H')$ mapping the subgroups of G' to subgroups of G containing N. Now $\Phi(\Psi(H')) = \varphi(\varphi^{-1}(H') = H'$ by (a), and $\Psi(\Phi(H)) = \varphi^{-1}(\varphi(H)) = NH = H$ by (a) and the assumption $N \subseteq H$, which implies that $NH = H$. Hence both compositions $\Phi \circ \Psi$ and $\Psi \circ \Phi$ give identities, so both are bijections between the corresponding sets $X_N(G)$ and $X(G')$ (see A.9.1). □

We use the following two results on groups in several exercises:

Proposition A.2.14. *Let G be a finite abelian group. Then the exponent of G, that is, the least positive integer m such that $g^m = e$ for each element $g \in G$ (see p. 68) equals the maximal order n of the elements of G. Moreover, m divides the order of G.*

Proof. First we note that if $g, h \in G$ are elements whose orders $o(g)$ and $o(h)$ are relatively prime, then $o(gh) = o(g)o(h)$. In fact, if r is the order of gh, then $(gh)^r = e$ gives $(gh)^{ro(h)} = g^{ro(h)} = e$. Hence $o(g) \mid ro(h)$ and by the assumption, $o(g) \mid r$. Similarly, $o(h) \mid r$, so $o(g)o(h) \mid r$. Since, of course, $(gh)^{o(g)o(h)} = e$, we have $r = o(g)o(h)$.

Now let n be the maximal order of elements in G and let g be an element of that order. Assume that h is an element whose order does not divide n. Then $\gcd(o(h), n) = d$ and $n = da$, $o(h) = db$, where $\gcd(a, b) = 1$ and $b > 1$ (if $b = 1$ then the order of h divides n). The order of g^d equals a, and the order of h^d equals b, so the order of $g^d h^d = (gh)^d$ equals ab. Thus the order of gh equals $dab = nb > n$. This is a contradiction, since n was the maximal order of the elements in G. Thus $b = 1$ and $o(h) \mid n$, so n is the exponent of G.

The last statement is clear, since the order of every element of G divides the order of this group, so in particular, the exponent divides this order. □

Proposition A.2.15. *A group G of order at least 2 has no nontrivial subgroups if and only if it is cyclic of prime order.*

Proof. If the order of G is a prime number p and H is its subgroup, then as the order of H divides p by Lagrange's theorem A.2.6, the order of H is 1 or p. Thus H is either the identity subgroup or the whole group G.

Now assume that G has at least one element different from the identity and no nontrivial subgroups. Take $g \in G$, $g \neq e$. The cyclic subgroup $\langle g \rangle$ of G is nontrivial, so $G = \langle g \rangle$. If the order of g is infinite, then G contains nontrivial subgroups, e.g. $\langle g^2 \rangle$. Thus the order of g is finite, say, n. If n is not a prime number, then $n = kl$, where $k > 1, l > 1$. Thus G contains a nontrivial subgroup, e.g. $\langle g^k \rangle$. Hence n must be a prime number. □

A group G without nontrivial normal subgroups is called **simple**. Simple groups are in some sense the building blocks of all finite groups, since a group G having a nontrivial normal subgroup N can be described in terms of N and G/N (even if such a description is usually not "simple"). The simple groups have been classified thanks to the efforts of many generations of mathematicians. The classification was completed around 1980; its proof still takes up several thousands of pages. Those simple groups which we use are the alternating groups A_n for $n \geq 5$. The groups from A.2.15 which do not contain any subgroups are of course simple. These are all simple abelian groups.

A trivial, but important, situation when a group is described in terms of its subgroups or quotients is the construction of products. If G_1 and G_2 are two groups, then we can form a new group $G_1 \times G_2$, called their **product** (or **direct product**), which consists of all pairs (g_1, g_2), where $g_1 \in G_1$, $g_2 \in G_2$ with coordinate-wise operation, that is, $(g_1, g_2)(g_1', g_2') = (g_1 g_1', g_2 g_2')$. It is not difficult to see how to embed the groups G_1, G_2 as (normal) subgroups of $G_1 \times G_2$ or to obtain them as quotients of the product.

There are two groups related to a given group G which in some sense give a measure of the non-commutativity of G. The first is the **center** $C(G)$ of G, consisting of those elements of G which commute with all elements in this group, that is, $C(G) = \{x \in G | \forall g \in G \; xg = gx\}$. Of course, the center is a normal subgroup of G and the group is abelian exactly when $C(G) = G$.

The second group is the so-called commutator group of G, which we define now. If $g_1, g_2 \in G$, then $g_1 g_2$ and $g_2 g_1$ differ by an element $c \in G$ such that $g_1 g_2 = c g_2 g_1$ (c "corrects" $g_2 g_1$ to be the same as $g_1 g_2$). Of course, we have $c = g_1 g_2 g_1^{-1} g_2^{-1}$. The element c is often denoted by $[g_1, g_2]$ and called the **commutator** of g_1, g_2. Notice that $[g_1, g_2] = e$ precisely when g_1, g_2 commute. The subgroup of G generated by all commutators of pairs $g_1, g_2 \in G$ is denoted by G' (sometimes by $[G, G]$) and called the **commutator group** of G. The group G' consists simply of all finite products of commutators, since the inverse of a commutator is a commutator: $[g_1, g_2]^{-1} = [g_2, g_1]$. Notice that the group G' reduces to e if and only if all commutators are trivial, that is, if the group G is abelian.

The commutator subgroup G' of G is normal. In fact, if $c = [g_1, g_2]$ is a commutator and $g \in G$, then $gcg^{-1} = [gg_1g^{-1}, gg_2g^{-1}]$ is also a commutator. Since conjugation by g is an automorphism, it maps any product of commutators onto a product of commutators. Thus, the commutator group G' is normal, since it is invariant with respect to conjugation. The quotient group G/G' is abelian, since $G'g_1G'g_2 = G'g_2G'g_1$ is equivalent to $[g_1, g_2] \in G'$. Moreover, every normal subgroup N of G such that G/N is abelian contains G', since $Ng_1Ng_2 = Ng_2Ng_1$ implies $[g_1, g_2] \in N$, that is, $G' \subseteq N$. This means that G/G' is the biggest abelian quotient group of G. In fact, if $G' \subseteq N$, then we have an imbedding of the pair $G' \subseteq G$ into the pair $N \subseteq G$ and as a consequence an induced surjection of the quotient G/G' onto the quotient G/N (see A.2.12). As a matter of fact, every subgroup N of G containing G' is normal, which immediately follows from A.2.13(a).

Rings

A **ring** is a set R with two operations usually called addition "+" and multiplication "·" such that R with respect to addition is an abelian group, multiplication is associative and addition is distributive with respect to multiplication, that is, $a(b + c) = ab + ac$ and $(b + c)a = ba + ca$ for all $a, b, c \in R$. In this text, we always assume that R is commutative, that is, $ab = ba$ for all $a, b \in R$. We say that R is a **ring with identity** if there exists an element $1 \in R$ such that $1a = a$ for all $a \in R$. It is easy to see that the identity 1 is unique if it exists. In the sequel, we only meet rings with identity. A **zero divisor** in a ring R is a nonzero element $a \in R$ such that $ab = 0$ for some nonzero $b \in R$. A ring without zero divisors is called an **integral domain**. We will always assume that an integral domain has an identity element 1. Thus, the ring $\mathbb{Z}$ is an integral domain, whereas the rings $\mathbb{Z}_n$ of integers modulo n (with the sum $a + b$ and the product ab of $a, b \in \mathbb{Z}_n$ taken as the residues of the integers $a + b, ab$ when divided by n—see A.2.9 and A.3.6 for a formal argument) are integral domains only when n is a prime number. For example, in $\mathbb{Z}_4 = \{0, 1, 2, 3\}$, we have $2 \cdot 2 = 0$.

A **subring** of a ring R is any nonempty subset R' which is also a ring with respect to the same addition and multiplication as in R. In practical terms, in order to check that a nonempty subset R' of R is a ring, it suffices to check that when $a, b \in R'$, then $a + b, ab, -a \in R'$. All other conditions in the definition of ring are automatically satisfied, since they are satisfied in R.

If a ring R is a subring of a ring S and $\alpha \in S$, then $R[\alpha]$ denotes the smallest subring of S containing R and α. It is easy to see that it consists of all sums $r_0 + r_1\alpha + \cdots + r_n\alpha^n$, where $r_i \in R$ and n is a nonnegative integer. For example, the ring of integers $\mathbb{Z}$ is a subring of the complex numbers $\mathbb{C}$ and taking $\alpha = i$, we get the ring $\mathbb{Z}[i]$. This ring consists of all $a + bi$, where $a, b \in \mathbb{Z}$, since the equality $i^2 = -1$ makes all powers of i with exponents bigger than 1 needless. Notice that $\mathbb{Z}[i]$ is called the ring of **Gaussian integers** (see Exercise 11.7). A similar construction gives $\mathbb{Z}[\varepsilon]$, where $\varepsilon^3 = 1, \varepsilon \neq 1$, which is called the ring of **Eisenstein integers** (see Exercise 11.6).

If R is a commutative ring with identity, then an element $a \in R$ is called a **unit** (or an **invertible element**) if there exists an $a' \in R$ such that $aa' = 1$. The element a' is called an **inverse** of a. All invertible elements form a group with respect to the multiplication in R, which is often denoted by R^*. For example, we have $\mathbb{Z}^* = \{\pm 1\}$, $\mathbb{Z}[i]^* = \{\pm 1, \pm i\}$. In the ring $\mathbb{Z}_n$, the invertible elements form the group $\mathbb{Z}_n^*$ consisting of all k such that $\gcd(k, n) = 1$. Its order is the value of Euler's function $\varphi(n)$ (for the definition of Euler's function, see p. 279). In fact, if $\gcd(k, n) = 1$, then $kx + ny = 1$ for some integers x, y, so $kx = 1$ in $\mathbb{Z}_n$, that is, k is invertible. Conversely, if k is invertible in $\mathbb{Z}_n$, then $kx = 1$ for some $x \in \mathbb{Z}_n$, so $n|kx - 1$, that is, $kx - 1 = ny$ for some $y \in \mathbb{Z}$. Hence, $\gcd(k, n) = 1$. Thus we have $|\mathbb{Z}_n^*| = \varphi(n)$.

A commutative ring with identity in which every nonzero element is invertible is called a **field**. Some facts about fields which we use in this book are discussed in section "Fields" of this Appendix.

If R_1 and R_2 are rings, then it is possible to form a new ring $R_1 \times R_2$ consisting of pairs (r_1, r_2), where $r_1 \in R_1$ and $r_2 \in R_2$, which are added and multiplied coordinate-wise. This ring is called the **product** (or **direct product**) of the two rings. Of course, it is possible to extend this definition to any finite number of rings. If both rings have an identity (both denoted by 1), then $(1, 1)$ is the identity of the product. For future use, observe that, in this case, (r_1, r_2) is invertible if and only if r_1 and r_2 are invertible and

$$(R_1 \times R_2)^* = R_1^* \times R_2^*, \tag{A.1}$$

where on the right-hand side, we have the product of groups.

If R, R' are rings, then a **ring homomorphism** is a function $\varphi : R \to R'$ which respects both addition and multiplication, that is, $\varphi(a + b) = \varphi(a) + \varphi(b)$ and $\varphi(ab) = \varphi(a)\varphi(b)$. If R, R' are rings with identity, we shall also assume that a ring homomorphism maps the identity in R onto the identity in R'. The **kernel** of φ is its kernel as a homomorphism of the additive group R into the additive group R', so $\mathrm{Ker}(\varphi) = \{a \in R | \varphi(a) = 0\}$. This set is not only a subgroup of the additive group R, but it is a subring with a stronger property: if $a \in \mathrm{Ker}(\varphi)$ and $b \in R$, then $ab \in \mathrm{Ker}(\varphi)$. Indeed, if $\varphi(a) = 0$, then $\varphi(ab) = \varphi(a)\varphi(b) = 0$ for any $b \in R$. A subring I of R having this strong property for multiplication, that is, $ab \in I$ whenever at least one of the factors $a, b \in R$ is in I, is called an **ideal** (it is called **proper** if $I \neq R$). Thus kernels of the homomorphisms of R are ideals. In fact, every ideal is also a kernel of a homomorphism, which will be clear in a moment. Notice that $\mathrm{Ker}(\varphi) = \{0\}$ if and only if $\varphi : R \to R'$ is an injective function (see A.2.10). An injective ring homomorphism $\varphi : R \to R'$ is often called an embedding. In the particular case when φ is injective and surjective (that is, $\varphi(R) = R'$), it is called a (ring) **isomorphism**.

In any ring, one can fix a finite number of elements $a_1, \ldots, a_k$ and consider all sums $r_1 a_1 + \cdots + r_k a_k$, where $r_1, \ldots, r_k \in R$. It is easy to check that all such sums of products form an ideal I in R. It is called the **ideal generated by** $a_1, \ldots, a_k$ and is denoted by $(a_1, \ldots, a_n)$. The elements $a_1, \ldots, a_k$ are called **generators** of I. An ideal may have many sets of generators. If there is one generator of an ideal I, that is, $I = (a)$ for some $a \in R$, then I is called **principal**. A ring R is called a **principal ideal ring** if every ideal in R is principal. If, moreover, R has an identity and there are no zero divisors in R (that is, R is an integral domain), then it is called a **principal ideal domain** (PID). The following examples of such rings are very important:

Proposition A.3.1. *The ring of integers $\mathbb{Z}$ and the polynomial rings $K[X]$, where K is a field, are principal ideal domains.*

Proof. Let I be an ideal in $\mathbb{Z}$. The zero ideal is always principal and generated by 0, so assume that I is nonzero. Let d be a nonzero integer in I with the least absolute value. If $a \in I$, we use the division algorithm and get $a = dq + r$, where $0 \leq r < |d|$. But $r = a - dq \in I$, so by the definition of d, we have $r = 0$. Hence each element of I is a multiple of d, that is, $I = (d)$ is principal. Notice that looking at I as a subgroup of the cyclic group $\mathbb{Z}$ (with addition and a generator 1), we have also proved that it is cyclic (and generated by some integer d). This shows that every subgroup of an infinite cyclic group is cyclic by A.2.3 and A.2.13.

Let now I be an ideal in $K[X]$. The zero ideal is always principal and generated by 0, so assume that I is nonzero. Let d be a nonzero polynomial in I of the least possible degree. If $f \in I$, we use the division algorithm and get $f = dq + r$, where $-1 \leq \deg r < \deg f$. But $r = f - dq \in I$, so by the definition of d, we have $r = 0$. Hence each element of I is a multiple of d, that is, $I = (d)$ is principal. □

In any ring R, it is possible to consider the notion of divisibility saying that $a \in R$ is **divisible** by $b \in R$ (or b **divides** a) if there is a $c \in R$ such that $a = bc$ (notation: $b \mid a$). However, such a notion only has meaningful properties when R is an integral domain (there are no zero divisors in R and $1 \in R$), which we will assume when discussing divisibility. Notice that $b \mid a$ is equivalent to the inclusion $(a) \subseteq (b)$. We have $(a) = (b)$ if and only if $a = b\varepsilon$ and $b = a\eta$ for some $\varepsilon, \eta \in R$. If R is an integral domain and $ab \neq 0$, then these equalities imply $\varepsilon\eta = 1$. Hence ε and η are units in R. In general, two elements of a ring which differ by a unit factor are called **associated**. We summarize:

Proposition A.3.2. *Two elements $a, b \in R$ in an integral domain R generate the same ideal, that is, $(a) = (b)$, if and only if they are associated. In particular, we have $(a) = R$ if and only if a is a unit.*

A nonzero element $a \in R$ is called **irreducible** if it is not a unit and it is not a product of two non-units. We say that a ring is a **unique factorization domain** (UFD) if it is an integral domain in which every nonzero element which is not a unit can be represented uniquely as a product of irreducible elements. Uniqueness here means that if $r \in R$ and

$$r = a_1 \cdots a_k = b_1 \cdots b_l,$$

where a_i, b_j are irreducible and $k, l \geq 1$, then $k = l$ and it is possible to order the factors in such a way that a_i and b_i are associated.

Notice that the irreducible elements in the ring of integers $\mathbb{Z}$ are exactly the prime numbers (up to sign). In the polynomial rings $K[X]$ over fields K, the irreducible elements are exactly the irreducible polynomials.

Proposition A.3.3. *Any principal ideal ring is a unique factorization domain. In particular, the ring of integers $\mathbb{Z}$ and the polynomial rings $K[X]$, where K is a field, are UFDs.*

We give a proof that polynomial rings $K[X]$ are UFDs in A.4.2 (similar arguments work for the ring of integers $\mathbb{Z}$). For a general proof, see [L], Chap. II, Thm. 5.2.

If $a, b \in R$, then we say that $d \in R$ is a **greatest common divisor** (gcd) of a, b if d divides both these elements and every common divisor of a and b divides d. If d exists, it is often denoted by (a, b) or $\gcd(a, b)$. If $ab \neq 0$ and d exists then it is defined only up to a unit in R. In fact, if d' is also a greatest common divisor of $a, b \in R$, $ab \neq 0$, then by the definition, we have $d \mid d'$ and $d' \mid d$, that is, $(d) = (d')$ (see A.3.2).

In some sense, a dual to the notion of greatest common divisor is the notion of **least common multiple** (denoted by $\operatorname{lcm}(a, b)$ or $[a, b]$ for $a, b \in R$). If R is an integral domain, then a least common multiple of two elements $a, b \in R$ is an element $m \in R$ which is divisible by both these elements and divides every element with this property. As for a greatest common divisor, if m exists, it is unique up to a unit.

Proposition A.3.4. *Let R be a principal ideal domain.*
(a) Any two elements $a, b \in R$ have a greatest common divisor and it is equal to any generator d of the ideal $(a, b) = (d)$.
(b) If $a, b \in R$ and $d = \gcd(a, b)$, then there exist $x, y \in R$ such that $d = ax + by$.
(c) Any two elements $a, b \in R$ have a least common multiple and it is equal to any generator m of the ideal $(a) \cap (b) = (m)$. Moreover, we have $(a, b)((a) \cap (b)) = (ab)$, so that

$$\operatorname{lcm}(a, b) \gcd(a, b) = (ab).$$

Proof. We prove (a) and (b) simultaneously. Consider the ideal (a, b) in R. Since it is principal, there exists d such that $(a, b) = (d)$. Of course, we have $d = ax + by$ for some $x, y \in R$, which proves (b) as soon as we show that $d = \gcd(a, b)$. But $a, b \in (a, b) = (d)$, so $d \mid a$ and $d \mid b$. On the other hand, if $d' \mid a$ and $d' \mid b$, then the equality $d = ax + by$ shows that $d' \mid d$. Hence d really is a greatest common divisor of a, b.
(c) The ideal $(a) \cap (b)$ is principal and if m is its generator, then $(m) \subseteq (a)$ and $(m) \subseteq (b)$, so $a \mid m$ and $b \mid m$. On the other hand, if $a \mid m'$ and $b \mid m'$ for some $m' \in R$, then $(m') \subseteq (a)$ and $(m') \subseteq (b)$, so $(m') \subseteq (a) \cap (b) = (m)$. This implies that $m \mid m'$. Thus, m is a least common multiple of a and b.

Now we prove a formula relating $\operatorname{lcm}(a, b)$ to $\gcd(a, b)$. Assume that $ab \neq 0$ (if $ab = 0$ the equality can be checked immediately) and let

$$d' = \frac{ab}{\operatorname{lcm}(a, b)}.$$

We want to show that $d' = d$ (up to an invertible element). The equality above implies that:

$$\frac{a}{d'} = \frac{\operatorname{lcm}(a,b)}{b} \quad \text{and} \quad \frac{b}{d'} = \frac{\operatorname{lcm}(a,b)}{a},$$

so d' divides both a and b (since on the right-hand sides are elements of R). Thus d' as a common divisor to a and b divides the greatest common divisor d. Conversely, both a and b divide $\frac{ab}{d}$, since it is both a multiple of a (as $a\frac{b}{d}$) and b (as $\frac{a}{d}b$). Thus $\operatorname{lcm}(a,b)$ divides $\frac{ab}{d}$, which means that the quotient $\frac{ab}{d}/\operatorname{lcm}(a,b) = \frac{d'}{d}$ is an element of R. Hence, we get that d divides d'. Since both $d' \mid d$ and $d \mid d'$, these two elements are associated. □

Sometimes, the elements x, y in A.3.4(b) may be found effectively. In particular, it is possible when R is the ring of integers or a polynomial ring $K[X]$ over a field K using the Euclidean algorithm. Since the last nonzero remainder in the Euclidean algorithm for $a, b \in R$ is a greatest common divisor of these two elements, it is possible to express it as a linear combination of a, b by tracing back all the steps of this procedure. In these two most common cases, that is, $R = \mathbb{Z}$ or $R = K[X]$, both $\gcd(a,b)$ and $\operatorname{lcm}(a,b)$ are chosen in a unique way. In $\mathbb{Z}$, the units are ± 1, and the standard choice is to define $\gcd(a,b)$ and $\operatorname{lcm}(a,b)$ as positive integers (if $ab \neq 0$). In $R = K[X]$ the units are all nonzero constants and in any class of associated elements, it is possible to choose monic polynomial (that is, with leading coefficient 1). Thus $\gcd(a,b)$ and $\operatorname{lcm}(a,b)$ are defined uniquely as monic polynomials (if $ab \neq 0$).

The following application of A.3.4(b) is in the case of $R = \mathbb{Z}$ sometimes called the Fundamental theorem of arithmetic (as it implies the uniqueness of factorization of the integers bigger than 1 as products of prime numbers):

Proposition A.3.5. *Let R be a principal ideal domain and $p \in R$ an irreducible element. If $p \mid ab$, then $p \mid a$ or $p \mid b$.*

Proof. Take $d = \gcd(a,p)$. Since $d \mid p$, we have $d = 1$ or $d = p$ (up to a unit factor). In the second case, the result is true, while in the first case, we have $1 = ax + py$ by A.3.4(b). Hence, we have $b = abx + pby$, which shows that $p \mid b$ (since $p \mid ab$). □

Notice that A.3.5 shows that irreducible elements in principal ideal domains have the fundamental property of prime numbers in the ring of integers (or the irreducible polynomials in the polynomial rings). This property says that if a prime number (an irreducible polynomial) divides a product, then it must divide some of the factors of this product.

We return to arbitrary rings and their ideals. Any ideal I in a ring R is a subgroup of the additive group R, which by definition is abelian. Therefore, we can form the quotient group R/I with respect to addition, since I is a normal subgroup of the additive group R. But R/I inherits a ring structure from R when multiplication of the cosets $a + I, b + I$, $a, b \in R$ is defined by the equality $(a + I)(b + I) = ab + I$. Here it is necessary to know that the result of multiplication of $a + I$ and $b + I$

does not depend on the representation of cosets by the representatives a, b. In fact, we may have $a + I = a' + I$ and $b + I = b' + I$ for some $a', b' \in R$. Then our multiplication is correctly defined if $ab + I = a'b' + I$. But $a + I = a' + I$ if and only if $a' - a \in I$ and similarly, $b + I = b' + I$ if and only if $b' - b \in I$ according to the properties of cosets in groups. Thus $a' = a + i$ and $b' = b + j$, where $i, j \in I$ and $a'b' - ab = ib + ja + ij \in I$, so $a'b' + I = ab + I$. The ring R/I is called the **quotient ring** of R modulo I. As for groups, we have a natural surjection $\iota : R \to R/I$ mapping an element $a \in R$ onto the coset $a + I \in R/I$. This mapping is not only a group homomorphism, but also a ring homomorphism, since $\iota(ab) = ab + I = (a + I)(b + I) = \iota(a)\iota(b)$. Its kernel is the set of $a \in R$ such that $\iota(a) = a + I = I$, that is, $a \in I$. Thus $\mathrm{Ker}(\iota) = I$, which shows that every ideal I really is a kernel of a ring homomorphism—namely, the kernel of the natural surjection $\iota : R \to R/I$.

Example A.3.6. If $R = \mathbb{Z}$ and $I = (n)$ is the ideal of all multiples of $n \in \mathbb{Z}$, then the quotient ring $\mathbb{Z}/(n)$ consists of the same elements (cosets) as the quotient group $\mathbb{Z}/\langle n \rangle$ and coincides with $\mathbb{Z}_n$ (see A.1.2 and A.2.9). Thus, $\mathbb{Z}_n$ also has a ring structure (addition and multiplication modulo n).

If $\varphi : R \to R'$ is any ring homomorphism and $I = \mathrm{Ker}(\varphi)$ its kernel, then there is a similar relation between the quotient R/I and the image $\varphi(R)$ (often denoted by $\mathrm{Im}(\varphi)$) as for group homomorphisms, since φ is a group homomorphism of R into R' considered as additive groups and I is its kernel. The only new ingredient is the presence of multiplication and the mapping $\varphi^* : R/I \to R'$, where $\varphi^*(a+I) = \varphi(a)$ is not only a group isomorphism of the quotient R/I onto the image $\varphi(R)$ but also a ring isomorphism. In order to check this, we only need to note that

$$\varphi^*((a + I)(b + I)) = \varphi^*(ab + I) = \varphi(ab) = \varphi(a)\varphi(b) = \varphi^*(a + I)\varphi^*(b + I).$$

Thus we can summarize these facts in what is called the **fundamental theorem on ring homomorphisms**:

A.3.7 Fundamental Theorem on Ring Homomorphisms. *If $\varphi : R \to R'$ is a ring homomorphism, then the kernel $I = \mathrm{Ker}(\varphi)$ is an ideal of R and the function $\varphi^*(a + I) = \varphi(a)$ is a ring isomorphism of $R/\mathrm{Ker}(\varphi)$ and $\mathrm{Im}(\varphi)$.*

As in the case of group homomorphisms, Theorem A.3.7 is a special case of a more general result in which a ring homomorphism induces a homomorphism of quotient rings. Let I be an ideal in a ring R, and I' an ideal in a ring R'. We say that a ring homomorphism $\varphi : R \to R'$ is a **homomorphism of pairs** if $\varphi(I) \subseteq I'$—the pair $I \subseteq R$ is mapped into the pair $I' \subseteq R'$. As in the case of groups A.2.12, we have:

A.3.8 Ring Homomorphism of Pairs. *Let $\varphi : R \to R'$ be a ring homomorphism of the pair $I \subseteq R$ into the pair $I' \subseteq R'$. Then there exists an induced homomorphism*

$$\varphi^* : R/I \to R'/I'$$

such that $\varphi^*(r + I) = \varphi(r) + I'$. Moreover, we have $\mathrm{Ker}(\varphi^*) = \varphi^{-1}(I')/I$ and $\mathrm{Im}(\varphi^*) = \varphi^*(R/I) = (\varphi(R) + I')/I'$.

The proof of this result follows exactly the proof of A.2.12 plus an easy argument concerning the behaviour of φ^* with respect to the multiplication in the rings R/I and R'/I'. The following property of ring homomorphisms is similar to the property of group homomorphisms considered in A.2.13 and its proof follows exactly the same pattern:

Proposition A.3.9. *Let $\varphi : R \to R'$ be a ring homomorphism with kernel $I = \mathrm{Ker}(\varphi)$.*
(a) The image $\varphi(S)$ of a subring S of R is a subring of R' and the inverse image $\varphi^{-1}(S')$ of any subring S' of R' is a subring of R containing I. We have $\varphi(\varphi^{-1}(S')) = S'$ and $\varphi^{-1}(\varphi(S)) = S + I$. The inverse image $\varphi^{-1}(I')$ is an ideal in R if I' is an ideal in R'.
(b) If φ is surjective (so $R' \cong R/I$), then there is a one-to-one correspondence between the ideals J' of R' and the ideals J of R containing I in which J' corresponds to the inverse image $\varphi^{-1}(J')$ and J to its image $\varphi(J)$.

If R is a ring and I, J its ideals, then it is possible to form their sum $I + J$ and product IJ. The sum is simply the set of all sums $i + j$, where $i \in I$ and $j \in J$. It is easy to check that all such sums form an ideal. The product IJ is defined as the set of all finite sums of products ij, where $i \in I$ and $j \in J$. Here it is also evident that such finite sums of products form an ideal. If R is a principal ideal ring and $I = (a)$, $J = (b)$, then $I + J = (a, b) = (\gcd(a, b))$ and $IJ = (ab)$.

Two ideals I, J in any ring are called **relatively prime** if $I+J = R$. This condition simply means that $1 = i+j$, where $i \in I$ and $j \in J$ (since every ideal which contains 1 is equal to R). In general, if $I + J = R$, then $I \cap J = IJ$. In fact, it is evident that $IJ \subseteq I \cap J$. But if $1 = i+j$ for some $i \in I, j \in J$ and $x \in I \cap J$, then $x = ix + xj \in IJ$, so we also have $I \cap J \subseteq IJ$.

Notice that if I is relatively prime to J and J', then it is also relatively prime to JJ'. In fact, $I+J = R$ and $I+J' = R$ imply that $i+j = 1$ and $i'+j' = 1$ for $i, i' \in I$, $j \in J, j' \in J'$. Thus, we have $ii' + ij' + i'j + jj' = 1$. Take $i'' = ii' + ij' + i'j \in I$ and $jj' \in JJ'$. Then $i'' + jj' = 1$, that is, $I + JJ' = R$. In several places, we use the following isomorphism:

Proposition A.3.10. *If I, J are relatively prime ideals in a ring R, then the homomorphism $\varphi : R \to R/I \times R/J$ such that $\varphi(r) = (r + I, r + J)$ induces an isomorphism $\varphi^* : R/(I \cap J) \to R/I \times R/J$.*

Proof. It is easy to check that φ is a homomorphism and that $\mathrm{Ker}(\varphi) = I \cap J$ ($\varphi(r) = (r + I, r + J) = (0, 0)$ if and only if $r + I = I$ and $r + J = J$, that is, $r \in I$ and $r \in J$).

We show that φ is surjective. Since $I + J = R$, we have $1 = i + j$ for some $i \in I$ and $j \in J$. Notice that $1 + I = j + I$ and $1 + J = i + J$. Hence, for $(x + I, y + J)$ in $R/I \times R/J$, we can take $r = xj + yi$ and then

$$\begin{aligned}\varphi(r) &= (xj + yi + I, xj + yi + J) = (xj + I, yi + J) = ((x + I)(j + I), (y + J)(i + J)) \\ &= ((x + I)(1 + I), (y + J)(1 + J)) = (x + I, y + J),\end{aligned}$$

so φ is surjective and by A.3.7, we get the induced isomorphism φ^*. □

Example A.3.11. Take $R = \mathbb{Z}, I = (a)$ and $J = (b)$ where a, b are relatively prime, so that $I + J = \mathbb{Z}$ by A.3.4(a). Thus, we have $I \cap J = IJ = (ab)$ and A.3.10 gives

$$\mathbb{Z}/(ab) \cong \mathbb{Z}/(a) \times \mathbb{Z}/(b),$$

where the isomorphism maps the residue of an integer n modulo ab to the pair of its residues (n_1, n_2) modulo a and b. In more compact notation, we write $\mathbb{Z}_{ab} \cong \mathbb{Z}_a \times \mathbb{Z}_b$. If $n = p_1^{k_1} \cdots p_r^{k_r}$ is a factorization of an integer into a product of different prime numbers $p_1, \ldots, p_k$, then applying the formula above several times, we get a very useful isomorphism of rings:

$$\mathbb{Z}_n \cong \mathbb{Z}_{p_1^{k_1}} \times \cdots \times \mathbb{Z}_{p_r^{k_r}}.$$

Taking units in these rings, we also have an isomorphism of groups, which we use several times (see A.1):

$$\mathbb{Z}_n^* \cong \mathbb{Z}_{p_1^{k_1}}^* \times \cdots \times \mathbb{Z}_{p_r^{k_r}}^*.$$

Notice that the groups $\mathbb{Z}_{p^k}^*$ are cyclic when p is an odd prime number, while $\mathbb{Z}_{2^k}^*$ is a product of a cyclic group of order 2^{k-2} by a cyclic group of order 2 when $k > 2$ (see [K], Prop. 1.4.2 and 1.4.3). Of course, the groups $\mathbb{Z}_{2^k}^*$ for $k = 1, 2$ are cyclic.

Similarly, if $R = K[X]$ is a polynomial ring, $I = (f(X))$ and $J = (g(X))$, where $f(X), g(X)$ are relatively prime polynomials, then as above, we have $I + J = K[X]$ and A.3.10 gives this time

$$K[X]/(fg) \cong K[X]/(f) \times K[X]/(g). \tag{A.2}$$

A.3.12 Characteristic of a Ring. If R is a commutative ring with identity, then the additive subgroup $\langle 1 \rangle$ generated by the identity consists of multiples $k \cdot 1$ for $k \in \mathbb{Z}$. If the order of this group is infinite, then it is isomorphic with the group of integers $\mathbb{Z}$. In this case, we say that the characteristic of R is zero. If 1 generates a finite group of order n, then $\langle 1 \rangle = \{0 \cdot 1, 1 \cdot 1, \ldots, (n-1) \cdot 1\}$ and $n \cdot 1 = 0$. This group is isomorphic to $\mathbb{Z}_n$, the group of residues modulo n. In this case, we say that R has characteristic n. Observe that n is the order of 1 in the additive group of the ring R in this case. The characteristic of a commutative ring R with identity is sometimes denoted by char(R). Notice that the multiples $k \cdot 1$ for $k \in \mathbb{Z}$ form a subring of R, so a commutative ring R with identity contains a subring isomorphic to the integers $\mathbb{Z}$ if its characteristic is 0, and a subring isomorphic to $\mathbb{Z}_n$ if its characteristic is n. In particular, the ring of integers $\mathbb{Z}$ has characteristic 0, and the ring $\mathbb{Z}_n = \{0, 1, \ldots, n-1\}$ of residues modulo n has characteristic n. For convenience of references, we record these facts:

Proposition A.3.13. *The characteristic of a commutative ring with identity is either* 0 *or a positive integer n. The identity* $1 \in R$ *generates a subring isomorphic to the integers $\mathbb{Z}$ in the first case, and a subring isomorphic to $\mathbb{Z}_n$ in the second case.*

Notice that if the number n is composite, that is, $n = kl$, where $k, l > 1$, then the ring $\mathbb{Z}_n$ has zero divisors, since $k \neq 0$ and $l \neq 0$ but $kl = 0$. If the number $n = p$ is a prime, then $\mathbb{Z}_p$ is without zero divisors, since $kl = 0$ in this ring means that $p \mid kl$, which implies that $p \mid k$ or $p \mid l$, that is, $k = 0$ or $l = 0$ in $\mathbb{Z}_p$.

Polynomial Rings

If R is a commutative ring, then the ring of polynomials with coefficients in R is the set of all expressions $a_0 + a_1X + a_2X^2 + \cdots$ where $a_i \in R$, almost all a_i are 0 and X is a symbol (called a variable), which are added and multiplied as polynomials with number coefficients. This ring is denoted by $R[X]$. The construction may be used inductively—we can construct a polynomial ring over $R[X]$ and so on. Thus $R[X][Y] = R[X, Y]$ is the polynomial ring in two variables over R. Continuing in this way, we get polynomial rings $R[X_1, \ldots, X_n]$ over R in n variables $X_1, \ldots, X_n$.

The most common rings in this book are polynomial rings over fields and, in particular, the polynomial rings $K[X]$ in one variable over fields K. In this case, we have the well-known division algorithm which for $f(X), g(X) \in K[X]$, with $g(X) \neq 0$, gives the unique quotient $q(X) \in K[X]$ and the remainder $r(X) \in K[X]$ such that $f(X) = g(X)q(X) + r(X)$ and $\deg r(X) < \deg g(X)$.

A.4.1 Factor Theorem. *Let $K \subseteq L$ be a field extension.*
(a) The remainder of $f \in K[X]$ divided by $X - a$, $a \in L$, is equal to $f(a)$.
(b) An element $a \in L$ is a zero of $f \in K[X]$ if and only if $X - a | f(X)$ (in $L[X]$).

Proof. (a) The division algorithm for polynomials gives

$$f(X) = (X - a)q(X) + r,$$

where $\deg r < 1$, that is, the remainder r is a constant polynomial. Thus, taking $X = a$, we get $f(a) = r$.
(b) Using (a), we have $f(a) = 0$ if and only if $r = f(a) = 0$. □

The polynomial rings $K[X]$, like the ring of integers $\mathbb{Z}$, are unique factorization domains (UFDs, see p. 248). This property is shared by all principal ideal domains (see A.3.3).

The unique factorization property also holds in the polynomial rings over the ring of integers and over fields, since in general it is true that if a ring R is a UFD, then the polynomial ring $R[X]$ is also a UFD (see [L], Chap. IV, Thm. 2.3). Below, we give a proof that the polynomial rings over fields are unique factorization domains:

Theorem A.4.2. *Let K be a field. Every polynomial of degree ≥ 1 in $K[X]$ is a product of irreducible polynomials. If*

$$f = p_1 \cdots p_k = p'_1 \cdots p'_l,$$

where p_i and p'_i are irreducible polynomials, then $k = l$ and with a suitable numbering of the factors p_i, p'_j, we have $p'_i = c_i p_i$, where $c_i \in K$.

Proof. First we prove by induction that every polynomial $f(X)$ of degree at least one is a product of irreducible polynomials. It is clear for polynomials of degree one (they are irreducible). Assume that we have proved that every polynomial of degree less than $n > 1$ is a product of irreducible polynomials. Take an arbitrary polynomial $f(X)$ of degree n. If $f(X)$ is irreducible, we have what we want. If $f(X)$ is reducible, then $f(X) = g(X)h(X)$, where $1 \leq \deg g < n$ and $1 \leq \deg h < n$, so both g, h are products of irreducible polynomials. Thus $f(X)$ is also such a product.

Now consider two factorizations of $f(X)$ given in the theorem:

$$p_1 \cdots p_k = p'_1 \cdots p'_l,$$

where all p_i, p'_j are irreducible. We prove the theorem by induction with respect to $m = k + l$. If $m = 2$, then we have one factor to the left and one to the right, so the claim that $k = l$ is true (and the factors are equal). Assume that the theorem is true when the number of factors p_i, p'_j is less than $m \geq 2$. Consider the case when the number of factors is m. The irreducible polynomial p_k divides the product on the right-hand side. Hence p_k must divide at least one of the factors of this product (see A.3.5). Say that $p_k | p'_l$ (we can change the numbering of the factors if necessary). But both these polynomials are irreducible, so $p'_l = c_k p_k$ for some constant c_k. We divide both sides by p_k and get

$$p_1 \cdots p_{k-1} = c_k p'_1 \cdots p'_{l-1},$$

so the number m is now $k + l - 2$ and by our inductive assumption the theorem is true. Thus $k - 1 = l - 1$, that is, $k = l$ and by a suitable numbering of the factors, we get $p'_i = c_i p_i$ for $i = 1, \ldots, k - 1$. □

Fields

A field is a commutative ring with identity in which nonzero elements form a group with respect to multiplication. Notice that zero divisors are absent in fields, so any field is an integral domain (if a is invertible with inverse a^{-1} and $ab = 0$, then $a^{-1}ab = b = 0$). Any field contains at least two elements: 0 and 1. A subring of a field K which is also a field (with respect to the same addition and multiplication) is called a **subfield** of K (it is called **proper** if it is not equal to K). In order to check that K' is a subfield of K it suffices to check that $a, b \in K'$ imply $a+b, ab, -a, a^{-1} \in K'$ (a^{-1} for $a \neq 0$).

Among all fields considered in this text, a special role is played by the field of rational numbers $\mathbb{Q}$ and the fields of residues modulo prime numbers $\mathbb{Z}_p$. These fields are very often denoted by $\mathbb{F}_p$ and we shall use this notation. The fact that

$\mathbb{F}_p$ is a field follows from a general property of finite integral domains—any finite integral domain is a field. In fact, if R is a finite integral domain and $a \in R$, $a \neq 0$, then the powers a^n for $n = 1, 2, \ldots$ are nonzero and cannot be different, so there exist exponents k, l such that $l > k$ and $a^l = a^k$. Hence, we have $a^k(a^{l-k} - 1) = 0$, which implies $a^{l-k} = 1$ as $a^k \neq 0$. Thus a has an inverse a^{l-k-1} in R. Notice that $\mathbb{Q}$ has characteristic 0 (the order of 1 in the additive group of $\mathbb{Q}$ is infinite) and $\mathbb{F}_p$ has characteristic p (the order of 1 with respect to addition is p)—see A.3.12 and A.3.13.

Theorem A.5.1. *(a) The characteristic of a field is* 0 *or a prime number.*
(b) Any field K of characteristic 0 *contains a unique subfield isomorphic to the rational numbers $\mathbb{Q}$, and any field of characteristic p contains a unique subfield isomorphic to $\mathbb{F}_p$.*

Proof. (a) The characteristic of any commutative ring with identity is either 0 or a positive integer n by A.3.13. If the characteristic of a field K is not 0, then it is n for some positive integer n and K contains a subring isomorphic to $\mathbb{Z}_n$, also by A.3.13. But if n is composite (that is, $n = kl, k < n, l < n$), then $\mathbb{Z}_n$ contains zero divisors ($kl = 0$, $k \neq 0, l \neq 0$), which are absent in fields (see p. 255). Thus n must be a prime.
(b) If a field K has characteristic 0, then the multiples $k \cdot 1$, $k \in \mathbb{Z}$ form a subring of K isomorphic to $\mathbb{Z}$. Since every quotient $(k \cdot 1)/(l \cdot 1)$, where $l \neq 0$, must belong to K, this field contains a subfield isomorphic to the rational numbers $\mathbb{Q}$. If K has characteristic p, then K contains a subfield isomorphic to $\mathbb{F}_p$. Such subfields of K are unique, since every subfield contains 1, so it must contain the subfield generated by this element. In fact, the subfield generated by 1 is the intersection of all subfields of K (it is contained in any subfield). □

The fields $\mathbb{Q}$ and $\mathbb{F}_p$ are called **prime**. They are the smallest ones in the sense that every field contains (an isomorphic copy) of exactly one of them and they do not contain any proper subfields.

The fields which we use to exemplify the Galois theory in this book are mainly number fields, that is, subfields of the field of complex numbers $\mathbb{C}$, finite fields (see Chap. 5) and fields of rational functions with coefficients in number fields or finite fields. The latter are fields of fractions of polynomial rings just as the rational numbers are the field of fractions of the ring of integers. Both these cases are covered by a very useful construction of the field of fractions of any integral domain. We describe this construction now.

Take any integral domain R (for example, the ring of integers $\mathbb{Z}$). Consider the set $K = R_0$ of all pairs (a, b) such that $a, b \in R, b \neq 0$. We think of a as numerator and b as denominator of a fraction. We say that (a, b) and (c, d) are equivalent if $ad = bc$ (both these pairs define the same fraction). This equivalence can be denoted in some way, for example as $(a, b) \sim (c, d)$. Denote by a/b the set of all pairs equivalent to (a, b). Such a class will be called a fraction. Now we define addition and multiplication of fractions in such a way that we get a field:

$$\frac{a}{b} + \frac{c}{d} = \frac{ad + bc}{bd}, \qquad \frac{a}{b}\frac{c}{d} = \frac{ac}{bd}.$$

Some minor formal calculations are needed to show that the addition and multiplication of fractions give the same result independently of the presentation of the involved fractions by the pairs of elements of R (for example, we have to show that if $(a, b) \sim (a', b')$ and $(c, d) \sim (c', d')$, then $(ad + bc, bd) \sim (a'd' + b'c', b'd')$). When this is done, we have to check that the fractions really form a field and this also needs some very simple (but also tedious) computations. Anyway, we check that we get a field in which R can be naturally embedded by mapping $a \in R$ onto the fraction $a/1$. In fact, in this way, we usually consider R as a subring of its **field of fractions** $K = R_0$. Of course, if $R = \mathbb{Z}$, then $\mathbb{Z}_0 = \mathbb{Q}$ is the field of rational numbers. If $R = K[X]$ is a polynomial ring over a field K, then $R_0 = K(X)$ is the field of rational functions with coefficients in K.

Notice that the field of fractions of R is very often called the quotient field of R (meaning the field of quotients a/b rather than fractions a/b). There is a little danger that using "quotient ring" in this sense may be confused with quotients of rings modulo ideals. We use the term "field of fractions" in order to avoid such misunderstandings.

In many places in the book, we consider finite subgroups of fields. The most important property of such subgroups is that they are cyclic:

Proposition A.5.2. *Any finite subgroup of the multiplicative group of a field K is cyclic.*

Proof. Let G be a subgroup of order n of the multiplicative group of a field K. Let the exponent of G be m (see A.2.14), that is, $x^m = 1$ for every element $x \in G$. Since the equation $X^m - 1 = 0$ has at most m solution in the field K and every element of G satisfies this equation, we have $n \leq m$. But as we know (see A.2.14), the exponent m is the maximal order of the elements of G and $m \mid n$. Hence $m = n$ and the group G has an element of order n, that is, G is cyclic. □

Notice that as a special case, the last result says that the groups of nonzero residues $(\mathbb{Z}/p\mathbb{Z})^*$ modulo a prime number p are cyclic as the groups of nonzero elements in the fields $\mathbb{F}_p = \mathbb{Z}/p\mathbb{Z}$ (the order of this group is $p - 1$). More generally, for any positive integer n, the group of all nth roots of unity in any field (see p. 59), that is, the group of all solutions of the equation $X^n = 1$, is cyclic, since it is a finite group. In the particular case of the field of complex numbers, all solutions of the equation $X^n = 1$, that is, all nth roots of unity, form a group of order n. This group (often denoted by U_n or C_n) consists of the numbers $\varepsilon_k = e^{\frac{2\pi ik}{n}} = \cos\frac{2\pi k}{n} + i\sin\frac{2\pi k}{n}$ for $k = 0, 1, \ldots, n-1$. This cyclic group has $\varphi(n)$ generators given by ε_k for k relatively prime to n (see A.2.2 and p. 279 for the definition of Euler's function $\varphi(n)$). The usual choice of generator is ε_1, for which we have $\varepsilon_k = \varepsilon_1^k$.

A.5.3 Maximal Ideals and Fields. Now, we will discuss ideals in rings for which the quotient rings R/I are fields. First notice the following simple fact:

Proposition A.5.4. *A ring with identity has only the trivial ideals* (0) *and R if and only if it is a field.*

Proof. If R is a field and I is a nonzero ideal, then there is an $a \in I$ such that $a \neq 0$. Thus a has an inverse, that is, there is an $a' \in R$ such that $aa' = 1$. Hence $1 \in I$, so $I = R$.

If R is a ring whose only ideals are (0) and R, then for any $a \in R$, $a \neq 0$, we can take the principal ideal Ra. This ideal is nonzero, since $a = a \cdot 1 \in I$. Hence $I = R$. This means that $1 = a'a$ for some $a' \in R$, so a has an inverse in R. Hence R is a field. □

An ideal I in a ring R is called **maximal** if it is not equal to R and if an ideal J is such that $I \subseteq J \subseteq R$, then $J = I$ or $J = R$. A very important property of maximal ideals, which we use on some occasions, is the following result:

Proposition A.5.5. *An ideal I in a ring R with identity is maximal if and only if R/I is a field.*

Proof. If R/I is a field, then there are no nontrivial ideals in this ring (see A.5.4), so the ideal I is maximal by A.3.9(b). Conversely, if I is a maximal ideal, then A.3.9(b) says that R/I is without proper ideals, so it is a field by A.5.4. □

We apply A.5.5 mostly when R is the ring of integers or the ring of polynomials over a field. As we know, all ideals in $\mathbb{Z}$ and $K[X]$ (K a field) are principal (see A.3.1). For such rings, we have:

Proposition A.5.6. *In a principal ideal ring, the maximal ideals are exactly those generated by irreducible elements.*

Proof. If R is a principal ideal ring and (a) an ideal generated by an irreducible element a (see the definition of irreducible elements on p. 248), then the ideal is maximal. In fact, if $(a) \subseteq (b) \subset R$, then $a \in (b)$ gives $a = bc$, where $c \in R$. Hence, we have that b or c must be a unit in R. But b is not a unit (otherwise $(b) = R$), so c is a unit, which means that $(a) = (b)$ (see A.3.2). Thus (a) is maximal.

Conversely, if (a) is maximal, then a must be irreducible, since otherwise $a = bc$, where both b and c are not units. Hence, we have $(a) \subset (b) \subset R$, since $(a) = (b)$ implies that c is a unit, whereas $(b) = R$ gives that b is a unit (see A.3.2). □

In the ring $\mathbb{Z}$, the irreducible elements are prime numbers (up to sign), so all maximal ideals are the ideals (p), where p is a prime number. In the polynomial rings $K[X]$, K a field, the irreducible elements are exactly all irreducible polynomials $f(X)$, so maximal ideals are exactly those generated by irreducible polynomials. Thus each quotient $\mathbb{Z}/(p)$ is a field (with p elements). Similarly, each quotient ring $K[X]/(f(X))$ is a field when $f(X)$ is an irreducible polynomial. We record this particular case as we refer to it several times:

Proposition A.5.7. *(a) The quotient ring $\mathbb{Z}/(n)$ is a field if and only if n is a prime number.*
(b) The quotient ring $K[X]/(f(X))$ is a field if and only if $f(X)$ is an irreducible polynomial.

If $K[X]$ is a polynomial ring over a field K and $I = (f(X))$ an ideal generated by a nonconstant polynomial $f(X) \in K[X]$, then we write $[g(X)] = g(X) + I$ to denote the class of $g(X)$ in the quotient ring $K[X]/(f(X))$. We have, as usual, $[g(X)] = [h(X)]$ if and only if $g(X)-h(X) \in I = (f(X))$, that is, $g(X)-h(X)$ is a multiple of $f(X)$. Thus, for every class $[g(X)]$, we have $[g(X)] = [r(X)]$, where $g(X) = f(X)q(X) + r(X)$ with a unique remainder $r(X)$ such that $\deg(r(X)) < \deg(f(X))$. If $f(X)$ has degree n, then $r(X)$ has degree at most $n-1$. Of course, we have $[f(X)] = [0]$.

If $a \in K$ is a constant polynomial, then its class $[a]$ may be simply identified with a, since for classes of the constant polynomials, we have $[a] = [b]$ if and only if $f(X)$ (a nonconstant polynomial) divides $a - b$ (a constant), so it must be $a = b$. We will always identify $[a]$ with a for constants so that K will be considered as a subring of the quotient $K[X]/(f(X))$. If we denote $[X]$ by α and if $f(X) = a_nX^n + \cdots + a_1X + a_0$ is a polynomial of degree n, then $0 = [f(X)] = a_n[X]^n + \cdots + a_1[X] + a_0 = a_n\alpha^n + \cdots + a_1\alpha + a_0$, that is, we have $f(\alpha) = 0$, which means that $f(X)$ has a zero α in the quotient $K[X]/(f(X))$. For any class $[r(X)]$ in $K[X]/(f(X))$, where $r(X) = b_{n-1}X^{n-1} + \cdots + b_1X + b_0$, we have $[r(X)] = b_{n-1}[X]^{n-1} + \cdots + b_1[X] + b_0 = b_{n-1}\alpha^{n-1} + \cdots + b_1\alpha + b_0$. We use the above description of the quotient $K[X]/(f(X))$ in several proofs, so we record it for reference:

Proposition A.5.8. *Let $f(X) = a_nX^n + \cdots + a_1X + a_0 \in K[X]$ be a nonconstant polynomial of degree n. Then every element of the quotient $K[X]/(f(X)) = K[\alpha]$ has a unique representation as a polynomial expression $b_{n-1}\alpha^{n-1} + \cdots + b_1\alpha + b_0 \in K[\alpha]$, where $\alpha = [X]$ is a zero of $f(X)$ and $b_i \in K$.*

As an example, we have $\mathbb{R}[X]/(X^2 + 1) = \mathbb{R}[\alpha]$, where $\alpha^2 + 1 = 0$ and every element has a unique representation as $a + b\alpha$, $a, b \in \mathbb{R}$. Thus the quotient is the field of complex numbers as $\alpha^2 = -1$.

In particular, when $K = \mathbb{F}$ is a finite field with q elements and $f(X)$ is a polynomial of degree n, then $\mathbb{F}[X]/(f(X)) = \mathbb{F}[\alpha]$ is a ring with q^n elements $b_{n-1}\alpha^{n-1} + \cdots + b_1\alpha + b_0$ (we have q possible choices for each $b_i \in \mathbb{F}$, $i = 0, 1, \ldots, n-1$). Of course, we have $f(\alpha) = 0$. As we know, this ring $\mathbb{F}[X]/(f(X))$ is a field exactly when $f(X)$ is irreducible in $\mathbb{F}[X]$.

Modules over Rings

If R is a ring and M is an abelian group, then we say that M is a left module over R if for every pair $(r, m) \in R \times M$, we have an element $rm \in M$ such that
(a) $r(m_1 + m_2) = rm_1 + rm_2$,
(b) $(r_1 + r_2)m = r_1m + r_2m$,
(c) $(r_1r_2)m = r_1(r_2m)$,
(d) $1m = m$,
where $r, r_1, r_2 \in R$ and $m, m_1, m_2 \in M$. A right module is defined in a similar way. If $R = K$ is a field, then K-modules are called vector spaces or linear spaces (over the field K) and their elements are called vectors. A **homomorphism** of modules over R is a function $\varphi : M \to M'$ satisfying:
(a) $\varphi(m_1 + m_2) = \varphi(m_1) + \varphi(m_2)$,
(b) $\varphi(rm) = r\varphi(m)$,

when $m, m_1, m_2 \in M$ and $r \in R$. The kernel of φ is its kernel as a homomorphism of abelian groups, that is, $\mathrm{Ker}(\varphi) = \{m \in M | \varphi(m) = 0\}$. The kernel is also a submodule of M, that is, if $r \in R$ and $m \in \mathrm{Ker}(\varphi)$, then $rm \in \mathrm{Ker}(\varphi)$, since $\varphi(rm) = r\varphi(m) = 0$ if $\varphi(m) = 0$.

If M_1, M_2 are R-modules, then the set of pairs (m_1, m_2), where $m_1 \in M_1$ and $m_2 \in M_2$, form a module when the addition of pairs and multiplication by elements of R are defined on coordinates: $(m_1, m_2) + (m_1', m_2') = (m_1 + m_1', m_2 + m_2')$ and $r(m_1, m_2) = (rm_1, rm_2)$. This new module is denoted by $M_1 \times M_2$ and called the **direct sum** (or **direct product**) of M_1, M_2. This definition can be extended to an arbitrary finite number of modules (it is also possible to consider infinite families but the definitions of sum and product are then different).

In this book, we need some knowledge of modules over fields, the ring of integers and polynomial rings. We say that a module M is finitely generated if there are elements $m_1, \ldots, m_k \in M$ such that every element $m \in M$ can be expressed as a linear combination of these elements, that is, if there are $r_1, \ldots, r_k \in R$ such that $m = r_1m_1 + \cdots + r_km_k$. We write $M = \langle m_1, \ldots, m_k\rangle$. If such a representation of m is unique, then $m_1, \ldots, m_k$ is called a **basis** of M. If there is one element $m \in M$ such that $M = \langle m\rangle$, then M is called a **cyclic module**. Notice that if $M = \langle m\rangle$ is cyclic, then we have a surjective homomorphism $\varphi : R \to M$ such that $\varphi(r) = rm$. If I denotes the kernel of φ, then $R/I \cong M$. Thus a cyclic module can also be defined as a module isomorphic to a module R/I. If $R = \mathbb{Z}$ is the ring of integers, than the cyclic modules are $\mathbb{Z}/(n) = \mathbb{Z}_n$, which are finite cyclic groups when $n > 0$ or the infinite cyclic group $\mathbb{Z}$ when $n = 0$. This explains the terminology.

If M is a left module over a ring R, then the **annihilator** of M is the set of all $r \in R$ such that $rM = 0$ (that is, $rm = 0$ for each $m \in M$). The annihilator of M is an ideal in R, which is denoted by $\mathrm{Ann}_R(M)$. In fact, if $r_1, r_2 \in \mathrm{Ann}_R(M)$, that is, $r_1M = r_2M = 0$, then $(r_1 - r_2)M = 0$, so $r_1 - r_2 \in \mathrm{Ann}_R(M)$. Similarly, if $r \in \mathrm{Ann}_R(M)$, that is, $rM = 0$, then for any $r' \in R$, we also have $(r'r)M = 0$, so $r'r \in \mathrm{Ann}_R(M)$. Notice that the annihilator of the R-module $M = R/(a)$ is equal to (a). In fact, it is clear that $aM = 0$, so $a \in \mathrm{Ann}_R(M)$. On the other hand, if $r \in \mathrm{Ann}_R(M)$, that is, $rM = 0$, then in particular $r(1 + (a)) = r + (a) = 0$, so $r \in (a)$. Hence $\mathrm{Ann}_R(M) = (a)$. In the particular case when $M = R = R/(0)$, we have $\mathrm{Ann}_R(R) = (0)$.

Theorem A.6.1. *If R is a principal ideal ring, then every finitely generated module M over R is a finite direct sum of cyclic R-modules. Moreover,*

$$M = R/(a_1) \times R/(a_2) \times \cdots \times R/(a_r),$$

where $a_1 \mid a_2 \mid \ldots \mid a_r$, $Ann(M) = (a_r)$ and the ideals $(a_1), (a_2) \ldots, (a_r)$ are uniquely defined by M.

We cannot give a proof of this result here, but only note that we use it in two particular situations, which we explain below (for a proof, see [L], Chap. III, Thm. 7.7). Notice that it may happen that some $a_i = 0$, which corresponds to $R/(a_i) = R$. In our notation, we follow the convention that $0 \mid a$ for all $a \in R$.

If $R = \mathbb{Z}$, then A.6.1 says that every abelian group is a direct sum of finitely many cyclic groups. These cyclic groups are either finite of the form $\mathbb{Z}/(a)$, $a > 0$, or infinite of the form $\mathbb{Z}$. If the group is finite, then there are only the summands of the first type. Each integer $a > 1$ can be factorized as a product of prime powers and each cyclic group $\mathbb{Z}/(a)$ can be split into a product of cyclic groups $\mathbb{Z}/(p^k)$ for prime numbers p and their exponents k such that $p^k \mid a$ and $p^{k+1} \nmid a$ according to A.3.11. The set of such cyclic groups is also uniquely determined by the isomorphism class of G (notice that the same prime p with the same exponent k may appear several times). We record this result as we refer to it occasionally:

A.6.2 Fundamental Theorem on Finite Abelian Groups. *Every finite abelian group G is a direct product*

$$G = \mathbb{Z}/(a_1) \times \mathbb{Z}/(a_2) \times \cdots \times \mathbb{Z}/(a_r),$$

where $a_1 \mid a_2 \mid \ldots \mid a_r$, $Ann(G) = (a_r)$ *and the ideals* $(a_1), (a_2) \ldots, (a_r)$ *are uniquely defined by G. The cyclic groups* $\mathbb{Z}/(a_i)$ $(a_i > 1)$ *can be represented as direct products of cyclic groups whose orders are prime powers. The number of such factors in the product and the orders of the cyclic groups in it are also uniquely determined by the group.*

The second important case when we use modules over principal ideal rings is the case of polynomial rings over fields. As we know, every ring $R = K[X]$, where K is a field, is a principal ideal domain (see A.3.1). Consequently, Theorem A.6.1 is true over R. The cyclic modules in this case are quotients $K[X]/(f(X))$, where $f(X)$ is a polynomial. If this is the zero polynomial, then the cyclic module is $K[X]$ itself. This module is infinite-dimensional as a vector space over K (a basis is $1, X, X^2, \ldots$). If $f(X)$ is a polynomial of degree $n > 0$, then by A.5.8, the quotient consists of all sums $b_0 + b_1\alpha + \cdots + b_{n-1}\alpha^{n-1}$, where $b_i \in K, f(\alpha) = 0$ ($\alpha = [X]$ is the class of X) and the representation of each element of $K[X]/(f(X))$ in such a form is unique, that is, $1, \alpha, \ldots, \alpha^{n-1}$ form a basis of $K[X]/(f(X))$ over K

Similarly to the case of integers, every polynomial $f(X)$ can be factored as a product of irreducible factors, so by (A.2), we can represent M uniquely as a product of cyclic modules $K[X]/(p(X)^k)$, where $p(X)$ are irreducible polynomials. Thus, for the polynomial rings $K[X]$, the general case A.6.1 leads to the following result:

Proposition A.6.3. *Every* $K[X]$*-module M which is finite-dimensional as a vector space over K is a direct product*

$$M = K[X]/(f_1) \times K[X]/(f_2) \times \cdots \times K[X]/(f_r),$$

where $f_1 \mid f_2 \mid \ldots \mid f_r$, $Ann(M) = (f_r)$ *and the ideals* $(f_1), (f_2) \ldots (f_r)$ *are uniquely defined by M. The cyclic modules* $K[X]/(f_i)$ $(\deg f_i > 0)$ *can be represented as direct products of cyclic modules* $K[X]/(p^k)$*, where p are irreducible polynomials in* $K[X]$*. The number of such factors in the product and the ideals* (p^k) *are also uniquely determined by the module M.*

Notice that if the polynomial f_r which generates the annihilator of M is separable (without multiple zeros), then $M = K[X]/(f_r)$.

A.6.4 Group Rings. If G is a finite group and R a ring, then it is possible to form a ring, denoted by $R[G]$ and called the group ring of G over R. Formally, this ring consists of all functions $\varphi : G \to R$. Such functions are added and multiplied in the following way: $(\varphi + \psi)(g) = \varphi(g) + \psi(g)$ and $\varphi\psi(g) = \sum_{h\in G} \varphi(h)\psi(h^{-1}g)$. In practice, we denote $\varphi(g)$ by r_g and write any element of $R[G]$ as a sum $\varphi = \sum_{g\in G} r_g g$. Such sums are added by adding the corresponding coefficients for $g \in G$ and multiplied in the "usual way" (according to the above definitions of addition and multiplication). This means that if $\psi = \sum_{g\in G} s_g g$, then the coefficient of g in the product $\varphi\psi$ is the sum of $r_{g'}s_{g''}$ such that $g'g'' = g$.

In Chap. 11, we defined the notion of a G-module. If A is such a G-module, which is an R-module for a ring R with the property $g(ra) = r(ga)$ for each $r \in R$ and $g \in G$, then it becomes a module over the group ring $R[G]$ if we define $(\sum_{g\in G} r_g g)a = \sum_{g\in G} r_g(ga)$. Conversely, each $R[G]$-module A is a G-module (as in Chap. 11) and at the same time an R-module (and both structures are related by $(rg)a = g(ra) = r(ga)$ when $r \in R$, $g \in G$ and $a \in A$).

The Chinese Remainder Theorem

The Chinese remainder theorem (CRT) says that if $d_1, \dots, d_k \in \mathbb{Z}$ are relatively prime positive integers and $r_1, \dots, r_k$ are any integers, then there is an integer a such that the remainder of a when divided by d_i is r_i for $i = 1, \dots, k$. When we say that the remainder of a when divided by d is r, we mean that d divides $a - r$, which is denoted by $d \mid a - r$ or $a - r \in (d)$ or $a \equiv r \pmod{d}$ (a is congruent to r modulo d). This result is true in much greater generality. Since we use it both for integers and polynomials, we give a formulation which covers these two cases:

A.7.1 Chinese Remainder Theorem. *Let R be an integral domain and let $d_1, \dots, d_k \in R$ be pairwise relatively prime elements, that is, the ideals (d_i, d_j) are equal to* R for $i \neq j$. *If $r_1, \dots, r_k \in R$, then there exists an $a \in R$ such that $a - r_i \in (d_i)$ for all $i = 1, \dots, k$.*

Proof. We use induction starting with $k = 1$ when the result is of course true. Notice now that it is true for $k = 2$, since by A.3.10, we have a surjection of R onto $R/(d_1) \times R/(d_2)$ mapping a onto $(a + (d_1), a + (d_2))$. But this means that if we take $(r_1 + (d_1), r_2 + (d_2))$, then we can find an $a \in R$ such that $a + (d_1) = r_1 + (d_1)$ and $a + (d_2) = r_2 + (d_2)$, that is, $a - r_1 \in (d_1)$ and $a - r_2 \in (d_2)$.

Assume now that the theorem is true for $k - 1$ elements of R and consider k (pairwise) relatively prime elements $d_1, \dots, d_k \in R$. Let $d = d_1 d_2 \cdots d_k$. First we note that every d_i ($i = 1, \dots, k$) and d/d_i are relatively prime. In fact, we noted earlier (see the text preceding A.3.10) that an ideal relatively prime with two other ideals is relatively prime with their product ((d/d_i) is the product of all (d_j) for

$j \neq i$). Now we use k times the case of two relatively prime elements: d_i and d/d_i, ($i = 1, \ldots, k$). We find $x_1, \ldots, x_k \in R$ such that

$$x_i - 1 \in (d_i) \quad \text{and} \quad x_i \in (d/d_i) \subseteq (d_j) \quad \text{for } j \neq i.$$

Take now $a = r_1 x_1 + \cdots + r_k x_k$. We check that $a - r_i \in (d_i)$ for all $i = 1, \ldots, k$. In fact, we have $a - r_i = r_1 x_1 + \cdots + r_i(x_i - 1) + \cdots + r_k x_k \in (d_i)$. □

Example A.7.2. Let p_1, p_2, p_3 be three prime numbers, for example, $p_1 = 3, p_2 = 5, p_3 = 11$. We want to find an integer a such that $a \equiv 2 \pmod{p_1}$, $a \equiv 4 \pmod{p_2}$ and $a \equiv 8 \pmod{p_3}$. Let $d = p_1 p_2 p_3$. By Fermat's Little Theorem (see Exercise 6.3(c)), we may choose $x_1 = (p_2 p_3)^{p_1 - 1} \equiv 1 \pmod{p_1}$, $x_2 = (p_1 p_3)^{p_2 - 1} \equiv 1 \pmod{p_2}$, $x_3 = (p_1 p_2)^{p_3 - 1} \equiv 1 \pmod{p_3}$, since $x_i \equiv 1 \pmod{p_i}$ and $x_i \equiv 0 \pmod{p_j}$ for $j \neq i$. According to the proof above, we have $a = 2x_1 + 4x_2 + 8x_3 = 2 \cdot (5 \cdot 11)^2 + 4 \cdot (3 \cdot 11)^4 + 8 \cdot (3 \cdot 5)^{10} \equiv 74 \pmod{3 \cdot 5 \cdot 11}$. The choice of x_i above is given by a general formula (using FLT), but in the exercises in Chap. 15 (and in this example), it is not difficult to find suitable x_i by simple computations (for example, we can consider the numbers $11k + 8$ and try to find k such that the number gives 2 modulo 3 and 4 modulo 5—this is $k = 6$). The command "mod" in Maple is very useful for computations of residues.

Permutations

If a set X is finite, the bijective functions on it are often called permutations of X. The most common case is $X = \{1, 2, \ldots, n\}$. All bijective functions on X form a group with respect to the composition of functions. This group is denoted by S_n and is called the **symmetric group**. The order of S_n is $n!$. Thus, a permutation of $1, \ldots, n$ is a bijective function $\sigma : \{1, \ldots, n\} \to \{1, \ldots, n\}$, which sometimes is denoted by

$$\sigma = \begin{pmatrix} 1 & \ldots & n \\ i_1 & \ldots & i_n \end{pmatrix},$$

when $\sigma(k) = i_k$ for $k = 1, \ldots, n$. Permutations can be conveniently written as a product of cycles. A **cycle** is a permutation such that for a subset $\{i_1, \ldots, i_k\}$ of $X_n = \{1, \ldots, n\}$, we have $\sigma(i_1) = i_2, \sigma(i_2) = i_3, \ldots, \sigma(i_k) = i_1$ and $\sigma(i) = i$ for $i \notin \{i_1, \ldots, i_k\}$. Such a cycle is denoted by $(i_1, i_2, \ldots, i_k)$. A **transposition** is a cycle of length 2, that is, a permutation (i, j) which shifts i and j. An arbitrary permutation σ can be written as a composition of disjoint cycles, since we can start with 1, take its image, the image of the image, and so on. Finally, we have to return to 1 (if we do not already have $\sigma(1) = 1$ at the beginning). In this way, we get a cycle starting with 1. If all $1, \ldots, n$ are in the cycle, then σ is one cycle. Otherwise, we choose the smallest number which is not in the cycle starting with 1, and construct a second cycle, which is disjoint from the first one. We continue the process so that every number is in a cycle. Notice that we usually omit the cycles of length 1, that is, those corresponding to $\sigma(i) = i$, but we denote by (1) the cycle corresponding to the identity (essentially, the identity is a composition of n cycles (i) for $i = 1, \ldots, n$). For example:

$$\sigma = \begin{pmatrix} 1\,2\,3\,4\,5\,6\,7\,8 \\ 4\,7\,8\,5\,1\,6\,2\,3 \end{pmatrix} = (1, 4, 5)(2, 7)(3, 8).$$

The cycle permutations have many pleasant properties which are important in the study of the permutation groups. We record the property proved above as well as the following useful facts:

Proposition A.8.1. (*a*) *Any permutation in* S_n *can be obtained as a composition of disjoint cycles.*
(*b*) *Any permutation in* S_n *can be obtained as a composition of transpositions* (*even of only two adjacent numbers*).
(*c*) *If* $\sigma \in S_n$ *and* $\tau = (i_1, \ldots, i_k) \in S_n$, *then* $\sigma\tau\sigma^{-1} = (\sigma(i_1), \ldots, \sigma(i_k))$.
(*d*) *Any two cycles in* S_n *having the same length are conjugate.*

Proof. (b) Any permutation $i_1, \ldots, i_n$ of $1, \ldots, n$ can be obtained from $1, \ldots, n$ by a chain of transpositions of two adjacent numbers. Using (a), it is sufficient to check it for cycles. We have the following evident equality:

$$(i_1, \ldots, i_k) = (i_1, \ldots, i_{k-1})(i_{k-1}, i_k),$$

so every cycle can be represented as a composition of transpositions repeatedly shortening the component cycles, replacing them by transpositions according the formula above. Observe that we compose ("read") from right to left as usual for functions on sets. As regards transpositions (i,j) (where $j > i + 1$), we have the following formula, which is easy to check:

$$(i,j) = (i, i + 1)(i + 1, i + 2)\cdots(j - 1, j)(j - 2, j - 1)\cdots(i, i + 1).$$

This formula shows that every transposition is a composition of transpositions of adjacent numbers.
(c) Let $\varrho = \sigma\tau\sigma^{-1}$. We have to show that $\varrho(\sigma(i_1)) = \sigma(i_2), \ldots, \varrho(\sigma(i_k)) = \sigma(i_1)$ and for every $i \neq \sigma(i_r)$, where $r = 1, \ldots, k$, we have $\varrho(i) = i$.
We check simply that $\varrho(\sigma(i_1)) = \sigma\tau\sigma^{-1}(\sigma(i_1)) = \sigma(\tau(i_1)) = \sigma(i_2)$ and, in general, $\varrho(\sigma(i_r)) = \sigma\tau\sigma^{-1}(\sigma(i_r)) = \sigma(\tau(i_r)) = \sigma(i_{r+1})$ for $r = 1, \ldots, k$, when the addition of these indices r is performed modulo k.
If $i \neq \sigma(i_r)$, then of course, $\sigma^{-1}(i) \neq i_r$ for all $r = 1, \ldots, k$, that is, τ does not move $\sigma^{-1}(i)$. Hence, we have $\varrho(\sigma(i)) = \sigma\tau\sigma^{-1}(i) = \sigma(\tau(\sigma^{-1}(i))) = \sigma(\sigma^{-1}(i)) = i$, that is, ϱ doesn't move i.
(d) Take two cycles of the same length $(i_1, \ldots, i_k)$ and $(j_1, \ldots, j_k)$. Choose a permutation $\sigma \in S_n$ such that $\sigma(i_r) = j_r$ for $r = 1, \ldots, k$ and σ is any bijection from the set $i \neq i_r$ on the set of $j \neq j_r$ Then, it follows from (c) that $\sigma(i_1, \ldots, i_k)\sigma^{-1} = (j_1, \ldots, j_k)$. □

A permutation σ of $X = \{1, 2, \ldots, n\}$ is called even if for an even number of pairs $i, j \in X$ such that $i < j$, we have $\sigma(i) > \sigma(j)$ (this number may be 0). The even permutations of X form a subgroup of S_n called the **alternating group**, which is denoted by A_n. Its index in S_n is 2, that is, its order is $n!/2$. Among many possible proofs of this, we choose an argument which is useful in Chap. 15.

As we know, the group S_n acts on the set $X = R[X_1, \ldots, X_n]$ (R any ring) of polynomials in variables $X_1, \ldots, X_n$ by the formula: $\sigma(f(X_1, \ldots, X_n)) = f(X_{\sigma(1)}, \ldots, X_{\sigma(n)})$.

Consider the following polynomial in $\mathbb{Z}[X_1, \ldots, X_n]$:

$$\Delta(X_1, \ldots, X_n) = \prod_{1 \leq i < j \leq n} (X_i - X_j).$$

It has as many factors as the number of pairs i, j such that $i < j$. For any permutation $\sigma \in S_n$, we have $\sigma(\Delta(X_1, \ldots, X_n)) = \Delta(X_{\sigma(1)}, \ldots, X_{\sigma(n)})$. This polynomial also has $X_i - X_j$ as its factors but each time $\sigma(i) > \sigma(j)$, the corresponding factor $X_{\sigma(i)} - X_{\sigma(j)}$ in $\sigma(\Delta(X_1, \ldots, X_n))$ differs by sign from the factor $X_{\sigma(j)} - X_{\sigma(i)}$, which appears in $\Delta(X_1, \ldots, X_n)$. Hence a permutation σ is even if and only if $\sigma(\Delta) = \Delta$ and odd if and only if $\sigma(\Delta) = -\Delta$. If σ, τ are two even permutations, then $\sigma\tau(\Delta) = \sigma(\tau(\Delta)) = \Delta$, which shows that the even permutations form a group. Recall that it is denoted by A_n. We also have that if both σ, τ are odd, then $\sigma\tau(\Delta) = \sigma(\tau(\Delta)) = \sigma(-\Delta) = \Delta$, so $\sigma\tau$ is an even permutation, that is, $\sigma\tau \in A_n$. This property implies that A_n has index 2 in S_n, that is, the subgroup A_n has two cosets in S_n: one is A_n and the second is σA_n for any odd permutation σ (for example $\sigma = (1, 2)$). In fact, if τ is also odd, then $\sigma A_n = \tau A_n$, since $\sigma^{-1}\tau \in A_n$ as a product of two odd permutations ($\sigma^{-1} \notin A_n$, since $\sigma \notin A_n$ as A_n is a subgroup). Thus the number of elements in A_n is half of the number of elements in S_n, that is, it equals $n!/2$.

Proposition A.8.2. *The group A_n is the only subgroup of S_n having index* 2.

Proof. Let H be a subgroup of index 2 in S_n. We claim that no transpositions (a, b) are in H. In fact, the subgroup H is normal in S_n as a subgroup of index 2 (see A.2.8). If a transposition $t = (a, b) \in H$, then every other transposition is also in H, since it is conjugate to t (see A.8.1(d)). But the transpositions generate the whole group S_n, so we would have $H = S_n$, which is not true. Now a product of two elements which are not in H is an element belonging to H, since H has index 2 in S_n. Thus any product of two transpositions is in H. But any even permutation is a product of an even number of transpositions, so it is an element of H. Thus $A_n \subseteq H$, which means that $H = A_n$, since these groups have the same order. □

A.8.3 Cayley's[1] Theorem. *Every group G of order n can be embedded into a permutation group S_N, where $N \geq n$. It is always possible to choose N to be a prime number.*

[1]Arthur Cayley, 16 August 1821–26 January 1895.

Proof. Let $G = \{g_1, \dots, g_n\}$, where $g_1 = e$ is the identity. For $g \in G$ denote by $\sigma_g \in S_n$ the permutation such that $\sigma_g(i) = j$ if $gg_i = g_j$, that is, $gg_i = g_{\sigma_g(i)}$. Mapping g onto the permutation σ_g gives an embedding of G into S_n. In fact, denote by $\Phi : G \to S_n$ the function $\Phi(g) = \sigma_g$. For every i, we have $(gg')g_i = g_{\sigma_{gg'}(i)}$ and at the same time, $g(g'g_i) = gg_{\sigma_{g'}(i)} = g_{\sigma_g(\sigma_{g'}(i))}$. Hence, we have $\sigma_{gg'}(i) = \sigma_g(\sigma_{g'}(i))$. Thus for every i, we have $\Phi(gg')(i) = \sigma_{gg'}(i) = \sigma_g(\sigma_{g'}(i)) = \Phi_g(\Phi_{g'}(i)) = \Phi_g \circ \Phi_{g'}(i)$. This shows that $\Phi(gg') = \Phi(g) \circ \Phi(g')$, that is, Φ is a group homomorphism. It is clear that Φ is an injection as $\Phi(g) = (1)$ if and only if $\sigma_g = g_1$ (see A.2.10).

Notice that the group S_n can be embedded into any group S_N, where $N \geq n$. For example S_n can be regarded as all permutations of the first n numbers among $1, \dots, n, \dots, N$. In particular, one may choose N to be a prime number (see an application of this in Exercise 9.17). □

Proposition A.8.4. *The group S_n is generated by the cycle $\sigma = (1, 2, \dots, n)$ and the transposition $\tau = (1, 2)$.*

Proof. Let H be a subgroup of S_n generated by σ and τ. For $1 \leq k \leq n-2$, we have $\sigma^k(1) = k+1$ and $\sigma^k(2) = k+2$. Hence $\sigma^k\tau\sigma^{-k} = (k+1, k+2)$. Thus the subgroup H contains all transpositions of any two consecutive numbers among $1, 2, \dots, n$. But any permutation in S_n can be obtained as a product of such transpositions (it is possible to get any permutation $a_1, \dots, a_n$ from $1, \dots, n$ by a chain of transpositions of two adjacent numbers—see A.8.1(b)). □

Proposition A.8.5. *Let G be a transitive subgroup of the group S_n containing the cycle $\sigma = (1, 2, \dots, n-1)$ and a transposition $\tau = (i, j)$ $(i < j)$. Then $G = S_n$*

Proof. First of all, we observe that G contains all transpositions (k, n), where $k = 1, \dots, n$. In fact, there is an $f \in G$ such that $f(j) = n$, since G acts transitively on $\{1, \dots, n\}$. Let $f(i) = k_0$. By A.8.1(c), we have $f(i,j)f^{-1} = (f(i), f(j)) = (k_0, n)$. Now we easily check that all conjugates $\sigma^r(k_0, n)\sigma^{-r} = (\sigma^r(k_0), n)$, where $r = 1, \dots, n-1$ give all transpositions (k, n) with $k = 1, \dots, n-1$ (see A.8.1(c)). The transpositions (k, n) generate the whole group S_n, since for any transposition (a, b), where $1 \leq a < b < n$, we have $(a, b) = (a, n)(b, n)(a, n)$. Thus G contains all transpositions of $1, \dots, n$, which proves that $G = S_n$ by A.8.1(b). □

Group Actions on Sets

We say that a group G acts on a set X if for every pair $(g, x) \in G \times X$, we have an element $gx \in X$ such that $(gg')x = g(g'x)$ and $ex = x$ (e is the identity in G). Thus an action of G on X is a function from $G \times X$ to X satisfying these two assumptions, which can be best understood in terms of the transformations of the set X. If we fix g, then we get a function $\sigma_g : x \mapsto gx$ for $x \in X$ (instead of the term function, we often use the term transformation in this case). The function σ_g is a bijection on the set X, since it has an inverse function $\sigma_{g^{-1}}$. In fact, we have $\sigma_g \circ \sigma_{g'}(x) = \sigma_g(\sigma_{g'}(x)) = \sigma_g(g'x) = g(g'x) = (gg')x = \sigma_{gg'}(x)$, that is,

$\sigma_g \circ \sigma_{g'} = \sigma_{gg'}$ and $\sigma_e(x) = ex = x$, that is, $\sigma_e = id_X$ is the identity mapping on X. Hence, $\sigma_g \circ \sigma_{g^{-1}} = \sigma_{gg^{-1}} = \sigma_e = id_X$ and similarly, $\sigma_{g^{-1}} \circ \sigma_g = id_X$, which shows that σ_g and $\sigma_{g^{-1}}$ are inverses of each other (see below A.9.1).

If X is a set, then all bijective functions $\sigma : X \to X$ form a group under composition of functions, that is, if $\tau : X \to X$ is also a bijection, then the composition $\tau\sigma(x) = \tau(\sigma(x))$ for $x \in X$ is a bijection on X. The composition of functions is associative, the identity is the identity function $id(x) = x$ for $x \in X$, and the inverse of σ is the inverse function σ^{-1}. Denoting by $\mathcal{B}(X)$ the group of all bijections on X, we define a **transformation group** of X as any subgroup of $\mathcal{B}(X)$. Notice that, for simple combinatorial reasons, if X is a finite set, then every injective or surjective function $\sigma : X \to X$ is automatically bijective (that is, an injective function must also be surjective, and a surjective function must be injective). This easy observation is often useful, as is the following general property of functions, which we often use:

Lemma A.9.1. *If $\sigma : X \to Y$ and $\tau : Y \to X$ are functions such that $\tau \circ \sigma = id_X$, then σ is injective and τ is surjective, so if also $\sigma \circ \tau = id_Y$, then both σ and τ are bijective.*

Proof. If $x, x' \in X$ and $\sigma(x) = \sigma(x')$, then $\tau \circ \sigma(x) = \tau \circ \sigma(x')$, so $x = x'$. Hence σ is injective (different x, x' have different images under σ). If $x \in X$, then $x = \tau(\sigma(x))$, so τ is surjective (every element x of X is an image of an element $\sigma(x)$ of Y). □

The observations above concerning actions of groups on sets can be now simply expressed by saying that any action of a group G on a set X defines a homomorphism from G to the transformation group $\mathcal{B}(X)$ such that $g \in G$ maps onto the transformation σ_g. Denoting such a homomorphism by $\Phi : G \to \mathcal{B}(X)$, we have $\Phi(g) = \sigma_g$. It is easy to see that also conversely, any homomorphism $\Phi : G \to \mathcal{B}(X)$ defines an action of G on X if we define $gx = \Phi(g)(x)$. The properties $\Phi(gg') = \Phi(g)\Phi(g')$ and $\Phi(e) = Id$ of the homomorphism Φ immediately translate to $(gg')x = g(g'x)$ and $ex = x$ for $g, g' \in G$ and $x \in X$.

Let a group G act on a set X. The **orbit** of $x \in X$, denoted by Gx, is the set of all images of x under the elements of G, that is, $Gx = \{gx, x \in X\}$. The **stabilizer** of $x \in X$ in G, denoted by G_x, is the subgroup of G consisting of those elements which map x to itself, that is, $G_x = \{g \in G | gx = x\}$. If $g \in G$, we denote by X^g the set of all elements in X fixed by g, that is, $X^g = \{x \in X | gx = x\}$. We say that G **acts transitively** on X if for each pair $x, x' \in X$ there is a $g \in G$ such that $x' = gx$.

Proposition A.9.2. *If G is a finite group acting on a finite set X, then:*
(a) The orbits of G on X are disjoint and cover the set X, that is, they form a partition of X. The number of elements in the orbit of $x \in X$ is equal to the index of the stabilizer of x in G, that is, $|Gx| = [G : G_x] = |G|/|G_x|$. Moreover, if $x' = hx$ for $h \in G$, then $G_{x'} = hG_xh^{-1}$. Thus, we have

$$X = \bigcup_x Gx \quad and \quad |X| = \sum_x \frac{|G|}{|G_x|},$$

where x represent different orbits of G on X.

(*b*) (**Burnside's Lemma**) *Denoting by* $|G/X|$ *the number of orbits of* G *on* X*, we have*

$$|G/X| = \frac{1}{|G|}\sum_{g\in G}|X^g|.$$

(*c*) *If* X *has* p *elements, where* p *is a prime and* G *acts transitively on* X*, then* p *divides the order of* G.

Proof. (a) It is clear that each element $x \in X$ belongs to an orbit (its own). Two different orbits are disjoint, since if $x \in Gx_1$ and $x \in Gx_2$ for some $x \in X$, then for any $x' \in Gx_1$, we have $x' = g'x_1$ and $x = g_1x_1 = g_2x_2$, that is, $x_1 = g_1^{-1}g_2x_2$ gives $x' = g'g_1^{-1}g_2x_2$, so $x' \in Gx_2$. Hence, we have $Gx_1 \subseteq Gx_2$. By symmetry, we also have the inclusion $Gx_2 \subseteq Gx_1$, which gives $Gx_1 = Gx_2$. Another argument is to use the fact that the orbits are equivalence classes of an equivalence relation $x \sim x'$ on X: we declare $x, x' \in X$ equivalent if and only if they belong to the same orbit. When we check that the relation is reflexive, symmetric and transitive, we get that different orbits are disjoint, since different equivalence classes of an equivalence relation are disjoint and form a partition of X (see A.1.1).
The number of elements in an orbit Gx is equal to the number of different elements gx, where $g \in G$. Now $gx = g'x$ if and only if $g^{-1}g'x = x$, that is, $g^{-1}g' \in G_x$, which is equivalent to $g'G_x = gG_x$. Thus, the number of different elements in Gx is equal to the number of cosets of G_x in G, that is, the index $[G : G_x] = |G|/|G_x|$ of G_x in G.
If $x' = hx$ for some $h \in G$, then

$$g \in G_{x'} \Leftrightarrow gx' = x' \Leftrightarrow ghx = hx \Leftrightarrow h^{-1}ghx = x \Leftrightarrow h^{-1}gh \in G_x \Leftrightarrow g \in hG_xh^{-1},$$

so $G_{x'} = hG_xh^{-1}$.
(b) First note that

$$\sum_{g\in G}|X^g| = \sum_{x\in X}|G_x|,$$

since both these numbers are equal to the number of pairs $(g, x) \in G \times X$ such that $gx = x$. In fact, imagine a "multiplication table" for multiplication of elements of G by the elements of X in which the intersection of the row g with the column x is gx and count the number of occurrences of $gx = x$ in two ways: by rows and by columns.
We have:

$$\sum_{g\in G}|X^g| = \sum_{x\in X}|G_x| = \sum_{x\in X}\frac{|G|}{|Gx|} = |G|\sum_{x\in X}\frac{1}{|Gx|}.$$

Now we observe that the last sum is simply the number of orbits, since each fraction $\frac{1}{|Gx|}$ appears as many times as the number of elements in the orbit of x, which means that each orbit contributes 1 to this sum (the elements of X are distributed among all orbits which are disjoint and cover X). Thus,

$$\sum_{g\in G} |X^g| = |G||G/X|,$$

which proves Burnside's Lemma (which in reality was proved already by Cauchy and somewhat later by Frobenius).

(c) By our assumption, we have only one orbit $Gx = X$ for any $x \in X$, so by (a), we have $|Gx||G_x| = |G|$. Hence $p = |Gx|$ divides $|G|$. □

The formula in A.9.2(a) is often used when $X = G$ and the action of the group G is given by conjugation (that is, $g \cdot x = gxg^{-1}$). In this case, the equality

$$|G| = \sum_x \frac{|G|}{|G_x|}$$

is often called the **class formula**, since $\frac{|G|}{|G_x|} = [G : G_x] = |Gx|$ is the number of elements in the group G which are conjugate to x. We use this formula in such a way in Exercise 12.6. Notice that classes with only one element, that is, $|Gx| = 1$, correspond to $G_x = G$, that is, they correspond to the case when $gxg^{-1} = x$ for every $g \in G$. Such elements x are exactly the elements of the center $C(G)$ of G (see p. 245). For an arbitrary element $x \in G$, the group G_x consists, in this case, of those elements $g \in G$ for which $gxg^{-1} = x$, that is, those which commute with x. It is usually called the **centralizer** of the element $x \in G$.

Very often, we consider $X = G$ and H is a subgroup of G acting by multiplication, say from the left. Then the orbit of $x \in G$ is the coset Hx and the formula of A.9.2(a) is simply the splitting of G into the right cosets of H in G. We give a few important applications, which we use on different occasions in the exercises.

A.9.3 Cauchy's[2] Theorem. *Let G be a finite group whose order is divisible by a prime number p. Then the group G contains an element of order p.*

Proof. Consider the set X of all p-tuples $(g_1, g_2, \ldots, g_p)$ such that $g_i \in G$ and $g_1g_2\cdots g_p = e$. The number of elements in X is of course n^{p-1}, where $n = |G|$ (one can choose $g_1, \ldots, g_{p-1}$ arbitrarily, in n different ways each, and compute uniquely the corresponding last factor g_p).

Let $H = \langle\sigma\rangle$ be the cyclic group of order p generated by the permutation $\sigma = (1, 2, \ldots, p)$ (the cycle of length p mapping $1 \mapsto 2 \mapsto \cdots \mapsto p \mapsto 1$). The group H acts on the set X, by shifting the indices of $g_1, g_2, \ldots, g_p$ one place to the right circularly (the last to the first): $\sigma(g_1, g_2, \ldots, g_p) = (g_{\sigma(1)}, g_{\sigma(2)}, \ldots, g_{\sigma(p)})$. In fact,

[2]Augustin-Louis Cauchy, 21 August 1789–23 May 1857.

if $g_1 g_2 \cdots g_p = e$, then $g_2 \cdots g_p g_1 = e$, since we can multiply the first equality by g_1^{-1} from the left and then by g_1 from the right.

If $x = (g_1, g_2, \dots, g_p) \in X$, then its orbit Hx contains p different elements unless all $g_1 = g_2 = \cdots = g_p$ when all shifts of x give the same element x, that is, the orbit Hx has only one element. We just want to show that there is an $x = (g, g, \dots, g) \in X$ with $g \neq e$, that is, that the number of elements of X whose orbit has only one element is bigger than 1. Denote by k the number of orbits Hx of length 1 and by s the number of orbits Hx consisting of p elements.

According to A.9.2(a), the number of elements in X (n^{p-1}) is equal to the sum of the numbers of elements in all orbits. Thus $n^{p-1} = k + ps$. Since p divides the order n of the group G, the last equality shows that p divides k, so we really have $k > 1$. □

We also use the formula A.9.2(a) in the proof of the following theorem of Sylow,[3] which has many applications in Galois theory. Let G be a group and p a prime number. If p divides the order of G and p^k is the highest power of p dividing it, then each subgroup of G of this order is called a **Sylow p-subgroup** of G. In general, a group whose order is a power of a prime p is called a **p-group**. The Sylow subgroups play an important role and the main result about them is usually formulated as three theorems (below (a), (b), (c)):

A.9.4 Sylow's Theorems. *Let G be a finite group and p a prime number dividing the order of G. Then*
(a) G contains Sylow p-subgroups;
(b) any two Sylow p-subgroups of G are conjugate;
(c) the number of Sylow p-subgroups of G divides their index and is equal to 1 *modulo p.*

Proof. (a) We use induction with respect to the order of G. If $|G| = 2$, then the group is cyclic of order 2 and the theorem is of course true. Assume that it is true for all groups of order less than the order of a given group G. We prove (a) for G. Assume that G has proper subgroups, since otherwise it is a cyclic group of a prime order p and the theorem is automatically true for G (see A.2.15). Let p^k be the highest power of a prime number dividing the order of G. If G has a proper subgroup H of order divisible by p^k, then the theorem is true by induction, since a subgroup of H of order p^k is a subgroup of G of this order. Thus, assume that p^k does not divide the orders of all proper subgroups of G. Consider now the action of G by conjugation (so $X = G$, $H = G$ and $g \cdot x = gxg^{-1}$). As we noted above, the stabilizer $G_x = \{g \in G \mid gxg^{-1} = x\}$ is the centralizer of the element $x \in G$. We know that $G_x = G$ (that is, G_x is not proper) if and only if x is in the center $C(G)$ of G. The class formula of A.9.2(a) gives

$$|G| = \sum_x \frac{|G|}{|G_x|} = |C(G)| + \sum_x \frac{|G|}{|G_x|},$$

[3] Peter Ludwig Mejdell Sylow, 12 December 1832–7 September 1918.

where in the sum to the right, we take only x representing the classes for which the group G_x is proper. Hence, the prime number p divides all terms $\frac{|G|}{|G_x|}$ for which G_x is a proper subgroup (since then $p^k \mid |G|$ and $p^k \nmid |G_x|$) and $\frac{|G|}{|G_x|} = 1$ each time x is in the center of G. Since p^k divides $|G|$ to the left, we get that p divides $|C(G)|$. Thus the order of $C(G)$ is divisible by p, which implies by Cauchy's theorem A.9.3 that there is an element $g \in C(G)$ of order p. Hence the subgroup $\langle g \rangle$ has order p and is normal in G, since g is in the center of this group. The quotient group $G/\langle g \rangle$ has order $|G|/p$ less than the order of G and the highest power of p dividing this order is p^{k-1}. By the inductive assumption, the group $G/\langle g \rangle$ contains a subgroup of order p^{k-1}. Now the inverse image of this subgroup in G (see A.2.13) has order p^k, so it is a Sylow p-subgroup of G.
(b) Let H_1 and H_2 be two Sylow p-subgroups of a group G. We want to show that there is an element $x \in G$ such that $H_1 = xH_2x^{-1}$. Once again, we use the class formula of A.9.2(a) but this time, we choose $X = G$ and we act on X by the subgroup $H_1 \times H_2$ of $G \times G$ in the following way $(h_1, h_2)x = h_1xh_2^{-1}$ for $(h_1, h_2) \in H_1 \times H_2$ and $x \in G$. It is easy to check that this definition really does define an action of $H_1 \times H_2$ on G:

$$((h_1, h_2)(h_1', h_2'))x = (h_1h_1', h_2h_2')x = (h_1h_1')x(h_2h_2')^{-1}$$

$$= h_1(h_1'xh_2'^{-1})h_2^{-1} = (h_1, h_2)((h_1', h_2')x)$$

for $(h_1, h_2), (h_1', h_2') \in H_1 \times H_2$, $x \in G$ and, of course, $(e, e)x = x$ for the identity $(e, e) \in H_1 \times H_2$. The orbit of $x \in G$ is H_1xH_2 (this set is called the double coset of H_1, H_2 in G). The isotropy group G_x of $x \in G$ consists of all (h_1, h_2) such that $h_1xh_2^{-1} = x$, that is $h_1 = xh_2x^{-1}$. Such pairs (h_1, h_2) are exactly those belonging to the intersection $H_1 \cap xH_2x^{-1}$. Hence the number of elements in the orbit H_1xH_2 is equal to

$$\frac{|H_1 \times H_2|}{|H_1 \cap xH_2x^{-1}|} = \frac{|H_1||H_2|}{|H_1 \cap xH_2x^{-1}|}$$

and the class formula of A.9.2(a) says that

$$|G| = \sum_x \frac{|H_1||H_2|}{|H_1 \cap xH_2x^{-1}|},$$

where x represent different orbits of $H_1 \times H_2$ on G (double cosets). Now $|H_1||H_2|$ is divisible by $p^kp^k = p^{2k}$ and $|H_1 \cap xH_2x^{-1}|$ as a subgroup of H_1 is divisible by at most p^k. If $H_1 \cap xH_2x^{-1}$ is a proper subgroup of H_1, then its order is at most p^{k-1}, which means that $\frac{|H_1||H_2|}{|H_1 \cap xH_2x^{-1}|}$ is divisible by at least $p^{2k-(k-1)} = p^{k+1}$. Since the order of G on the left is divisible exactly by p^k, we get a contradiction if all terms to the right are divisible by p^{k+1}. Hence it must be a term in which $H_1 \cap xH_2x^{-1}$ is not a proper subgroup of H_1, that is, there must exist an $x \in G$ such that $H_1 = xH_2x^{-1}$.

(c) Since all Sylow p-subgroups H of G are conjugate, the number n_p of different such groups is equal to the index of the normalizer $\mathcal{N}(H)$ in G (for the definition of $\mathcal{N}(H)$, see p. 241) In fact, we have $xHx^{-1} = x'Hx'^{-1}$ if and only if $x^{-1}x'H = Hx^{-1}x'$, that is, $x^{-1}x' \in \mathcal{N}(H)$, which is equivalent to $x'\mathcal{N}(H) = x\mathcal{N}(\mathcal{H})$. Thus the number of Sylow p-subgroups divides the order of G. Now in order to prove that this number is equal to 1 modulo p, we use the same action on G as in (b) with $H_1 = H_2 = H$. Since we want to find the number of Sylow p-subgroups, which is equal to the index of $\mathcal{N}(H)$ in G, we have to show that the quotient $n_p = |G|/|\mathcal{N}(H)|$ is equal to 1 modulo p. According to the class formula in (b), we have:

$$|G| = \sum_x \frac{|H||H|}{|H \cap xHx^{-1}|},$$

and we split the sum into those terms which correspond to $x \in \mathcal{N}(H)$ and those for which $x \notin \mathcal{N}(H)$ (notice that $H \subseteq \mathcal{N}(H) \subseteq G$). In the first case, we have $xHx^{-1} = H$, so the term is equal to $|H|$. The number of such terms is equal to the index of H in $\mathcal{N}(H)$, so this first sum is simply the order $|\mathcal{N}(H)|$. As in (b), each term $\frac{|H||H|}{|H \cap xHx^{-1}|}$ of the second type is divisible by p^{k+1} (since $H \cap xHx^{-1}$ is proper subgroup of H, its order is at most p^{k-1} and the numerator is $|H|^2 = p^{2k}$). Hence the sum of the terms of the second type is $p^{k+1}m$ for an integer m. Thus $|G| = |\mathcal{N}(H)| + p^{k+1}m$, which gives the number of Sylow p-subgroups:

$$n_p = \frac{|G|}{|\mathcal{N}(H)|} = 1 + \frac{p^{k+1}m}{|\mathcal{N}(H)|} = 1 + pm'$$

for an integer m', since the number $\frac{p^{k+1}m}{|\mathcal{N}(H)|}$ is an integer (as a difference of two integers) and it must be divisible by p, since the order of $\mathcal{N}(H)$ as a subgroup of G is at most divisible by p^k. This proves that the number of Sylow p-subgroups of G is equal to 1 modulo p. □

Symmetric Polynomials

A polynomial $f(X_1, \ldots, X_n) \in R[X_1, \ldots, X_n]$ over a ring R is called **symmetric** if it is unchanged by any permutation of its variables $X_1, \ldots, X_n$. This can be expressed in terms of the group action of the symmetric group S_n on the ring of polynomials: $\sigma(f(X_1, \ldots, X_n)) = f(X_{\sigma(1)} \ldots, X_{\sigma(n)})$. The polynomial f is symmetric if and only if $\sigma(f) = f$ for all $\sigma \in S_n$. The **elementary symmetric polynomials** are the polynomials $s_1 = X_1 + \cdots + X_n$, $s_2 = X_1X_2 + X_1X_3 + \cdots + X_{n-1}X_n, \ldots, s_n = X_1X_2 \cdots X_n$. They appear in the coefficients of the general polynomial of degree n (see p. 78):

$$g(X) = \prod_{i=1}^{n}(X - X_i) = X^n - s_1X^{n-1} + s_2X^{n-2} + \cdots + (-1)^{n-1}s_{n-1}X + (-1)^n s_n.$$

The following result has many applications:

A.10.1 Fundamental Theorem on Symmetric Polynomials. *If* R *is a (commutative) ring and* $f(X_1, \dots, X_n) \in R[X_1, \dots, X_n]$ *is a symmetric polynomial, then there is a polynomial* $h \in R[X_1, \dots, X_n]$ *such that* $f(X_1, \dots, X_n) = h(s_1, \dots, s_n)$.

Thus, the fundamental theorem on symmetric polynomials says that every symmetric polynomial is a polynomial of the elementary symmetric ones. The idea of the usual proof of this statement, using mathematical induction with respect to the degree of the polynomial, is very simple, but its formulation needs the careful definition of an ordering of monomials, which we omit here (see e.g. [L], Chap. IV, §6). Such a proof gives a way to effectively find a polynomial h for a given polynomial f.

A.10.2 Discriminant of a Polynomial. Let $\alpha_1, \dots, \alpha_n$ be the zeros of a polynomial $f(X) = a_nX^n + a_{n-1}X^{n-1} + \dots + a_1X + a_0 \in K[X]$ in a splitting field L of $f(X)$ over a field K ($a_n \neq 0$). The **discriminant** of $f(X)$ is then

$$\Delta(f) = \prod_{1 \leq i < j \leq n} (\alpha_i - \alpha_j)^2.$$

It is easy to see that the permutations of $\alpha_1, \dots, \alpha_n$ do not affect the discriminant—it is fixed by all permutations of the zeros (as it is fixed by all transpositions—see A.8.1). Since the Galois group $G(L/K)$ can be considered as a subgroup of all permutations of the zeros, this property of $\Delta(f)$ implies that it is an element of the field K (see **T.9.1**(b), Exercises 9.25 and 15.2).

It is also possible to prove that $\Delta(f) \in K$ using a somewhat different argument applied to the "general discriminant"

$$\Delta(g(X)) = \prod_{1 \leq i < j \leq n} (X_i - X_j)^2 \in \mathbb{Z}[X_1, \dots, X_n]$$

defined by the general equation above. The discriminant $\Delta(g(X))$ is a symmetric polynomial of $X_1, \dots, X_n$. According to the fundamantal theorem on symmetric polynomials, there is a polynomial $h \in \mathbb{Z}[X_1, \dots, X_n]$ such that $\Delta(g(X)) = h(s_1, \dots, s_n)$. If now $f(X) \in K[X]$ and $\alpha_i \in L$ are the zeros of $f(X)$ in its splitting field L, then we replace X_i by α_i in the equality $\Delta(g(X)) = h(s_1, \dots, s_n)$ Formally, we take a homomorphism of $\mathbb{Z}[X_1, \dots, X_n]$ mapping X_i onto the zeros α_i of the polynomial $f(X)$ in L ($\mathbb{Z}$ is mapped into the prime field in K when 1 in $\mathbb{Z}$ is mapped to 1 in K). Then $\Delta(g(X))$ maps to $\Delta(f(X))$ and s_i to $(-1)^i a_{n-i}/a_n$ for $i = 1, \dots, n$, according to Vieta's formulae when $s_1, \dots, s_n$ are evaluated for $X_1 = \alpha_1, \dots, X_n = \alpha_n$. Thus, we get $\Delta(f(X)) = h(s_1(\alpha_1, \dots, \alpha_n), \dots, s_n(\alpha_1, \dots, \alpha_n)) \in K$.

In this book, we use the polynomial h mainly for $n = 2, 3, 4$. For $n = 2$, see p. 3, and for $n = 3$ see p. 177. Already for $n = 4$, it is a little laborious to compute. In Maple, it is possible to get the discriminant using the command[4] `discrim(f(T),T)`, where T is the variable in the polynomial $f(T)$. For $n = 4$, we gave the expression of $\Delta(f)$ in terms of the coefficients on p. 89.

Transitive Subgroups of Permutation Groups

A permutation group $G \subseteq S_n$ is called **transitive** if for any pair $i, j \in \{1, 2, \ldots, n\}$ there is a permutation $\sigma \in G$ such that $\sigma(i) = j$. Transitive permutation groups appear when we study Galois groups of irreducible polynomials. In fact, we know that if $f(X) \in K[X]$ is an irreducible polynomial of degree n, then the Galois group $G(K_f/K)$ acts transitively on its n zeros. We met this property in several exercises. We know that every element of the Galois group is a permutation of the zeros of $f(X)$ and that there is always an automorphism of K_f over K which maps a given zero onto any other zero of this polynomial (see Exercises 7.4(a), 9.3 and 9.4). When we want to classify the Galois groups of polynomials over fields, it is necessary to know transitive subgroups of permutation groups. This knowledge is used in computer programs in order to find Galois groups of (irreducible) polynomials. For example, in Pari/GP, there is such a possibility for irreducible polynomials up to degree 11, and in Maple up to degree 9 (in 2017).

The symmetric group S_3 of all permutations of $1, 2, 3$ consists of $3! = 6$ elements. We can represent each permutation as an isometry of the plane mapping an equilateral triangle with vertices at three points 1, 2, 3 to itself. If $\{a, b, c\} = \{1, 2, 3\}$, then there are three rotations: the identity (1), the rotations $\pm 120^0$: $(1, 2, 3)$, $(1, 3, 2)$ and three symmetries in the three heights of the triangle: $(1, 2), (2, 3), (1, 3)$. There are only two transitive subgroups of S_3: the group S_3 itself and the subgroup of the rotations (all even permutations) $A_3 = \{(1), (1, 2, 3), (1, 3, 2)\}$. These facts are very easy to check (e.g. by listing all the subgroups of S_3).

The symmetric group S_4 of all permutations of $1, 2, 3, 4$ consists of $4! = 24$ elements. We can represent each permutation as an isometry of the space mapping a regular tetrahedron with vertices at four points 1, 2, 3, 4 to itself. If $\{a, b, c, d\} = \{1, 2, 3, 4\}$, then each non-identity permutation can be written as a cycle or a composition of them:

six symmetries (a, b) of order 2: in the planes through the vertices c, d and the mid-point of the side between a and b;

eight rotations (a, b, c) of order 3: around the axis through the vertex d perpendicular to the plane through the vertices a, b, c;

three rotations $(a, b)(c, d)$ of order 2: 180° around the axis through the mid-points of the sides a, b and c, d. Notice that these rotations together with the identity (1)

[4]The discriminant of a polynomial is sometimes defined with a factor, which is a power of the leading coefficient with suitable sign. Check, in particular, the definition in Maple if you want to use it!

form a transitive group of order 4. It is a transitive permutation representation of the Klein four-group, which we denote by V_4, that is,

$$V_4 = \{(1), (1,2)(3,4), (1,3)(2,4), (1,4)(2,3)\}.$$

Six cycles (a, b, c, d) of order 4: compositions of the rotation (a, b, c) with the symmetry (c, d).

Together with the identity (1), we get 24 possible isometric mappings of the tetrahedron to itself. Notice that the even permutations are exactly the 12 rotations, and the odd permutations are the 6 symmetries and the 6 compositions of a rotation of order 3 and a symmetry.

Now observe that every transitive subgroup of S_4 has at least four elements. As its order divides 24, the allowed orders are 4, 6, 8, 12, 24. Of course, S_4 and A_4 are transitive. By A.8.2, the group of even permutations A_4 is the only subgroup of S_4 of order 12.

In order to describe all transitive subgroups of S_4, notice that in any subgroup G, which contains at least one odd permutation σ, half of the elements are odd permutations, and the other half are even. In fact, if G_0 denotes all even permutations in G, then G_0 is a (normal) subgroup of G and $G = G_0 \cup \sigma G_0$, since every permutation τ in G is either even or, if it is odd, then $\sigma^{-1}\tau$ is even (that is, in G_0).

The group S_4 has three subgroups of order 8. All are isomorphic to the square group D_4 and are transitive. In order to prove this, assume that G is a subgroup of order 8. Then it must consist of both even and odd permutations—they cannot all be even, since a group of order 8 cannot be a subgroup of a group of order 12. The subgroup G_0 of G consisting of the even permutations (that is, rotations of the tetrahedron) must be $G_0 = V_4$, since all the remaining rotations have order 3 (cannot belong to a group of order 8). The odd permutations in G cannot all be of order 2. Those are exactly the symmetries of the tetrahedron. Among four such symmetries, there are at least two which shift the same vertex (of four possible), that is, they have form (a, b) and (a, c). Then the group G contains $(a, b)(a, c) = (a, c, b)$, which as an element of order 3. This order is not allowed by G. Thus G must contain an odd permutation σ of order 4. An easy direct computation shows that there are exactly three possibilities for σV_4, which give three possibilities for G:

$$D_4 = V_4 \cup \{(1,2,4,3), (1,3,4,2), (1,4), (2,3)\},$$

$$D_4' = V_4 \cup \{(1,2,3,4), (1,4,3,2), (1,3), (2,4)\},$$

$$D_4'' = V_4 \cup \{(1,3,2,4), (1,4,2,3), (1,2), (3,4)\}.$$

It is clear that these groups are transitive. Notice that the groups D_4, D_4', D_4'' are all isomorphic. They consist of all isometries of a square corresponding to a numbering of its vertices a, b, c, d according to the rotations given by the elements of order 4 belonging to it. Recall that such a group contains three subgroups of order 4: V_4 (the rectangle group), the cyclic group of the rotations of the square:

$$C_4 = \{(1), (a, b, c, d), (a, d, c, b), (a, c)(b, d)\}$$

and a non-transitive representation of the Klein four-group (the rhombus group):

$$V_4' = \{(1), (a, b), (c, d), (a, b)(c, d)\}.$$

There are no transitive subgroups G of S_4 of order 6. In fact, such a subgroup cannot be cyclic, since there are no elements of order 6 in S_4. Thus it must be isomorphic to S_3. As we know, such a group has three elements of order 2 and two elements of order 3. Among the elements of order 2 at least two must be symmetries (otherwise V_4 is a subgroup of G, which is impossible). They must shift a common vertex a. Otherwise, they are of the form $(a, b), (c, d)$ with different a, b, c, d. Then $(1), (a, b), (c, d), (a, b)(c, d)$ is a subgroup of G, which is impossible. If (a, b) and (a, c) are in G, then it is easy to check that G is the group of all permutations of a, b, c. It is not transitive on the set $\{1, 2, 3, 4\}$.

Any subgroup of order 4 is either cyclic or isomorphic to the Klein four-group. A cyclic group of order 4 is generated by an element of order 4, which is a cycle (a, b, c, d). Of course, such a subgroup is transitive (since there are six such cycles and any cyclic group of order 4 has two of them, there are three cyclic subgroups of order 4).

Finally, there is only one transitive subgroup of order 4—the group V_4. In fact, a non-cyclic subgroup of order 4 must contain three elements of order 2. A similar argument to that given above in connection with the subgroups of order 6 shows that it is impossible to get a transitive group with two symmetries. Thus we can only have the rotations of order 2, giving V_4.

Summarizing, we have the following list of isomorphism types of transitive subgroups of S_4:

$$S_4, A_4, D_4, C_4, V_4.$$

We do not prove that the group S_5 has following five types of transitive subgroups (described below):

$$S_5, A_5, G_5, D_5, C_5.$$

The order of every transitive subgroup must be divisible by 5 according to A.9.2(d), so the allowed orders of transitive subgroups are 5, 10, 15, 20, 30, 60, 120. As we know by A.8.2, the subgroup A_5 is the only one of order 60. It is not too difficult to exclude the orders 15 and 30 as orders of subgroups of S_5. The subgroups of order 5 are the cyclic groups of this order generated by cycles (a, b, c, d, e) of length 5. These are of course transitive and often denoted by C_5 (there are six such subgroups). Numbering the vertices of a regular pentagon and taking all the isometries of it, we get a subgroup of S_5 of order 10, which is usually denoted by D_5 and called the dihedral group (of order 10)—see A.2.4. These are all subgroups of order 10 in S_5 and they are transitive (each contains four cycles of length 5 and there are six different subgroups of this type). Finally, there are transitive subgroups of order 20 denoted by G_5 (sometimes by $GA(1,5)$, or F_{20}, and called general affine group of order 20—see Exercise 12.4 for $n = 5$). They can be represented over the field $\mathbb{F}_5$ as affine transformations $f(x) = ax + b$, $a \in \mathbb{F}_5^*$, $b \in \mathbb{F}_5$ (replacing 0 by 5 in $\mathbb{F}_5 = \{0, 1, 2, 3, 4\}$, we get a corresponding permutation in S_5, for example, the mapping $f(x) = x + 1$ gives $(1, 2, 3, 4, 5)$ and $g(x) = 2x$ gives $(1, 2, 4, 3)$). There are six subgroups of S_5 of this type (each contains one of the subgroups of type D_5, and this one contains one of type C_5).

Some Arithmetical Functions

An **arithmetical function** is any function from the positive integers $\mathbb{N}$ to the complex numbers. In this book, there are two functions which appear in many contexts: Euler's (totient) function φ and the Möbius function μ.

Arithmetical functions form a ring under the usual addition and multiplication of functions: $(f + g)(n) = f(n) + g(n)$ and $(fg)(n) = f(n)g(n)$. But much more interesting is another ring structure on the set of arithmetical functions, which takes into consideration the divisibility relation in the ring $\mathbb{Z}$. It is called **Dirichlet convolution** and is defined in the following way:

$$(f \star g)(n) = \sum_{d|n} f(d)g\left(\frac{n}{d}\right) = \sum_{ab=n} f(a)g(b),$$

where the first sum is over all positive divisors of n, and the second, over all pairs a, b of positive divisors of n such that $ab = n$ (the last equality is only a change of the notation $d = a, b = n/d$). It is clear that the convolution is commutative, but the Reader may wish to write down the formulae showing the associativity and the distributivity of multiplication with respect to addition. Both are easy to check and the associativity follows immediately by showing that both $((f \star g) \star h)(n)$ and $(f \star (g \star h))(n)$ are equal to $\sum_{abc=n} f(a)g(b)h(c)$, where the sum is over all triples (a, b, c) of positive integers such that $abc = n$. The ring of arithmetical functions with addition of functions and the Dirichlet convolution as multiplication is often called the **Dirichlet ring**. It has an identity ε defined by $\varepsilon(1) = 1$ and $\varepsilon(n) = 0$ when $n \neq 1$. In fact, $(\varepsilon \star f)(n) = \sum_{ab=n} \varepsilon(a)f(b) = f(n)$, since only the term corresponding to $a = 1, b = n$ matters. We will denote by Id the function such that $Id(n) = n$ for every n. It is not the identity in the Dirichlet ring. Notice that an arithmetical function f has an inverse in the Dirichlet ring, that is, there is an arithmetical function g such that $f \star g = \varepsilon$ if and only if $f(1) \neq 0$. In fact, if $f \star g = \varepsilon$, then $(f \star g)(1) = f(1)g(1) = 1$, which shows that $f(1) \neq 0$. Conversely, if $f(1) \neq 0$, then we can find the function g inductively in the following way (it is unique, if it exists, since the units in every ring form a group). First we find $g(1)$ by solving the equation $f(1)g(1) = 1$. When this is done and we already have $g(k)$ for $k < n$, then we take the required equality $(f \star g)(n) = \sum_{ab=n} f(a)g(b) = \varepsilon(n) = 0$ when $n > 1$. In this equality, we know all the values of g for $b < n$ and the only value we have to find is the one corresponding to the term $f(1)g(n)$. Since $f(1) \neq 0$, we can compute $g(n)$ using this equality.

An arithmetical function f is called **multiplicative** if $f(ab) = f(a)f(b)$ when a, b are relatively prime, that is, the only common positive divisor of both these numbers is 1. In other words, the greatest common divisor $\gcd(a, b) = 1$.

Recall that the **Möbius function** $\mu(n)$ is the arithmetical function defined in the following way: $\mu(1) = 1$, $\mu(n) = 0$ if n is divisible by a square of a prime number and $\mu(n) = (-1)^k$, when n is a product of k different prime numbers. Since $\mu(1) = 1$, the Möbius function has an inverse with respect to the convolution, that is, there is a unique arithmetic function f such that $\mu \star f = \varepsilon$. Computing a few values of f (as we did in general case above), there is an evident guess that $f = \mathbf{1}$, where $\mathbf{1}$ is defined by $\mathbf{1}(n) = 1$ for every positive integer n. We record this and give a proof:

Proposition A.11.1. *(a) The Möbius function is multiplicative.*
(b) We have $\mathbf{1} \star \mu = \varepsilon$, that is,

$$\mathbf{1} \star \mu(1) = \mu(1) = 1 \quad \textit{and} \quad (\mathbf{1} \star \mu)(n) = \sum_{d|n} \mu(d) = 0 \quad \textit{for } n > 1.$$

Proof. (a) If a is divisible by k different primes and b by l different primes, then ab is divisible by $k + l$ primes. Hence, if $\gcd(a, b) = 1$, then $\mu(ab) = \mu(a)\mu(b)$, since both sides are equal to either 0 (if a or b is divisible by a square of a prime) or $(-1)^{k+l}$ (if both a and b are square free).
(b) We check immediately that $(\mathbf{1} \star \mu)(1) = 1$ and for $n > 1$, we obtain:

$$(\mathbf{1} \star \mu)(n) = \sum_{d|n} \mu(d) = 1 + \sum_{p_{i_1} \cdots p_{i_k} | n} (-1)^k = 1 + \sum_{k=1}^{r} (-1)^k \binom{r}{k} = (1-1)^r = 0,$$

where $p_1 < \ldots < p_r$ are all different prime numbers dividing n and the sum is over all possible products of k of these primes for $k = 0, 1, \ldots, r$ with $i_1 < \ldots < i_k$. The term 1 corresponds to $d = 1$. Any remaining d contains at least one prime number. It is possible to argue in many different ways in order to check that $\mathbf{1} \star \mu(n) = 0$ when $n > 1$. For example, we note that there are r summands containing only one prime and they contribute r summands equal to -1. Then there are $\binom{r}{2}$ products of 2 primes (contributing as many summands 1), $\binom{r}{3}$ products of 3 primes (contributing as many summands -1) and so on. This gives the expression of the sum of $(-1)^k$ as a sum of binomial coefficients with shifting signs depending on the number of primes in a divisor of n—even numbers of primes give 1, and odd numbers of primes in d give -1. □

Using the Möbius function, it is possible to prove one of the fundamental properties of arithmetic functions, which we use on several occasions in the exercises:

A.11.2 Möbius Inversion Formula. *If f is an arithmetic function and*

$$g(n) = \sum_{d|n} f(d),$$

then

$$f(n) = \sum_{d|n} \mu(d) g\left(\frac{n}{d}\right).$$

Proof. The theorem says that if $g = \mathbf{1} \star f$, then $f = \mu \star g$, but this is evident, since $\mu \star \mathbf{1} = \varepsilon$, so $f = \varepsilon \star f = \mu \star \mathbf{1} \star f = \mu \star g$. □

In Chap. 10, we use a multiplicative version of the Möbius formula. Such a form sometimes follows immediately from A.11.2 by replacing $f(n)$ and $g(n)$ by $\log f(n)$ and $\log g(n)$ (when these numbers are defined). But what we really need is a more general form of Möbius inversion:

A.11.3 Multiplicative Möbius Inversion Formula. *If $f : \mathbb{N} \to G$ is a function, where G is an abelian group (in multiplicative notation), then*

$$g(n) = \prod_{d|n} f(d)$$

implies

$$f(n) = \prod_{d|n} g\left(\frac{n}{d}\right)^{\mu(d)}.$$

Proof. In the proof, we prefer the additive notation in G (for typographical reasons). We have:

$$\sum_{d|n} \mu(d) g\left(\frac{n}{d}\right) = \sum_{d|n} \mu(d) \sum_{m|\frac{n}{d}} f(m) = \sum_{md|n} \mu(d) f(m) = \sum_{m|n} f(m) \sum_{d|\frac{n}{m}} \mu(d).$$

But the last sum is 0 if $n/m > 1$ according to A.11.1 and it is equal to 1 when $n/m = 1$, that is, when $m = n$, so the right-hand side is equal to $f(n)$. □

Recall that **Euler's totient** (or **phi**) **function** (or simply **Euler's function**, when only this one is considered, as in this book) φ is an arithmetic function such that $\varphi(n)$ equals the number of $1 \leq k \leq n$ such that $\gcd(k, n) = 1$. Those properties of Euler's function which we use are gathered in the following theorem:

Proposition A.11.4. (*a*) *Euler's function is multiplicative. More precisely, we have*

$$\varphi([a, b])\varphi((a, b)) = \varphi(a)\varphi(b),$$

so in particular, $\varphi(ab) = \varphi(a)\varphi(b)$ *if* $(a, b) = 1$ *(and consequently,* $[a, b] = ab$*).*
(b) If $n = p_1^{a_1} \cdots p_k^{a_k}$ *is presentation of* n *as a product of prime numbers* p_i *for* $i = 1, \ldots, k$*, then*

$$\varphi(n) = n\left(1 - \frac{1}{p_1}\right)\cdots\left(1 - \frac{1}{p_k}\right).$$

(c) We have $\sum_{d|n} \varphi(d) = n$*, that is,* $\mathbf{1} \star \varphi = Id$*.*

Proof. (a) If $(a, b) = 1$, then by A.3.11, we have a ring isomorphism $\mathbb{Z}_{ab} \cong \mathbb{Z}_a \times \mathbb{Z}_b$ (the residue of n modulo ab is mapped onto the pair of residues of n modulo a and b). Taking the invertible elements in the rings on both sides, we get an isomorphism $\mathbb{Z}_{ab}^* \cong \mathbb{Z}_a^* \times \mathbb{Z}_b^*$ (see (A.1)). As we know (see p. 247), $\varphi(n) = |\mathbb{Z}_n^*|$ is the number of invertible elements in the ring $\mathbb{Z}_n$. Hence, we have $\varphi(ab) = |\mathbb{Z}_{ab}^*|$, $\varphi(a) = |\mathbb{Z}_a^*|$, $\varphi(b) = |\mathbb{Z}_b^*|$, so $\varphi(ab) = \varphi(a)\varphi(b)$. The generalized formula follows easily from (b) below (we return to it).
(b) Since φ is multiplicative, it is sufficient to show that $\varphi(p^a) = p^a(1 - 1/p) = p^a - p^{a-1}$ when p is a prime. But among p^a numbers from 1 to p^a, those which are not relatively prime to p^a are exactly those divisible by p and their number is $p^a/p = p^{a-1}$. Thus, the number of those which are relatively prime to p is $p^a - p^{a-1}$. The formula in (a) in its generalized version follows immediately from the formula just proved by noting that in $\varphi((a, b))$ one takes the product over the primes p such that $p \mid a$ and $p \mid b$, while in $\varphi([a, b])$ the product is over those primes p for which $p \mid a$ or $p \mid b$.
(c) There are many different proofs of the equality. We use an argument related to convolution on arithmetic functions. The formula in (b) can be easily expressed in terms of the Möbius function, since multiplying the factors $1 - 1/p_i$ on the right-hand side, we get all possible terms of the type $\frac{\mu(d)}{d}$, where $d = p_{i_1} \cdots p_{i_r}$ is a divisor of n containing r different prime numbers dividing n for $r = 0, 1, \ldots, k$. Only such factors contribute nonzero terms to the sum on the right-hand side in

$$\frac{\varphi(n)}{n} = \left(1 - \frac{1}{p_1}\right)\cdots\left(1 - \frac{1}{p_k}\right) = \sum_{d|n} \frac{\mu(d)}{d}.$$

But this equality says simply that $\varphi(n) = \sum_{d|n} \frac{n}{d}\mu(d) = (Id \star \mu)(n)$. Since $\mathbf{1}$ is the inverse of μ, we multiply both sides of the equality $\varphi = Id \star \mu$ by $\mathbf{1}$ and obtain the required equality $\mathbf{1} \star \varphi = Id$. □

Characters and Pairing

The facts gathered in this section are only used in Chap. 11 in the proof of theorem **T.11.4** on Kummer extensions.

A **character** of a group G is a homomorphism $\chi : G \to \mathbb{C}^*$. All characters of a group G form a group $\widehat{G}$ when the multiplication of characters of G is defined in a natural way: if $\chi_1, \chi_2 \in \widehat{G}$, then $\chi_1\chi_2(g) = \chi_1(g)\chi_2(g)$ for $g \in G$.

Theorem A.12.1. *Let G be a finite abelian group and $\widehat{G}$ its group of characters. Then G and $\widehat{G}$ are isomorphic. In particular, we have $|\widehat{G}| = |G|$.*

Proof. If $G = \langle g \rangle$ is a cyclic group of order n with a generator g, then each character $\chi \in \widehat{G}$ is uniquely determined by its value $\chi(g)$ on g. Since this value must be an nth root of unity (as $g^n = 1$ in G), we have n different characters given by $\chi_r(g) = e^{\frac{2\pi ri}{n}}$ for $r = 1, \ldots, n$. The group $\widehat{G}$ is also cyclic and generated by e.g. χ_1. Thus, the groups G and $\widehat{G}$ are isomorphic.

Now, we prove the theorem for arbitrary finite abelian groups using induction and A.6.2, that is, the fact that each finite abelian group is a product of finite cyclic groups. If G is a finite abelian group, which is not cyclic, then according to A.6.2, it has a cyclic subgroup H such that $G = H \times H'$ and H' is a finite abelian group whose order is less than the order of G. It is an easy exercise to show that $\widehat{G} = \widehat{H} \times \widehat{H'}$ (every character on G is a unique product of characters on H and H', since the value of a character on H is uniquely determined by its value on a generator of H). As we know, the groups H and $\widehat{H}$ are isomorphic, and the groups H' and $\widehat{H'}$ are isomorphic as the order of H' is less than the order of G. Thus, we have an isomorphism of G and $\widehat{G}$. In particular, these groups have the same order. □

Let G, G' be abelian groups for which we use multiplicative notations. By a **pairing**, we mean a bimultiplicative function $\Phi : G \times G' \to \mathbb{C}^*$, that is,

$$\Phi(g_1g_2, g') = \Phi(g_1, g')\Phi(g_2, g'), \qquad \Phi(g, g_1'g_2') = \Phi(g, g_1')\Phi(g, g_2')$$

for $g, g_1, g_2 \in G$ and $g', g_1', g_2' \in G'$.

An example of such a pairing Φ is the function:

$$\Phi : G \times \widehat{G} \to \mathbb{C}^*,$$

where $\Phi(g, \chi) = \chi(g)$ for $g \in G$ and $\chi \in \widehat{G}$.

By the **left kernel** of Φ, we mean the subgroup of G:

$$G_0 = \{g \in G | \varphi(g, G') = 1\}.$$

The **right kernel** G_0' of Φ is defined similarly

$$G_0' = \{g' \in G' | \varphi(G, g') = 1\}$$

as a subgroup of G'.

Lemma A.12.2. *Let $\Phi : G \times G' \to \mathbb{C}^*$ be a bimultiplicative function such that G/G_0 or G'/G_0' is finite. Then both these groups are finite and each is isomorphic to the dual group of the other.*

Proof. For every element $g \in G$, we have a homomorphism $\Phi_g : G' \to \mathbb{C}^*$ such that $\Phi_g(g') = \Phi(g, g')$, so that $\Phi_g \in \widehat{G'}$. Mapping g onto Φ_g, we get a homomorphism $G \to \widehat{G'}$. But Φ_g maps G'_0 onto 1, so Φ_g defines a homomorphism $\overline{\Phi}_g : G'/G'_0 \to \mathbb{C}^*$ (see A.2.12). Thus, we have a homomorphism $G \to \widehat{G'/G'_0}$ mapping g onto $\overline{\Phi}_g$. The kernel of this homomorphism consists of all $g \in G$ such that $\Phi_g(g') = \Phi(g, g') = 1$ for every $g' \in G'$, which is exactly the left kernel G_0 of Φ. Thus according to A.2.11, the homomorphism $G \to \widehat{G'/G'_0}$, defines an injection:

$$G/G_0 \to \widehat{G'/G'_0}. \tag{A.3}$$

Interchanging the roles of G and G' and repeating exactly the same argument, we get an injection:

$$G'/G'_0 \to \widehat{G/G_0}. \tag{A.4}$$

Since we have injections (A.3) and (A.4), it is clear that if one of the groups G/G_0 or G'/G'_0 is finite, then the other is also finite. Moreover, the same injections together with A.12.1 show that

$$|G/G_0| \leq |\widehat{G'/G'_0}| = |G'/G'_0| \quad \text{and} \quad |G'/G'_0| \leq |\widehat{G/G_0}| = |G/G_0|,$$

so $|G/G_0| = |\widehat{G'/G'_0}|$ and $|G'/G'_0| = |\widehat{G/G_0}|$. Thus, both injections (A.3) and (A.4) are in fact isomorphisms. □

Zorn's Lemma

A relation $\leq$ on a set X (see p. 235) is called a **partial ordering** if for any $x, y, z \in X$, we have
(a) $x \leq x$;
(b) $x \leq y$ and $y \leq z$ imply $x \leq z$;
(c) $x \leq y$ and $y \leq x$ imply $x = y$.

When $x \leq y$, then we also write $y \geq x$.

A typical example is the set of real numbers $\mathbb{R}$ (or any of its subsets) with the usual relation $\leq$. Another common example is the set $\mathcal{S}(M)$ of all subsets of a set M with the inclusion $\subseteq$ as the relation $\leq$.

If X is a set with a partial ordering $\leq$, then we say that an element $x^* \in X$ is **maximal** if the relation $x^* \leq x$ for $x \in X$ implies that $x = x^*$. An element $y_0 \in X$ is called an upper bound for a subset Y of X if $y \leq y_0$ for all $y \in Y$. A subset Y of X is called a **chain** if for any two elements $y_1, y_2 \in Y$, we have $y_1 \leq y_2$ or $y_2 \leq y_1$.

A.13.1 Zorn's[5] Lemma. *Let X be a nonempty partially ordered set. If every chain in X has an upper bound, then there exists a maximal element in X. More precisely, for each $x \in X$, there exists a maximal element x^* such that $x \leq x^*$.*

For a proof of Zorn's Lemma and a discussion of its different applications in mathematics see [Ci], Chapter 4.

As an example and an important tool, which we use in order to prove the existence of algebraic closure of any field, we prove the following result:

Proposition A.13.2. *Every proper ideal in a commutative ring with identity is contained in a maximal ideal.*

Proof. Let R be a ring and I_0 a proper ideal (that is, $I_0 \neq R$). Let X be the set of all ideals in R which contain I_0. The set X is nonempty, since $I_0 \in X$. We consider X with inclusion as a partially ordered set. Let Y be a chain in X and let $J = \bigcup_{I \in Y} I$. We claim that J is a proper ideal. Let $r_1, r_2 \in J$, so that $r_1 \in I_1 \in Y$ and $r_2 \in I_2 \in Y$, where $I_1 \subseteq I_2$ or $I_2 \subseteq I_1$. Thus $r_1 - r_2 \in I_1$ or $r_1 - r_2 \in I_2$, which gives $r_1 - r_2 \in J$. If $r \in R$ and $r' \in J$, that is, $r' \in I$ for some $I \in Y$, then $rr' \in I \subseteq J$. Thus J is an ideal that contains I_0 (since $I_0 \subseteq I \in Y$). The ideal J is proper, since $1 \notin J$. Moreover, the ideal J is an upper bound for Y. By Zorn's Lemma, there exists a maximal element I^* in X, which means that I^* is a maximal ideal containing I_0. □

[5]Max August Zorn, 6 June 1906–9 March 1993.

References

[A] E. Artin, Galois Theory, Dover Publications, 1997.

[B] G.M. Bergman, Exercises supplementing those in Ian Stewart's "Galois Theory", 3rd Edition, https://math.berkeley.edu/~gbergman/ug.hndts/#m114_IStwrt_GT

[BR] T.R. Berger, I. Reiner, A proof of the normal basis theorem, Am. Math. Monthly, 82 (1975), 915–918.

[Ca] J.S. Calcut, Rationality and the Tangent Function, http://www.oberlin.edu/faculty/jcalcut/tanpap.pdf

[Ci] K. Ciesielski, Set Theory for the Working Mathematician, London Mathematical Society Student Texts, Cambridge University Press, 1997.

[C] P.M. Cohn, Algebra, vol. 3, Second Edition, John Wiley and Sons, 1991.

[D] R.A. Dean, A rational polynomial whose group is the quaternions, Am. Math. Monthly, 88(1981), 42–45.

[E] H.M. Edwards, Essays in constructive mathematics, Springer, 2005.

[F] E. Formanek, Rational function fields. Noether's problem and related questions. Journal of Pure and Applied Algebra, 31(1984), 28–36.

[G] K. Girstmair, Hippocrates' Lunes and Transcendence, Expo. Math., 21 (2003), 179–183.

[FK] E. Fried, J. Kollár, Authomorphism Groups of Algebraic Number Fields, Math. Z., 163 (1978), 121–123.

[GV] D. Gay, W.Y. Vélez, On the degree of the splitting field of an irreducible binomial, Pacific J. Math. 78 (1978), 117–120.

[J] N. Jacobson, Basic Algebra II, Second Edition, Dover Books on Mathematics (W.H. Freeman and Company), 1989.

[K] H. Koch, Number Theory, Algebraic Numbers and Functions, Graduate Studies in Mathematics vol. 24, AMS, 2000.

[L] S. Lang, Algebra, Third Edition, Addison-Wesley, 1993.

[LS] H.W. Lenstra and R.J. Schoof, Primitive normal bases for finite fields, Mathematics of Computation, 48 (1987), 217–231.

[MM] G. Malle, B.H. Matzat, Inverse Galois Theory, Springer, 1999.

[MP] The Five Squarable Lunes, http://www.mathpages.com/home/kmath171/kmath171.htm

[R] S. Roman, Field Theory, Second Edition, Graduate Texts in Mathematics vol. 158, Springer, 2006.

[S1] J.-P. Serre, Galois Cohomology, Springer, 1997.

[S2] J.-P. Serre, A Course in Arithmetic, Graduate Texts in Mathematics vol. 7, Springer, 2006.

J. Brzeziński, *Galois Theory Through Exercises*, Springer Undergraduate Mathematics Series, https://doi.org/10.1007/978-3-319-72326-6

[Sch] A. Schinzel, Polynomials with Special Regard to Reducibility, Encyclopedia of Mathematics and its Applications 77, Cambridge University Press, 2000.
[Tsch] N.G. Tschebotaröw, Grundzüge der Galois'schen Theorie, H. Schwerdtfeger (ed.), P. Noordhoff, 1950.
[V] J.K. Verma, Exercises in field theory and Galois theory, http://www.math.iitb.ac.in/~jkv/algebra2/prob.pdf

List of Notation

J. Brzeziński, *Galois Theory Through Exercises*, Springer Undergraduate Mathematics Series, https://doi.org/10.1007/978-3-319-72326-6

Index

J. Brzeziński, *Galois Theory Through Exercises*, Springer Undergraduate Mathematics Series, https://doi.org/10.1007/978-3-319-72326-6

Printed in the United States
By Bookmasters